Creating Balance?

Stephan Kaiser · Max Josef Ringlstetter ·
Doris Ruth Eikhof · Miguel Pina e Cunha
Editors

Creating Balance?

International Perspectives on the Work-Life Integration of Professionals

Springer

Editors
Professor Stephan Kaiser
Universität der Bundeswehr München
Werner-Heisenberg-Weg 39
85577 Neubiberg
Germany
stephan.kaiser@unibw.de

Professor Max Josef Ringlstetter
Katholische Universität
Eichstätt-Ingolstadt
Auf der Schanz 49
85049 Ingolstadt
Germany
max.ringlstetter@kueichstaett.de

Dr. Doris Ruth Eikhof
Institute for Socio-Management
Stirling Management School
University of Stirling
d.r.eikhof@stir.ac.uk

Professor Miguel Pina e Cunha
University Nova de Lisboa
Campus de Campolide
1099-032 Lisboa
Portugal
mpc@fe.unl.pt

ISBN 978-3-642-16198-8 e-ISBN 978-3-642-16199-5
DOI 10.1007/978-3-642-16199-5
Springer Heidelberg Dordrecht London New York

Cover design: WMXDesign GmbH, Heidelberg

Printed on acid-free paper

Springer is part of Springer Science+Business Media (www.springer.com)

Foreword

Although highly topical, research on the work-life integration of professionals has previously not been published in a comprehensive volume. The aim of our book project was therefore to provide a platform for distinguished researchers from across the world to share experiences and research findings and to generate new knowledge on the work-life integration of professionals. The result is a compendium of cutting-edge research that is relevant and accessible to both academics and practitioners. We hope that this compendium provides a source of knowledge and inspiration for academic and business audiences interested in work-life integration issues – in particular of professional workers – and advances our understanding and discussions in this area.

This compendium was written for researchers and teachers in the areas of management, education, psychology, sociology and related fields, as well as for executives and managers in the areas of business strategy, organization and HRM, and for students of business and management, human resource management, education, organizational psychology and related fields. We appreciate any feedback on the book or the general topic.

In this foreword we would like to acknowledge the people and institutions that made this book happen. We are grateful to our 50 international authors for their excellent contributions to the book and for their collaboration. We would also like to thank those authors who submitted proposals and whose contributions we were not able to fit in. Special thanks go to Cornelia Reindl. Not only did she coordinate all contributions and take responsibility for the organization of this project. She also contributed essentially to the content and direction of this book and without her support we would not hold this book in our hands today. The book is part of the research project "Work-Life Balance in Consulting Firms" located at the Catholic

University Eichstätt-Ingolstadt, Ingolstadt School of Management. Last but not least we would like to express our thanks to the funders of this project, the German Ministry for Education and Science (BMBF), the German Aerospace Centre (DLR) and the European Social Fund (ESF).

Stephan Kaiser, Max Josef Ringlstetter, Doris Ruth Eikhof, Miguel Pina e Cunha

GEFÖRDERT VOM

Bundesministerium für Bildung und Forschung

ESF Europäischer Sozialfonds für Deutschland

EUROPÄISCHE UNION

Projektträger im DLR Deutsches Zentrum für Luft- und Raumfahrt e.V.

Förderschwerpunkt Innovationsstrategien jenseits traditionellen Managements

IPOB

Introduction

Work-life balance and work-life integration are prominent buzzwords in today's management practice and theory, at practitioner and academic conferences, in lifestyle magazines and general media. Corporate websites try to impress with presentations of companies' work-life balance friendly workplaces and programs. Companies employing "extreme workers" (Hewlett and Luce, 2006), i.e., employees in highly demanding work environments are no exception: the websites of virtually all large consulting firms dutifully inform readers and potential applicants about how the companies address the challenge of balancing or integrating work with other life domains.

Key issues in the increasingly extensive discussion of work-life issues are work hours and work intensification, the intensifying competitive pressure on the job market, changed and still changing work and family-related values amongst young executives and the increasing career orientation of women. More than ever, organizations are expected to have policies and procedures in place that enable employees to work efficiently as well as integrate work and private life. Particularly in high performance work environments, where employees work long hours and perform complex tasks, work-life initiatives that address these challenges have gained both importance and prevalence. Nevertheless, in organizational reality attempts to reconcile work and life are often not very successful.

Work-life integration is a multifaceted and complex issue. Numerous antecedents on the individual, organizational, environmental, societal and cultural level influence individual work-life experiences (cf. Byron, 2005; Kossek et al., 1999; Mesmer-Magnus and Viswesvaran, 2007). In addition, it is not only work that impacts private life, but also vice versa (cf. Carlson and Kacmar, 2000; Eby et al., 2005; Kossek and Ozeki, 1998). Also, while the quality of the impact may be negative in the sense of conflict, positive impacts through, for instance, positive mood spillover or useful synergies exist as well (Carlson et al., 2006; Greenhaus and Powell, 2006; Hill et al., 2007; Wayne et al., 2007). Finally, for an increasing amount of workers life and work are not distinct spheres anymore. Boundaries between work and private life have become blurred – both intentionally and unintentionally.

In a number of ways, the work-life integration of professionals is particularly interesting. Whereas working hours for many groups of workers have decreased or remained constant, professionals' working hours have increased (cf. Brett and Stroh, 2003; Wharton and Blair-Loy, 2006). In addition, professionals across the world experience an intensification of work. They are required to perform more or more demanding tasks in the same time. Helped by recent technological developments, professional work has extended into times and spaces previously reserved for private matters (cf. Ahuja et al., 2007; Chesley, 2005; Valcour and Hunter, 2005). Thanks to Blackberries, mobile phones and wireless internet, professional workers can and do work during holidays, weekends and evenings as well as on commuter trains, in family homes, cafés and children's playgrounds. While this extensification can be both blessing and curse, professional workers should, in theory, be comparatively well placed to integrate work and life according to their individual preferences: they tend to have more autonomy and control over their work content and schedules and can often afford to buy in childcare or other domestic services. In reality though, the work-life integration of professionals remains prominently precarious.

Against this backdrop, this edited collection brings together current research on the work-life integration of professionals. The book is structured in five parts. The first part gives an overview of different approaches and concepts of work-life research. The remaining four parts deal with specific issues of the work-life integration of professionals. At the individual level, studies focus on, for instance, the meaning of work-life integration for managers and other highly-qualified employees, gender differences in the perception of work-life conflict, personal values and personality traits. Analysis of the use, implementation, and effectiveness of corporate work-life balance policies provides insight into the structural context of work-life integration at organizational level. Studies into (the blurring of) work-life boundaries, the spatial organization of work and the role of mobile technology in integrating work and life investigate the social embeddedness of professionals' work-life integration.

Part I: Perspectives on the Work-Life Integration of Professionals

Part I of the book gives an overview of the different concepts and approaches in work-life research. The first chapter, "Integrating Professional Work and Life: Conditions, Outcomes and Resources" (C. U. Reindl, S. Kaiser, M. L. Stolz), discusses the challenges of work-life integration for professionals. It highlights the conditions under which professionals work and explores the respective implications for work-life integrations. The authors then discuss individual and organizational benefits of successful work-life integration and the means of facilitating it. The second chapter, "Introducing Theoretical Approaches to Work-Life Balance and Testing a New Typology Among Professionals" (J. Rantanen, U. Kinnunen, S. Mauno, K. Tillemann), explores alternative approaches to conceptualising work-life balance. The authors introduce a new work-life balance typology of based on the synthesis of earlier role (conflict) theories and recent views on work-life balance. Four types of work-life balance – beneficial, harmful, active and passive – are proposed that differ with respect to psychological functioning and role engagement. This typology is tested empirically among three samples of professionals. The third chapter, "From 'Balancers' to 'Integrators'? Young Professionals Talk About 'Work' and the Rest of 'Life' in the U.K." (S. Lewis, U. Chatrakul Na Ayudhya), takes a more critical approach to the topic. It contrasts previous conceptions of work-life balance with the work-life perspectives and expectations of young British and Asian adults during their transition from university to professional life. The authors argue that a lifecourse perspective needs to be adopted in order to contextualize and understand individuals' perceptions and expectations of work-life balance. The final chapter of this part, "Redefining Success: The Foundation for Creating Work-Life Balance" (J.L. Whittington, R. Maellaro, T. Galpin), takes a closer look at the conceptualization of personal success. The authors propose a Whole-Life Model as the basis for a more comprehensive view on well-being that places career and work in the context of overall life development. They explore strategies for developing well-being, work, career and life with the aim of achieving both work-life balance and success in a broader sense.

Part II: Antecedents and Consequences Related to Work-Life Integration

Part II of the book investigates antecedents and consequences of work-life integration from different perspectives and with varying foci. The first chapter, "Personality and Work-Life Integration" (J. Michel, M. Clark), explores how personality impacts the work-life integration of professionals. After providing a typology of primary theories of work and life, major conceptions of personality are linked to the work-life integration of professionals. Thirdly, the authors provide a qualitative review of personality and work-life integration constructs followed by suggestions for future research, such as a cognitive-affective approach to work and life. The next chapter, "Potential Antecedents and Consequences of Work-Family Conflict: A Three Country Study" (R. J. Burke, L. Fiksenbaum, G. Ek-Kot, M. Koyuncu, M. Jing), extends findings about the bi-directional measure of work-family conflict and its three dimensions by presenting research findings from three different countries and cultures: Turkey, China and Egypt. Lastly, A. C. Peng, R. Ilies and N. Dimotakis in their chapter "Work-Family Balance, Role Integration and Employee Well-Being" examine the antecedents and consequences of work-family role integration. They suggest that for understanding the various effects of work-family integration the direction of work-family integration (i.e., integrating work into family or integrating family into work) is as important as the level of work-family integration. The chapter also provides an in-depth discussion of the effects of work-family integration on employee well-being and the practical implications of different role boundary management strategies for professionals and their employers.

Part III: Mothers, Fathers and Couples: Challenging Conventional Wisdom

Part III of the book takes a closer look at professionals in household roles that make work-life integration particularly salient: mothers, fathers and couples. The first chapter, "What to Expect When She's Expecting: Work-Life and Identity Integration Challenges and Opportunities of 'Soon-to-be' Working Professional Mothers" (J.J. Ladge, D. Greenberg, J. A. Clair), examines work-life and identity integration issues for professional women. Specifically, the authors investigate how expectant mothers' experiences and professional self-conceptualizations change as they begin to integrate new maternity-centered identities into their established identities as working professionals. S. Burnett, C. Catrell, C. Cooper and P. Sparrow in their chapter "Fatherhood and Flexible Working: A Contradiction in Terms?" take a male-centred perspective on work-life integration. They explore modern fatherhood in the context of professional employment, flexible working practices and gender. In particular, they examine the uptake of flexible working practices by predominantly Western, employed, white-collar, middle-class fathers and the family and workplace circumstances that affect it. The final chapter of this part, "Dual-Career Couples: Specific Challenges for Work-Life Integration" (A. Abele, J. Volmer), brings the gender perspectives together by shedding light on the situation of professional couples. The authors provide a comprehensive overview of work-life research into dual career couples. Against this backdrop they then discuss empirical findings regarding stress and well-being, career and mobility issues, marital satisfaction and partnership stability, parenthood and gender issues.

Part IV: Corporate Strategies Facilitating Work-Life Integration

Part IV of the book focuses on how organizations impact, facilitate and hinder successful work-life integration. A. Beauregard in her chapter "Corporate Work-Life Balance Initiatives: Use and Effectiveness" examines antecedents and outcomes of the uptake of corporate work-life balance initiative for professionals. She explains that work-life balance initiatives have the potential to ease the pressures of combining challenging, time-consuming jobs with family responsibilities and personal commitments outside of work, but that equity concerns and "backlash" among non-users of such initiatives also have to be taken into account. She also emphasises the role of corporate culture for the successful implementation of such policies. The second chapter, "Temporal Workplace Flexibility and Associated Work-Life Outcomes for Professionals" (E.J. Hill, S.J. Carroll, B.L. Jones, L.A. Buswell, T.A. Fackrell, A.M. Galovan), discusses the relationship between temporal workplace flexibility and work-life integration for professional employees. It shows that reduced work hours options, including regular part-time employment, job sharing, transition from leave and phased retirement, enable professionals to reduce the stress associated with overwork. The authors argue that temporal workplace flexibility is an important component in a strategy to enhance work-life integration for professional employees. The following chapter, “Causes and Consequences of the Utilization of Work-Life Policies by Professionals: ‘Unconditional Supervisor Support Required’” (B. Peper, J.S.E. Dikkers, M.L. Van Engen, C.J. Vinkenburg), takes a closer look at how supervisor and management support influence the work-life integration challenges of professionals. The authors analyze the utilization of work-life policies in association with work-life culture, work-life enrichment and career advancement with a qualitative as well as quantitative approach.

Part V: Boundary Perspectives on Professionals’ Work-Life Integration

Part V or the book focuses on an ever more pressing issue: the blurring of boundaries between work and non-work life. C. Baldry and J. Hyman in their chapter “The Pressures of Commitment: Taking Software Home” show how for a group of “new professionals”, software workers, work often permeates non-work lives both quantitatively (through the length and pattern of working hours) and qualitatively (for example due to exhaustion, stress or continuously thinking and talking about work when not in the workplace). Their case study shows that many software workers exhibit a voluntary extension of work into non-work time, facilitated by professional identities, commitment and membership in the software occupational community. However, this extension of work into non-work spheres is also partly imposed by managerial expectations and the intensification of project targets. The second chapter, “‘Myth of Separate Worlds’: An Exploration of How Mobile Technology Has Redefined Work-Life Balance” (L. Duxbury, R. Smart), discusses the role of mobile technologies in work-life integration and boundary management. The authors explore how mobile technology impacts the balance between work and family for professional employees, more explicitly how mobile technologies impact boundaries in time and space and why, given these issues, professional employees chose to adopt such technologies. The third chapter, “Working in Polycontextual Environments: An Empirical Analysis of Flexpatriates’ Lifestyles” (H. Mayerhofer, B. Müller, A. Schmidt), discusses boundary issues and their impact on work-life integration in a specific group of chronically flexible workers, flexpatriates. Focusing on the nature of flexpatriate lifestyle, the authors present four empirically grounded types – tough travellers, enjoyers, cosmopolitans and contactors. Workers of these four types vary with respect to their flexpatriate lifestyles and also require different human resource management approaches. The final chapter of the book, “Spatial Aspects of Professionals’ Work-Life Integration” (D. Anderson, C. Kelliher), presents a case study of a large UK company that implemented major changes in the spatial allocation of work. Their findings indicate that work-life integration of the professionals employed in this company had improved in many ways; however there were also many examples of a less positive impact. Because of the increased blurring of boundaries between the work and non-work, individuals experienced greater intrusion into their home lives, often in unanticipated, and unwelcome, ways.

The above chapters provide a wide-ranging overview of the specific issues and challenges that professional workers face in their attempt to integrate work-life domains and to satisfactorily fulfil role demands of both work and private life. However, they discuss only a selection of issues and do not give an exhaustive account of the research or the issues on the work-life integration of professionals. Work-life integration of professionals is and will continue to be a matter of great interest for research and organizational practice. We hope that this compendium contributes to a much-needed and fruitful debate of this issue.

References

Ahuja MK, McKnight DH, Chudoba KM, George JF, Kacmar CJ (2007) IT road warriors: balancing work–family conflict, job autonomy, and work overload to mitigate turnover intentions. MIS Q 31(1):1–17

Brett JM, Stroh LK (2003) Working 61 plus hours a week: why do managers do it?. J Appl Psychol 88(1):67–78

Byron K (2005) A meta-analytic review of work-family conflict and its antecedents. J Vocat Behav 67:169–198

Carlson DS, Kacmar KM, Wayne JH, Grzywacz JG (2006) Measuring the positive side of the work-family interface. Development and validation of a work-family enrichment scale. J Vocat Behav 68(1):131–164

Carlson DS, Kacmar KM, Williams LJ (2000) Construction and initial validation of a multidimensional measure of work-family conflict. J Vocat Behav 56(2):249–276

Chesley N (2005) Blurring boundaries? Linking technology use, spillover, individual distress, and family satisfaction. J Marriage Fam 67(5):1237–1248

Eby L, Casper WJ, Lockwood A, Bordeaux C, Brinley A (2005) Work and family research in IO/OB. Content analysis and review of the literature (1980–2002). J Vocat Behav 66(1):124–197

Edwards JR, Cooper CL (1990) The person-environment fit approach to stress: recurring problems and some suggested solutions. J Organ Behav 11(4):293–307

Greenhaus JH, Powell GN (2006) When work and family are allies. A theory of work-family enrichment. Acad Manage Rev 31(1):72–92

Hewlett SA, Luce CB (2006) Extreme jobs. The dangerous allure of the 70-hour workweek. Harv Bus Rev 49–59

Hill EJ, Allen S, Jacob J, Bair AF, Bikhazi SL, van Langeveld A (2007) Work-family facilitation. Expanding theoretical understanding through qualitative exploration. Adv Dev Hum Resour 9(4):507–526

Kossek EE, Noe RA, DeMarr BJ (1999) Work-family role synthesis. Individual and organizational determinants. Int J Confl Manage 10(2):102–129

Mesmer-Magnus JR, Viswesvaran C (2007) How family-friendly work environments affect work/family conflict. A meta-analytic examination. J Lab Res 27(4):555–574

Valcour MP, Hunter LW (2005). Technology, organizations, and work-life integration. In: Kossek EE, Lambert SJ eds Work and life integration. Organizational, cultural, and individual perspectives. Lawrence Erlbaum Associates, Mahwah, NJ, pp 61–84.

Wayne JH, Grzywacz JG, Carlson DS, Kacmar KM (2007) Work-family facilitation: a theoretical explanation and model of primary antecedents and consequences. Hum Resour Manage Rev 17(1):63–76

Wharton AS, Blair-Loy M (2006) Long work hours and family life: a cross-national study of employees' concerns. J Fam Issues 27(3):415–436

Contents

Contributors

Andrea E. Abele University of Erlangen-Nuremberg, Erlangen, Germany, abele@phil.uni-erlangen.de

Deirdre Anderson Cranfield School of Management, Bedford, UK, deirdre.anderson@cranfield.ac.uk

Chris Baldry University of Stirling, Stirling, UK, c.j.baldry@stir.ac.uk

T. Alexandra Beauregard London School of Economics, London, UK, a.beauregard@lse.ac.uk

Ronald J. Burke Schulich School of Business, York University, Toronto, ON, Canada, rburke@schulich.yorku.ca

Simon Burnett Lancaster University Management School, Lancaster, UK, s.burnett1@lancaster.ac.uk

Lydia A. Buswell School of Family Life, Brigham Young University, Provo, UT, USA, lideeuh@gmail.com

Sarah June Carroll School of Family Life, Brigham Young University, Provo, UT, USA, sarahjunecarroll@byu.edu

Uracha Chatrakul Na Ayudhya Middlesex University Business School, London, UK, u.chatrakul@mdx.ac.uk

Judith A. Clair Boston College, Chestnut Hill, MA, USA, clairju@bc.edu

Malissa A. Clark Wayne State University, Detroit, MI, USA, malissa@wayne.edu

Cary Cooper Lancaster University Management School, Lancaster, UK, c.cooper1@lancaster.ac.uk

Josje Dikkers VU University Amsterdam, Amsterdam, Netherlands, jdikkers@feweb.vu.nl

Nikos Dimotakis Michigan State University, East Lansing, MI, USA, dimotaki@msu.edu

Linda Duxbury Carleton University, Ottawa, ON, Canada, linda_duxbury@carleton.ca

Doris Ruth Eikhof University of Stirling, Stirling, UK, d.r.eikhof@stir.ac.uk

Ghada El-Kot Arab Academy for Science and Technology and Maritime Transport, Cairo, Egypt, ghada.elkot@plymouth.ac.uk

Tamara A. Fackrell School of Family Life, Brigham Young University, Provo, UT, USA, myattorneymediator@gmail.com

Lisa Fiksenbaum Department of Psychology, York University, Toronto, ON, Canada, lisafix@yorku.ca

Adam M. Galovan School of Family Life, Brigham Young University, Provo, UT, USA, agalovan@hotmail.com

Timothy Galpin University of Dallas, Dallas, TX, USA, tgalpin@gsm.udallas.edu

Caroline Gatrell Lancaster University Management School, Lancaster, UK, c.gatrell@lancaster.ac.uk

Danna Greenberg Babson College, Wellesley, MA, USA, dgreenberg@babson.edu

E. Jeffrey Hill School of Family Life, Brigham Young University, Provo, UT, USA, jeff_hill@byu.edu

Jeff Hyman University of Aberdeen, Aberdeen, UK, j.hyman@abdn.ac.uk

Remus Ilies Michigan State University, East Lansing, MI, USA, ilies@msu.edu

Weng Jing School of Hospitality and Tourism, Beijing Union University, Beijing, China, wjing7510@hotmail.com

Blake L. Jones School of Family Life, Brigham Young University, Provo, UT, USA, blakejones13@gmail.com

Stephan Kaiser Universität der Bundeswehr München, München, Germany, stephan.kaiser@unibw.de

Clare Kelliher Cranfield School of Management, Bedford, UK, clare.kelliher@cranfield.ac.uk

Ulla Kinnunen Department of Psychology, University of Tampere, Tampere, Finland, ulla.kinnunen@uta.fi

Mustafa Koyuncu Faculty of Commerce and Tourism Education, Nevsehir University, Nevsehir, Turkey, mkoyuncu34@gmail.com

Jamie J. Ladge Northeastern University, Boston, MA, USA, j.ladge@neu.edu

Suzan Lewis Middlesex University Business School, London, UK, s.lewis@mdx.ac.uk

Rosemary Maellaro University of Dallas, Dallas, TX, USA, rmaellaro@gsm.udallas.edu

Saija Mauno University of Jyväskylä, Jyväskylä, Finland, saija.mauno@psyka.jyu.fi

Helene Mayerhofer WU Vienna University of Economics and Business, Vienna, Austria, helene.mayerhofer@wu.ac.at

Jesse S. Michel Florida International University, Miami, FL, USA, jmichel@fiu.edu

Barbara Müller Austria, barbara.mueller@jku.at

Ann Chunyan Peng Michigan State University, East Lansing, MI, USA, peng@bus.msu.edu

Bram Peper Erasmus University Rotterdam, Rotterdam, Netherlands, peper@fsw.eur.nl

Miguel Pina e Cunha University Nova de, Lisboa, Portugal, mpc@fe.unl.pt

Johanna Rantanen University of Tampere, University of Jyväskylä, Jyväskylä, Finland, k.johanna.rantanen@jyu.fi

Cornelia U. Reindl Catholic University Eichstaett, Ingolstadt, Germany, cornelia.reindl@ku-eichstaett.de

Max Josef Ringlstetter Catholic University, Eichstätt-Ingolstadt, Ingolstadt, Germany, max.ringlstetter@ku-eichstaett.de

Angelika Schmidt WU Vienna University of Economics and Business, Vienna, Austria, angelika.schmidt@wu.ac.at

Rob Smart Carleton University, Ottawa, ON, Canada, robert_smart@carleton.ca

Paul Sparrow Lancaster University Management School, Lancaster, UK, p.sparrow@lancaster.ac.uk

Martin L. Stolz Catholic University Eichstaett-Ingolstadt, Ingolstadt, Germany, martin.stolz@ku-eichstaett.de

Kati Tillemann Estonian Business School, Tallinn University, Tallinn, Estonia, kati.tillemann@ebs.ee

Marloes van Engen Tilburg University, Tilburg, Netherlands, m.l.vengen@uvt.nl

Claartje Vinkenburg VU University Amsterdam, Amsterdam, Netherlands, cvinkenburg@feweb.vu.nl

Judith Volmer Department of Psychology, University of Erlangen-Nuremberg, Erlangen, Germany, judith.volmer@sozpsy.phil.uni-erlangen.de

J. Lee Whittington University of Dallas, Dallas, TX, USA, jleewhitt@aol.com

Part I
Perspectives on the Work-Life Integration of Professionals

Chapter 1
Integrating Professional Work and Life: Conditions, Outcomes and Resources

Cornelia U. Reindl, Stephan Kaiser, and Martin L. Stolz

1.1 Introduction

In the last 30 years a remarkable amount of research has been dedicated to the work–life interface, investigating work–life conflict and enrichment from many different angles and in different populations. A group that deserves particular attention and has received increasing interest in work–life research in the last few years is what we call professionals. These are usually hard working individuals in highly demanding work environments characteristically facing long working hours, high workload and pressure to succeed, and the need to travel frequently. On the other hand they usually have above-average incomes and high standards of living together with jobs they enjoy (Brett and Stroh, 2003; Litrico and Lee, 2008; Perlow and Porter, 2009; Robertson and Swan, 2003; Wallace, 1997). Professionals thus work and live in a situation which can be both impeding and enhancing work–life integration, i.e., *a sense of having achieved a satisfactory resolution of the multiple demands of work and non-work domains* (Higgins et al., 2000, p. 19). The group of professional workers faces specific challenges regarding work–life integration but also opportunities and resources to which other workers do not have access.

Work plays a dominant role in professionals' lives, most individuals who work in professional positions have a very high professional and organizational commitment and the importance attributed to the work role is high. With such a highly salient work role and accompanying benefits the question may arise whether work–life integration is an issue for professionals at all. We argue that it is particularly professionals who face challenges in balancing their life domains. Due to their qualifications, usually at college level and higher, the age of entering the job market is their mid to late twenties. Graduates usually start a family in their early thirties, i.e., in the phase of employment where they are still working their way up the institutional career ladder. Many professionals thus undergo the two most critical life phases within often overlapping time frames: entering professional employment and starting a family. During this critical phase, generally in high performance

C.U. Reindl (✉)
Catholic University Eichstaett, Ingolstadt, Germany
e-mail: cornelia.reindl@ku-eichstaett.de

S. Kaiser et al. (eds.), *Creating Balance?*, DOI 10.1007/978-3-642-16199-5_1,

work environments heavy work involvement is essential for organizational performance and career success (Perlow and Porter, 2009; Wharton and Blair-Loy, 2006). Corporate work cultures reflect this ethic leaving little space for individual work–life concerns (Dikkers et al., 2007; van Engen et al., 2009). Professionals themselves might view their self-concept as being much focused on the work domain as their work is very often highly meaningful and significant to them (Alvesson, 2000). However, anecdotal evidence suggests that life domains other than work do seem to play an important role for many professionals but integrating them with their extensive work role yields major challenges (Bagger et al., 2008; Cinamon and Rich, 2002; Friedman, 2006).

In this chapter after giving a brief idea about what we understand by the term "professionals" we present some thoughts and ideas on three major issues:

- Which chances and challenges do professionals face for successful work–life integration?
- (How) Do professionals benefit from work–life integration (initiatives)? How do organizations benefit from individual work–life integration?
- What can professionals and their organizations do to leverage work–life integration?

1.1.1 "Professionals" – Who Are They?

When talking about professionals we have in mind an idea as to which attributes characterize this kind of person and his/her job. However, in research many different groups are named as "professionals" and there exists no widely accepted definition (von Nordenflycht, 2010). Some groups of employees are called professionals without reference to professional norms or a validation of this label "profession" – which is often the case for managers. In many studies the sample is described as consisting of "professionals and managers" etc. Others are indeed professionals according to traditional definitions without being referred to as such but are rather described with their professional field (medics, lawyers, professors, etc.). Again others are in the process of professionalization, attempting to establish and institutionalize their status as professionals (cf. Greiner and Ennsfellner, 2010 for this discussion about management consultants).

Attempting to define what we mean by "professionals" as a specific part of the workforce, the starting point is a sociological approach. Freidson (1987) describes "professions" as

> (...) occupations especially distinguished by their orientation to serving the needs of the public through the schooled application of their unusually esoteric knowledge and complex skill. (p. 19)

This description underlines the fact that professionals are characterized as holding expertise in a specific field (von Nordenflycht, 2010). Traditionally members of long-established professions are called professionals such as judges and jurists, university professors, clergymen, physicians etc. but also dentists, architects, engineers and in recent decades also accountants (Bürger, 2005). Løwendahl (2005) gives a list of qualitative attributes which characterize professionals in a broader sense (p. 28):

- Members of a highly professionalized occupational group
- Higher education
- Emphasis on application and improvement of knowledge
- Respect for professional norms of behavior, including altruistic problem solving for the client, affective neutrality and the limitations of professional expertise
- Respect for and willingness to participate in peer reviews and sanctions.

Greiner and Ennsfellner (2010) describe the professional as

> any employee engaged in work predominantly intellectual and varied in character as opposed to routine mental, manual mechanical, or physical work; involving the consistent exercise of discretion and judgement in its performance; . . . requiring knowledge of an advanced type . . . customarily acquired by a course of specialized intellectual instruction and study in an institution of higher learning (p. 73 f.)

The authors also describe high standards of excellence – professionals being experts – and advanced education and training as features of professional workers. May et al. (2002) add another important aspect: professionals are usually highly committed to their tasks and to the solution of complex problems so that their work itself seems to be rewarding. Along with this goes their pursuit of personal development and ongoing professional challenge (Bürger, 2005).

These characteristics lead to a specific picture of professional workers. Regardless of their specific branch, professionals can be characterized as an occupational group with *high education, expertise and specialization* and *high professional and organizational commitment* (Løwendahl, 2005; von May et al., 2002; Nordenflycht, 2010) usually related to high work satisfaction, high work involvement and long working hours. This job profile has several implications for professionals' work organization and demands, some of which are impeding and some enhancing the satisfactory integration of an individual's work with other life domains such as family, friends, time for self etc (Friedman, 2008).

1.2 Conditions of Professionals' Work–Life Integration

In times of increasingly flexible careers and high job demands, many employees may find it challenging to integrate work with the rest of their lives. However, we affirm that professionals face specific conditions which imply specific barriers and also some advantages concerning work–life integration. Before examining some of these specific conditions we cast a very brief look at the status quo of research on the work–life conflict and integration of professionals.

A number of studies specifically investigate the work–life interface of professionals or those job profiles that we categorize as such. Most studies find that professionals face moderate to high work–family/work–life conflict (Ahuja et al., 2007; Bacharach et al., 1991; Bedeian et al., 1988; Casper et al., 2002; Greenhaus et al., 2003; Kasper et al., 2005; Kossek et al., 2006; Leiter and Durup, 1996; Lyness and Brumit Kropf, 2005; Taylor et al., 2009; Thomas and Ganster, 1995; Thompson et al., 1999; Wharton and Blair-Loy 2006 etc.). Usually it is found that the work-to-life conflict dimension is stronger than the life-to-work conflict dimension. Further factors that influence the perception of role conflict are typically gender (Duxbury and Higgins, 1991; Perrone, 2009) but also individual characteristics such as cultural background (Wharton and Blair-Loy, 2006) or personality factors (Beauregard, 2006). Not surprisingly dual-career couples with children are most affected by work–family conflict (Duxbury and Higgins, 1991).

To our knowledge no study explicitly compares professionals with non-professionals in terms of work–life issues. Frone et al. (1992) may present an exception here reporting differences in the work–family conflict of blue- vs. white-collar workers. According to the study, white-collar workers' job involvement was positively related to work–family conflict in contrast to blue-collar workers. For blue-collar workers work–family conflict was positively related to family distress, not so for white-collar workers. Nevertheless we cannot infer from the study whether we can categorize white-collar workers as professionals as for example administrative jobs also fall into the white-collar category. Moore (2007) compared workers and managers in a multi-national company and found that achieving work–life balance is easier for workers, although most of the company's work–life balance initiatives focus on the managers and the managers display greater loyalty to the company.

On the basis of these few comparative data we cannot infer that professionals generally perceive a stronger conflict between work and non-work domains. We do, however, argue that the integration of work and life is indeed a pressing issue for professionals and their organizations because of the conditions that characterize their work and the ideas and priorities professionals have.

1.2.1 Work Organization and the Meaningfulness of Work

The work demands and work organization of professionals are probably the greatest challenge for work–life integration together with the highly salient work role of these individuals. Due to their high commitment to the organization and even more to their profession (May et al., 2002) on the one hand and high work load and responsibility on the other, professionals usually work longer than the average employee. Work hours of 50–60 or more per week appear to be commonplace (Blair-Loy, 2009; Brett and Stroh, 2003; Hewlett and Luce, 2006; Perlow and Porter, 2009; Schor, 1998), often with the tendency to increase rather than decrease. With the development of the 24/7 economy, working times of professionals do not fit into the "classic" 9–5 pattern (Burke, 2009). Often professionals work overtime, refrain from using their vacation time (or work during the holiday), and work during the lunch break (Brett and Stroh, 2003; Burke, 2009; Perlow and Porter, 2009).

If not as a function of their high commitment, professionals also perceive long working hours as a necessity for their career advancement with "face-time" still appearing to be the primary indicator of performance (Milliken and Dunn-Jensen, 2005; Brett and Stroh 2003). The fact that individual performance and career dedication is associated primarily with presence at the workplace is anchored in the implicit culture of numerous organizations. Many companies do tread a new path here (Perlow and Porter 2009), yet still organizational cultures not only set great store on employee presence but even sanction employees who pursue successful work–life integration: employees who make use of policies supporting the integration of work with private life responsibilities such as leaves of absence or flexible working hours often experience a negative impact on performance ratings and promotions (Blair-Loy and Wharton, 2004; Dikkers et al., 2007; Judiesch and Lyness, 1999; Thompson et al., 1999). Beyond the negative career consequences generally imposed top-down the role of peer support should not be underestimated (Kossek et al., 1999). Working in environments that epitomize an *always-on* ethic (Perlow and Porter, 2009, p. 102) professionals frequently have a hard time "admitting" an interest in anything other than work. Therefore many professionals refrain from using supportive initiatives in favor of their career advancement and social status quo (Brett and Stroh, 2003).

However, despite all the pressure imposed on professionals, very often these individuals work long hours because they want to. As already mentioned professionals are usually highly committed to their profession. Hochschild (2000) raised the issue that work life can be more satisfying than home life in her bestselling book "The Time Bind". There is empirical evidence that supports this notion (Brett and Stroh, 2003; Hewlett and Luce, 2006). Reasons for this phenomenon are challenging tasks, rewards, opportunities for individual development and a sense of accomplishment. The workplace has somewhat become a place of intrinsic reward and social recognition, particularly if work tasks are non-routine, complex tasks requiring highly sophisticated skills and expertise, as is the case in most professional work, leading to a feeling of uniqueness and irreplaceability. Especially for women, as Hewlett and Luce (2006) found, receiving recognition for their work seems to be a major factor for "loving their job" as they put it. But also power and status associated with the job leads to this highly salient role of work in professional people's lives.

Professionals seem to find themselves in a dilemma between the external pressure put on them and the self-dedication to their work as well as high values in the work and non-work domain (Cinamon and Rich, 2002). The feeling of having such a high status in the organization and, more generally, the highly salient role of the work is likely to increase the work-to-life conflict and spillover as, for example, the nature of work makes it harder to cut down work hours or use work–life policies (Hewlett and Luce, 2006; Brett and Stroh, 2003). Similarly individuals with high responsibility jobs often find it hard to reduce working hours in practical terms: *Employees who are already putting in long hours have few degrees of freedom available to make these adjustments* (Blair-Loy and Wharton, 2004, p. 260). Part of this dilemma is that professionals perceive their work as part of their self-concept and of "who they are". Professionals' work–life integration – often perceived as vocation or destiny –

thus can in many respects not be treated with the same standards of work–life research and practice as in other occupations (Wharton and Blair-Loy, 2006).

1.2.2 Changing Mobility of Work

Another characteristic of professional work which can be both a curse and a blessing for work–life integration is the changing mobility of the workplace, i.e., changes in the physical place where work is done as well as the increasing use of mobile technology.

Rather than working in a regular office many professionals work from a "mobile workplace" often meaning a notebook in a hotel room for 3 days per week or more. Many "flexpatriates" (e.g., Mayerhofer et al., 2008) associate their work organization with positive experiences such as novelty of work situations, challenges, lifestyle and personal development as well as negative outcomes such as stress, periods of separation from the family, health complaints, psychological strain etc. and, often as a consequence, with cutbacks on behalf of the family and social life as a whole (Mayerhofer et al., 2008; Welch and Worm, 2006). The fact that professional work is not necessarily bound to one physical workplace frequently leads to a blurring of the invisible boundaries between work and non-work domains (Kossek et al., 2006). Work-related tasks are taken home to be completed after dinner with the family even more so because professionals perceive their work not only as a functional (economic) necessity or obligation but as a pursuit of personal interests (Brett and Stroh, 2003). Consequently professionals often deal with work–life issues in an integrative manner, i.e., boundaries between work and other life domains are rather permeable: work tasks are completed at home; family responsibilities can be organized during the workday. This can be both a blessing and a curse: many professionals, especially knowledge workers, have the opportunity of choosing where and when to work and thus can take this as an advantage in their work–life management. On the other hand it becomes more difficult to draw the line between life domains and concentrate 100% on the responsibilities of one domain. This might also mean that, as work from anywhere is possible, availability for work matters is extended to the non-work domains.

A facet of the increasingly mobile work organization that affects boundary management is the use of mobile technology (Duxbury et al., 1996; Evans and Bartolomé, 1986; Major and Germano, 2006; Mayerhofer et al., 2008; Valcour and Hunter, 2005). Mobile (communication) technology can be beneficial for integrating work with private life as Jackson (2005) states: *Just as work and home are becoming portable, so are relationships becoming more virtual and 'anytime anywhere'* (p. 140). Although mobile communication technology can have positive effects also for private life, e.g., for flexpatriates' communication with their home, it primarily affects the professional's availability for work-related issues. Managers and professionals are expected to be accessible 24/7 for their employer and clients (Major and Germano, 2006) contributing again to blurred boundaries between work and non-work and potential work–life conflict (Jackson, 2005).

1.2.3 Individual Values and Life Concepts

Aside from the nature of professional work, two developments are also relevant to the work–life integration of professionals: life concepts and life role values (beyond work).

Unlike past generations the typical family constellation is shifting more and more from single-earner towards dual-earner couples and even more so dual-career couples, i.e., partnerships where both partners pursue an ambitious career path usually working full-time (Rapoport and Rapoport, 1969). Whereas men's focus has shifted more towards engaging in family and private responsibilities women demonstrate a higher interest and investment in the work role (Barnett, 2004b; Bond et al., 2002). In the last 40 years women have achieved increasingly higher educational levels and have progressively entered professional occupations in greater numbers. With the parallel tendency for educationally equal partnerships, a dual-career couple work–life constellation is particularly a challenge that many professionals face. Individuals but also employing organizations have to react to this challenge and build ways of creating a satisfactory work–life interface in order to recruit and retain highly educated employees – especially women, as it appears (Barnett, 2004a; Tiedje, 2004) – ideally with a win–win outcome.

To some extent a side effect of the increasing focus on a dual agenda is that both the work role and the roles of family and private life are highly salient to these individuals: *Some individuals, tired of the fast-paced corporate grind, are getting off the fast track or declining promotions in order to spend more time with friends and family or to focus on self-development* (Sullivan and Mainiero, 2007 p. 239). With this societal change in values – a shift towards higher prioritization of family and private life – research has started to examine the role of individual priorities and role salience for work–life integration (Bagger et al., 2008; Carlson et al., 2000; Cinamon and Rich, 2002; Perrewe et al., 1999). The results indicate that the perception of work–life conflict is influenced by the value attributed to life roles. For example Cinamon and Rich (2002) showed that work–family conflict is highest for those with both high work and family importance attributes. This trend of changing values affects the organization in away that it has to consider employee needs in order to maintain employer attractiveness and retain employees. Burke (2009) puts it as follows:

> Effective and high performing organizations meet the needs of their people and the needs of the business simultaneously. Healthy employees in healthy organizations are more likely to achieve peak performance. (p. 170)

Although the majority of employees will face work–life integration questions at some point in their lives, professionals face specific challenges some of which are externally imposed and others are due to their own decisions. How individuals and organizations alike can benefit from satisfactory work–life integration and what can be done from both perspectives to increase it will be summarized in the following two sections.

1.3 Individual and Organizational Benefits of Successful Work–Life Integration

In the face of the challenges put upon professionals, organizations have recognized the necessity to introduce arrangements that satisfy both the organization's and individual's needs in order to maximize productivity and effectiveness. However, in times of tight HR budgets organizations are also under the pressure of legitimizing work–life integration arrangements.

Successful work–life integration has been shown to be related to several benefits for the individual, usually as a function of reduced conflict between work and non-work domains and/or positive role enhancement. Individual benefits can be reduced physical and psychological health problems, domain-specific and overall satisfaction and happiness with life, reduced stress and individual performance in work and non-work domains.

Organizations do profit from these individual outcomes in terms of intangible benefits, i.e., outcomes desired by the organization that are hard to measure and hard to influence and tangible benefits, i.e., desired organizational outcomes that can be expressed in facts and figures (Fig. 1.1). Satisfactory physical and psychological health, reduced stress and life satisfaction have been shown to yield intangible positive effects for the organization in terms of lower employee turnover intention, higher organizational commitment and higher individual performance. Independent of individual benefits the organization might experience positive effects concerning external communication and reputation by introducing work–life integration initiatives, for example by means of audits for work–family friendliness (Becker, 2004; Schneider et al., 2007). A number of studies have also taken a look at the business case of fostering employee work–life integration. Tangible organizational benefits such as organizational performance, reduced absenteeism and employee retention can result from successful work–life integration and related individual benefits. Beyond this, organizations that support employees in their work–life integration benefit from this in terms of employer branding and recruitment advantages (Harrington and Ladge, 2009).

In the following sections the individual and organizational benefits of successful individual work–life integration will be outlined briefly.

1.3.1 Individual Benefits

Although some studies suggest that work–life integration is not necessarily sought – see literature about work being an emotional refuge from home (Brett and Stroh, 2003; Edwards and Rothbard, 2000; Hochschild, 2000) – based on the much larger portion of studies it can be assumed that for the majority of individuals the achievement of work–life integration is desired and yields positive outcomes for the individual. Particularly in the context of changing life role values and the increasing frequency of dual-earner and dual-career relationships, work–life integration becomes both an objective and a necessity, even more so for professionals.

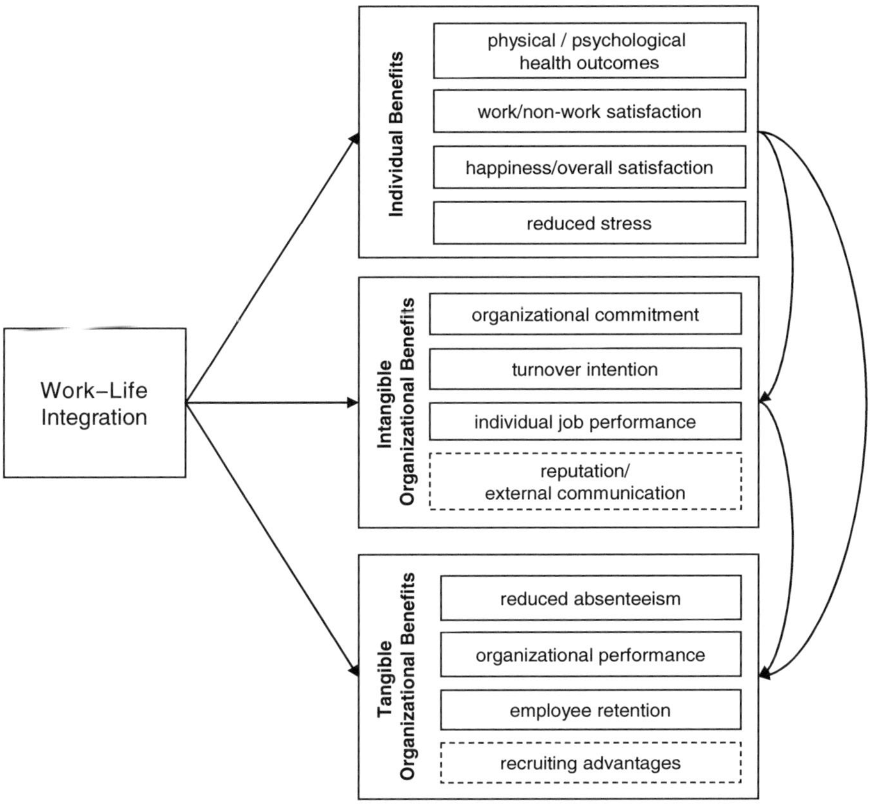

Fig. 1.1 Conceptual outcomes of successful work–life integration: individual benefits and organizational (intangible and tangible) benefits of work–life integration

Numerous studies examined the impact of work–life integration and balance on personal work and non-work outcomes and stress and health outcomes. Successful work–life integration is associated with positive outcomes for physical and psychological health, domain-specific and overall satisfaction as well as reduced stress.

1.3.1.1 Physical and Psychological Health Outcomes

Most research on the relationship between work–life integration and health outcomes focuses on the presence or absence of work–life conflict. Among others Jacobshagen et al. (2005) examined top managers' work–life conflict and found that it mediates the relationship between job stressors (overload, insecurity, work hours) and psychosomatic complaints. Burke and his colleagues pointed out in a number of publications how long work hours and work intensity negatively affect individuals' health (Burke, 2003; Burke, 2004; Burke and Cooper, 2008). Successful work–life integration is related to reduced physical and psychological complaints such as illnesses, burnout (Bianchi et al., 2005; Peeters et al., 2005) and depression but also reduced substance abuse (drugs, alcohol). In a study of employed parents Frone et al. (1997) found longitudinal evidence for the link of work-to-life conflict with depression, physical health complaints and hypertension and the link between family-to-work conflict and greater alcohol consumption.

It is also shown in several meta-analyses (Allen et al., 2000; Eby et al., 2005) that work–life conflict is very often related to negative health outcomes. In other words successful work–life integration should, to some extent, prevent individuals from severe illnesses that result from working long hours and experiencing work stress and the stress that results from struggling with integrating work and life.

1.3.1.2 Domain-Specific Satisfaction, Role Quality and Overall Satisfaction

Usually successful work–life integration is related to higher life, marital, and family satisfaction (Mauno and Kinnunen, 1999). Aryee et al. (1999) showed that lower and middle managers perceived conflict between life roles as related to lower life and family satisfaction (cf. also Bedeian et al., 1988). Perlow and Porter (2009) report similar positive results from an investigation at an international consulting firm. Management consultants experienced higher work and non-work role quality after more predictable work schedules and days off work were introduced into the previously very unpredictable working schedules. Hewlett and Luce (2006) show in their study of extreme workers that professionals who reduced their work time or took some time off were more satisfied with their life domains although reducing hours was perceived as very difficult. Similar results that indicate that work–life integration is linked to job, life and overall satisfaction can be found for different groups of professionals (health professionals: Thomas and Ganster, 1995; executives: Judge et al., 1994; dual-career professionals: Duxbury et al., 1996; IT personnel: Chang, 2008). The two meta-analyses of Allen et al. (2000) and Kossek and Ozeki (1998) also support this evidence.

Greenhaus et al. (2003) examined professional accountants' work–family balance and quality of life. They found that individuals who spent comparatively more time on family than work experienced higher quality of life than individuals who engaged in both life domains equally and who engaged more in the work domain. Bedeian et al. (1988) examined a similar group of employees with comparable results: work–family conflict was related to lower overall life satisfaction and marital satisfaction.

We can see that achieving the status of satisfactory engagement in work and non-work domains yields outcomes for individual job satisfaction, marital and family satisfaction and overall life satisfaction and happiness. These outcomes, particularly job satisfaction and overall life satisfaction will, in turn, have positive effects on organizational outcomes such as commitment or intent to leave the organization.

1.3.1.3 Reduced Stress

The perception of work–life conflict is very often associated with stress which in turn results in negative physical and psychological health outcomes. A large body of literature suggests that reducing work–life/work–family conflict results in decreased stress perception and thus avoidance of negative health outcomes, not only but even more so for professionals (Allen et al., 2000).

Several studies exist which specifically examine professionals' link between stress and work–life conflict. The results of a study of "IT road warriors" (Ahuja et al., 2007) – IT professionals who spend most of their workweek away from home at client sites – describe the positive relationship of work–family conflict with (work) exhaustion, i.e., achieved work–life integration should be related to less work exhaustion. The results suggest that work–family conflict is a key source of stress among IT road warriors because they have to juggle family and job duties as they work at distant client sites during the week. Data from married professional women in dual-career families in Singapore (Aryee, 1992) as well as other international data (Kinnunen and Mauno, 1998; Lu et al., 2009) widely confirm the positive effects of successful work–life integration on reducing stress factors and increasing well-being. Hewlett and Luce (2006) showed in a more descriptive approach that a lack of work–life integration is linked to stress and related outcomes.

That successful work-life integration yields benefits for the individual is a well-researched phenomenon. Individual characteristics, personality and life stage do influence the perception of work–life conflict and related outcomes but in general every individual strives for a – however shaped – solution of integrating work and non-work roles to fit his/her individual needs. If this is successfully achieved, individuals usually perceive a high quality of life in all domains, avoid negative health outcomes and feel less stressed.

1.3.2 Intangible Organizational Benefits

Although there is some literature emerging about the "business case" of work–life integration and related organizational initiatives the monetary or tangible impact of fostering individual work–life integration is still difficult to measure. Among the well-researched intangible benefits for the organization are desired behaviors such as low employee turnover intention and employee attitudes such as high organizational commitment.

1.3.2.1 Organizational Commitment, Professional Commitment, and Employee Turnover Intention

As professionals are highly specialized employees with expertise in a certain area they are not easily substituted. An organization thus has a major interest in retaining its employees, i.e., upholding high organizational commitment and reducing the intent to leave the organization. Organizational employee turnover is associated with time and costs that must be invested in an adequate substitute and is therefore a generally undesirable outcome for organizations. Even professional organizations that operate under an up-or-out practice will try to avoid losing employees that hold important knowledge and competencies and close client relations in order to retain their competitive advantage (Connor et al., 1999; Wright and Kehoe, 2009).

Professional work is, as we have learned earlier, associated with a demanding work organization and a certain lifestyle. Often when life circumstances change it is the professional commitment that is lowered because the work done is associated with limitations imposed on other life domains. Thus professional commitment, i.e., *[...] the relative strength of identification with and involvement in one's profession* (Morrow and Wirth, 1989) is also an issue for work–life integration questions.

The relationship between work–life conflict and organizational commitment has been researched intensively. In a sample of IT professionals Ahuja et al. (2007) were able to show a significant negative correlation between the two constructs (–0.32, $p < 0.01$), but were unable to prove causal links, i.e., that an increased work–life conflict leads to a reduced (affective) organizational commitment. Allen (2001), who interviewed 522 employees from different branches and organizations, was also able to support the postulated negative relationship. Similar results supporting the link between lower work–life conflict and increased organizational commitment were obtained by other studies (Aryee et al., 2005; Casper et al., 2002; Chang, 2008; Dex and Scheibl, 2001; Smith and Gardner, 2007) as well as the meta-analysis of Kossek and Ozeki (1999). Carr et al. (2008) found role centrality to be a moderator between work–life conflict and commitment. Their results indicate that if individuals view work as being more central to their lives, the negative relationships between work–family conflict and organizational attitudes and employee turnover intention is suppressed.

Concerning professional commitment, many employees leave their employer because their profession has become incompatible with their life priorities. Anecdotal evidence suggests that professional commitment and professional turnover are related to work–life conflict and work–life integration success. There are studies about the link between professional commitment and stress as outcomes (Kibeom et al., 2000; Lu et al., 2007; Sawada, 2009; Smith and Hall, 2008; Yongqing, 2006) and, for example, educational level and job autonomy (Giffords, 2009; Nogueras, 2006). However, there are no studies that specifically investigated this link. Although there is not much that organizations can do about professional commitment and employee turnover on the individual level, this is a phenomenon that organizations should monitor in order to keep track of such developments and to retain their own competitive position for recruiting employees.

1.3.2.2 Employee Performance

If an organization fosters individual work–life integration a central expectation is a positive impact on individual work performance. Professionals experiencing time stress or psychological stress because of their high workload and maybe because of the demands of private life that they cannot fulfill satisfactorily, will be likely to suffer from low concentration and in the end lower productivity. Thompson et al. (1999) put it this way:

> By not fostering a more balanced work–family life for employees, organizations are contributing to tensions in the employees' personal lives, the repercussions of which affect employees' ability to concentrate and be productive and creative on the job.

There is empirical evidence that supports this notion. Galinsky and Stein (1990) present indicators in their literature review that work–life balance initiatives improve employee productivity. Bloom and van Reenan (2006) found that work–life balance outcomes are significantly associated with better management, i.e., well-run firms are both more productive and offer better conditions to their employees. Better work–life balance practices are linked to higher individual productivity but this relationship becomes insignificant when compared with the standard for the overall quality of management. Others also report support for the relationship of work–life integration and performance. Based on social exchange theory (Muse et al., 2008), providing work–life benefits that employees use and/or value was positively associated with employees' feelings of perceived organizational support and affective commitment to the organization as well as reciprocation in the form of higher levels of task and contextual performance behaviors. These results occur regardless of the actual use of these programs, sometimes perception of their availability seems to suffice (cf. also Smith and Gardner, 2007).

Performance and productivity are very hard to measure, particularly in professional work as the output is very often intangible. However, if we think in terms of increased concentration and focus, as well as reduced stress and related illness symptoms and lower absenteeism, we can easily imagine how satisfactory work–life integration positively impacts professional employees' performance.

1.3.2.3 Organizational Reputation and External Communication

There is not much empirical evidence about the influence of organizational work–life integration-friendliness and organizational reputation and external communication. However we can observe a strong trend towards organizational attempts to present themselves as family-friendly or in more general terms of being considerate with respect to employees' work–life balance. Particularly in firms which demand extraordinary performance and engagement, for example management consultancies, work–life issues appear on the web sites of almost every large company. We argue that these organizations have recognized the trend towards increased attention for employees' work–life balance aspects, not only in order to improve recruitment but also to communicate a general message of organizational consideration of employees' private life needs.

Additionally we can see that family-friendly audits and programs enjoy increasing attention and usage (Becker, 2004; Schneider et al., 2007). These programs cannot only be seen as instruments for improved recruitment of employees but also as a general form of communication and increasing organizational reputation.

1.3.3 Tangible Organizational Benefits

Among the more tangible and measurable benefits of an organization that pays attention to individual work–life integration are the issues of absenteeism, organizational performance, employee retention as a function of high organizational commitment and low employee turnover intention (see above) and better recruitment opportunities in days of an increasing “war for talents”.

1.3.3.1 Reduced Absenteeism

A natural consequence of reduced employee stress, reduced physical and psychological health problems and high job and life satisfaction is reduced absenteeism. Obviously, organizations want to keep their absenteeism as low as possible in favor of performance. Several research works report that a work–life balance friendly firm that offers respective initiatives and epitomizes a supportive culture notices reduced absenteeism among employees (Boyar et al., 2005; Dex and Scheibl, 1999; Galinsky and Stein, 1990; Halpern, 2005). Van Steenbergen and Ellemers (2009) supported this link in a longitudinal study, work–life integration being positively linked to lower sickness absence as well as better physical health and higher objective job performance.

There is no specific evidence in the field of professionals and in an *always-on* culture it may be questioned whether employees will afford absenteeism because of work–life conflict and related stress symptoms. However, because health consequences of work–life integration problems and related stress are often long-term (burnout, depression, etc.), organizations should correspondingly follow a long-term oriented strategy and avoid absenteeism as a the first stage of employee turnover.

1.3.3.2 Organizational Performance

Organizational performance can again be seen as a function of several factors on the individual level (performance, commitment, job satisfaction, etc.) and the organizational level (low absenteeism, employee retention, organizational reputation). Most publications about the business case for work–life support of employees focus on these qualitative aspects because the quantitative factors are hard to encompass in the relationship with individual work–life integration (Beauregard and Henry, 2009; Connor et al., 1999; Konrad and Mangel, 2000; Perry-Smith and Blum, 2000). Beauregard and Henry (2009) present a comprehensive model of the mechanisms that link work–life integration and organizational performance arguing that reduced work–life conflict improves productivity which in turn saves costs. Further contributing factors are organizational citizenship behavior, improved job-related attitudes, improved retention and recruitment. Particularly in firms where organizational performance is difficult to measure as a construct this seems to be a valuable approach to visualize why organizations should care about employee work–life integration.

1.3.3.3 Employee Retention and Improved Recruiting

As already mentioned before successful work–life integration, or in other words, the state of not having to worry too much about other things while at work, also lowers employee turnover intention. Employee turnover is associated with high tangible and intangible costs for the organization. There will be costs from the time the person leaves (overtime rates of temporary employees to fill the vacant position etc.) until the time that the new employee actually provides a return on investment (recruitment costs, training costs, etc.) for an organization. This also means the longer an employee stays in the firm, the higher is the return on the investment in training of this employee (Dex and Scheibl, 1999). Beyond that in knowledge-intensive work, which most professional work is, knowledge and competencies are lost (Dex and Scheibl, 1999) and often client relations as is the case in professional service firms (Connor et al., 1999). This leads to a potential competitive disadvantage. Therefore even organizations that operate on an up-or-out principle will have some interest in the retention of highly-qualified and experienced employees.

With regards to recruitment, work–life integration issues seem to become a more and more prominent factor in employer choice, particularly in the face of the "war for talents" and demographic change (Connor et al., 1999; Johnson, 2004). Dex and Scheibl (2001) find recruitment advantages among the reasons for implementing work–life balance initiatives. Numerous inquiries report that work-life balance is a very important criterion for employer choice among students (Anderson and Kelliher, 2009; Human Resources Development Canada, 2002; Johnson, 2004; Scheltwort, 2004). With changing individual values of wanting both work and private life, and life concepts shifting towards dual-earner and dual-career couples, facilitating work–life integration has become, and will become, more and more a criteria for employee recruitment and retention.

1.4 Leveraging Professionals' Work–Life Integration: Increasing Person–Environment Fit

Now that we have taken a look at the individual and organizational benefits of satisfactory work–life integration in general and in the professional context in particular, the resulting question is: how can individuals and organizations create conditions that leverage work–life balance? What needs to be done from the individual perspective and the organizational perspective to maximize individual and organizational outcomes?

Achieving work–life integration is an issue where individuals need support on the one hand but which cannot be the sole responsibility of an employing organization on the other. Cooperation of the organization and the individual and mutual compliance are required for finding the right solutions for the right persons.

Drawing on previous research and anecdotal evidence of professionals we want to summarize some suggestions of how individuals and organizations in general and in professional work in particular can leverage work–life integration (cf. Fig. 1.2):

- Communicating individual needs is essential for successful work–life integration. Individual differences are so diverse that one-size-fits-all work–life initiatives do not cover a person's needs satisfactorily. Individuals must communicate their

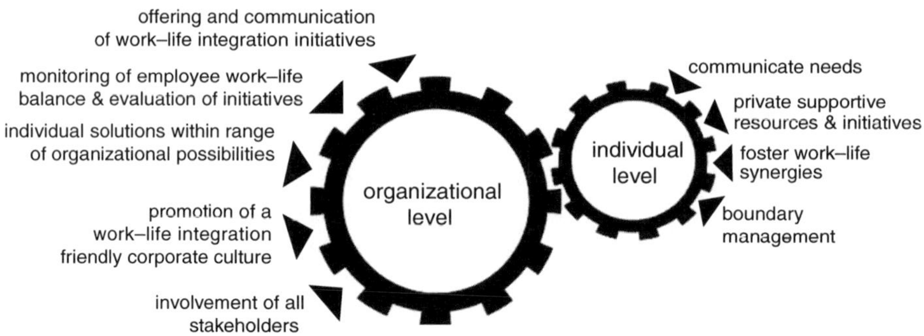

Fig. 1.2 Leveraging work–life integration on behalf of the organizational and individual level

specific situational needs in order to achieve the best possible work–life integration which the organization also benefits from, as we have seen. As Edmondson and Detert (2005) show individuals have to speak up in order to find consideration for their work–life integration concerns.

- The offer of work–life initiatives is not a new approach to dealing with work–life issues of employees. However, if such initiatives are offered very often their availability is not well communicated. Research has shown that the mere availability of work–life initiatives levers job satisfaction, organizational commitment etc. (Smith and Gardner, 2007). Several studies also give evidence that work–life policies such as flexible working times or part-time working are indeed successful in the professional context (Hill et al., 2004; Hill et al., 2006; Lee et al., 2002; Perlow and Porter, 2009).
- Organizations should monitor the work–life balance of their employees, particularly in highly demanding work environments where work–life conflict can cause serious health problems. Additionally, organizations should evaluate their work–life initiatives in terms of usage and effectiveness (Hill and Weiner, 2003). This increases the usefulness for employees and decreases costs for the organization. Moreover the promotion of pseudo work–life balance initiatives, i.e., organizational practices that do not necessarily increase work–life integration such as "Office Fridays", or "Home Office" in the case of management consultants, can be avoided and employers can focus on what their professionals really need.
- Instead of offering standard work–life integration initiatives organizations should, after evaluating the needs of their employees, tailor individual, flexible solutions within the range of the organizational framework. Work–life initiatives should also go beyond facilitating the care for dependants but consider individual needs of all employees (Casper et al., 2007; Lobel and Kossek, 2000).
- Individuals can do their share by using private initiatives and supportive services (household support, private childcare, cleaning services, etc.) in addition to organizational initiatives in order to increase the quality of their private life and decrease stress.

- Growing research evidence supports the notion that a work–life integration friendly corporate culture is necessary to make individual work–life integration possible and to allow employees to use respective initiatives (Casper et al., 2007; Connor et al., 1999; Dikkers et al., 2007; Peper et al., 2005; Thompson et al., 1999; Smith and Gardner, 2007). Particularly in highly demanding work environments the use of initiatives is often associated with negative impacts on career advancement and social recognition in the organization, extremely long work hours often seem to be part of the professional identity. Changes in the heads of the management, supervisors and co-workers are necessary to successfully implement work–life integration policies that are usable and actually used.
- Research indicates that besides the stress component of work–life conflict there is also a positive perspective to the interaction of work and non-work domains: work–life synergies and positive spillover (Ilies et al., 2009; van Steenbergen et al., 2007). Although not much evidence exists on how to actively foster this positive component we suggest that employees and particularly professionals should focus on tangible and intangible advantages that work offers for non-work domains and vice versa.
- Particularly in professional service firms where clients expect 24/7 availability of the professional, it is important to involve all stakeholders in work–life integration communications to reduce costs and still to satisfy all stakeholders' needs (Hyland and Jackson, 2006; Perlow and Porter, 2009). Part-time projects or job-sharing should be options to be communicated to the client as well as teleworking options to save on time lost with travel.
- Finally, individuals should become aware of their boundary management strategy, i.e., their preference either to keep work and non-work domains strictly separate or to integrate domains as tasks and responsibilities arise (Ashforth et al., 2000; Bulger et al., 2007; Kirchmeyer, 1995; Rothbard and Dumas, 2006). Once their preference on the integration–segmentation continuum is decided they should elaborate on respective organizational resources. Kirchmeyer (1995) found for example, that for the managers studied a strategy of flexible boundaries between work and private life was most effective. An effective boundary management strategy is particularly important for professionals who are obliged to use communication technologies extensively in their work (Chesley, 2005; Valcour and Hunter, 2005).

1.5 Conclusion

In this chapter we were trying to highlight the issue of work–life integration of professional workers. We understand professionals as an occupational group with high education, expertise and specialization and high professional and organizational commitment whose members work long hours and are highly involved in their work. Usually professionals work in highly demanding and highly competitive work environments that, besides some benefits and a certain lifestyle, are very often associated with cutbacks in other life domains. Therefore professionals are considered to be individuals for whom successful work–life integration is a particularly challenging endeavor.

We pointed out several conditions that professionals face for integrating their work and life domains as well as benefits for the individual and the organization if work–life integration is achieved.

Drawing general conclusions on the work–life integration of professionals is difficult because "professionals" are, as already mentioned at the beginning of the chapter, a very diverse group with high variations in job demands, job autonomy, working hours, work values etc. Accordingly there can be no overall implications for the group of professional workers as a whole. We would like to encourage further research to delve deeper into work–life research among workers with extreme jobs and to investigate similarities and disparities among professions.

References

Ahuja MK, McKnight DH, Chudoba KM, George JF, Kacmar CJ (2007) IT road warriors: balancing work–family conflict, job autonomy, and work overload to mitigate turnover intentions. MIS Q 31(1):1–17

Allen TD (2001) Family-supportive work environments. The role of organizational perceptions. J Vocat Behav 58(3):414–435

Allen TD, Herst D, Bruck CS, Sutton M (2000) Consequences associated with work-to-family conflict: a review and agenda for future research. J Occup Health Psychol 5(2): 278–308

Alvesson M (2000) Social identity and the problem of loyalty in knowledge-intensive companies. J Manage Stud 37(8):1101–1124

Anderson D, Kelliher C (2009) Flexible working and engagement. The importance of choice. Strateg HR Rev 8(2):13–18

Aryee S (1992) Antecedents and outcomes of work–family conflict among married professional women. Evidence from Singapore. Hum Relat 45(8):813–837

Aryee S, Fields D, Luk V (1999) A cross-cultural test of a model of the work–family interface. J Manage 25(4):491–511

Aryee S, Srinivas E, Tan HH (2005) Rhythms of life. Antecedents and outcomes of work–family balance in employed parents. J Appl Psychol 90(1):132–146

Ashforth BE, Kreiner GE, Fugate M (2000) All in a day's work. Boundaries and micro role transitions. Acad Manage Rev 25(3):472–491

Bacharach SB, Bamberger P, Conley S (1991) Work–home conflict among nurses and engineers. Mediating the impact of role stress on burnout and satisfaction at work. J Organ Behav 12(1):39–53

Bagger J, Li A, Gutek BA (2008) How much do you value your family and does it matter? The joint effects of family identity salience, family-interference-with-work, and gender. Hum Relat 61(2):187–211

Barnett R (2004a) Women and multiple roles. Myths and reality. Harv Rev Psychiatry 12(3): 158–164

Barnett RC (2004b) Women and work. Where are we, where did we come from, and where are we going?. J Soc Issues 60(4):667–664

Beauregard TA (2006) Predicting interference between work and home. A comparison of dispositional and situational antecedents. J Managerial Psychol 21(3):244–264

Beauregard TA, Henry LC (2009) Making the link between work–life balance practices and organizational performance. Hum Resour Manage Rev 19(1):9–22

Becker SJ (2004) Audit Beruf & Familie. Familienbewusste Personalpolitik durch freiwillige Unternehmensinitiativen. [Work & family audit: family-friendly HR politics by means of voluntary corporate initiatives]. In Badura H, Schellschmidt B, Vetter C, Bäcker G, (eds) Fehlzeiten-Report 2003. Zahlen, Daten, Analysen aus allen Branchen der Wirtschaft. Wettbewerbsfaktor Work–life-Balance. Betriebliche Strategien zur Vereinbarkeit von Beruf, Familie und Privatleben. Springer, Berlin, pp 161–169

Bedeian AG, Burke BG, Moffett RG (1988) Outcomes of work–family conflict among married male and female professionals. J Manage 14(3):475–491

Bianchi SM, Casper LM, Berkowitz King R (eds) (2005) Work, family, health, and well-being. Lawrence Erlbaum Associates, Mahwah, NJ

Blair-Loy M (2009) Work without end?: scheduling flexibility and work-to-family conflict among stockbrokers. Work Occup 36(4):279–317

Blair-Loy M, Wharton AS (2004) Organizational commitment and constraints on work–family policy use: corporate flexibility policies in a global firm. Sociol Perspect 47(3): 243–267

Bloom N, van Reenan J (2006) Management practices, work–life balance, and productivity. A review of some recent evidence. Oxf Rev Econ Policy 22(4):457–482

Bond JT, Thompson C, Galinsky E, Prottas D (2002) Highlights of the national study of the changing workforce: executive summary. Fam Work Inst 3:1–6

Boyar SL, Maertz Jr, CP, Pearson AW (2005) The effects of work–family conflict and family–work conflict on nonattendance behaviors. J Bus Res 58:919–925

Brett JM, Stroh LK (2003) Working 61 plus hours a week: why do managers do it?. J Appl Psychol 88(1):67–78

Bulger CA, Matthews RA, Hoffman ME (2007) Work and personal life boundary management: boundary strength, work/personal life balance, and the segmentation–integration continuum. J Occup Health Psychol 12(4):365–375

Bürger B (2005) Aspekte der Führung und der strategischen Entwicklung von Professional Service Firms: Der Leverage von Ressourcen als Ausgangspunkt einer differenzierten Betrachtung [Leadership issues and aspects of strategic development of professional service firms: resource leverage as a starting point of a differentiated examination]. Dt. Univ.-Verl, Wiesbaden

Burke RJ (2003) Work experiences, stress and health among managerial women. Research and practice. In Schabracq MJ, Winnubst JAM, Cooper CL (eds) The handbook of work and health psychology. Wiley, Chichester, pp 259–278

Burke RJ (2004) Work and personal life integration. Int J Stress Manage 11(4):299–304

Burke R (2009) Working to live or living to work. Should individuals and organizations care?. J Bus Ethics 84(0):167–172

Burke RJ, Cooper CL (eds) (2008) The long work hours culture. Causes, consequences and choices. Emerald, Bingley

Carlson DS, Kacmar KM, Williams LJ (2000) Work–family conflict in the organization. Do life role values make a difference?. J Manage 26(5):1031–1054

Carr JC, Boyar SL, Gregory BT (2008) The moderating effect of work–family centrality on work–family conflict, organizational attitudes, and turnover behavior. J Manage 34(2):244–262

Casper WJ, Martin JA, Buffardi LC, Erdwins CJ (2002) Work–family conflict, perceived organizational support, and organizational commitment among employed mothers. J Occup Health Psychol 7(2):99–108

Casper WJ, Weltman D, Kwesiga E (2007) Beyond family-friendly. The construct and measurement of singles-friendly work culture. J Vocat Behav 70(3):478–501

Chang S-i (2008) Work role stressors and turnover intentions: a study of IT personnel in South Korea. Zeitschrift für Personalforschung 22(3):272–290
Chesley N (2005) Blurring boundaries? Linking technology use, spillover, individual distress, and family satisfaction. J Marriage Fam 67(5):1237–1248
Cinamon RG, Rich Y (2002) Profiles of attribution of importance to life roles and their implications for the work–family conflict. J Couns Psychol 49(2):212–220
Connor M, Hooks K, McGuire T (1999) Gaining legitimacy for flexible work arrangements and career paths. The business case for public accounting and professional service firms. In Parasuraman S, Greenhaus JH (eds) Integrating work and family. Challenges and choices for a changing world. Praeger, Westport, CT, pp 154–166
Dex S, Scheibl F (1999) Business performance and family-friendly policies. J Gen Manage 24(4):22–37
Dex S, Scheibl F (2001) Flexible and family-friendly working arrangements in UK-based SMEs: business cases. Br J Ind Relat 39(3):411
Dikkers JS, Geurts SA, den Dulk L, Peper B, Taris TW, Kompier MA (2007) Dimensions of work–home culture and their relations with the use of work-home arrangements and work–home interaction. Work Stress 21(2):155–172
Duxbury LE, Higgins CA (1991) Gender differences in work–family conflict. J Appl Psychol 76(1):60–74
Duxbury LE, Higgins CA, Thomas DR (1996) Work and family environments and the adoption of computer-supported supplemental work-at-home. J Vocat Behav 49(1):1–23
Eby L, Casper WJ, Lockwood A, Bordeaux C, Brinley A (2005) Work and family research in IO/OB. Content analysis and review of the literature (1980–2002). J Vocat Behav 66(1): 124–197
Edmondson AC, Detert JR (2005) The role of speaking up in work–life balancing. In Kossek EE, Lambert SJ (eds) Work and life integration. Organizational, cultural, and individual perspectives. Lawrence Erlbaum Associates, Mahwah, NJ, pp 401–427
Edwards JR, Rothbard NP (2000) Mechanisms linking work and family. Clarifying the relationship between work and family constructs. Acad Manage Rev 25(1):178–199
Evans P, Bartolomé F (1986) The dynamics of work–family relationships in managerial lives. Int Rev Appl Psychol 35(3):371–395
Freidson E (1987) The future of the profession. J Dent Educ 51(3):140–144
Friedman SD (2006) Learning to lead in all domains of life. Am Behav Sci 49(9):1270–1297
Friedman SE (2008) Total leadership. Be a better leader. Have a richer life. Harvard Business Press, Boston, MA
Frone MR, Russell M, Cooper M (1992) Antecedents and outcomes of work–family conflict. Testing a model of the work–family interface. J Appl Psychol 77(1):65–78
Frone MR, Russell M, Cooper M (1997) Relation of work–family conflict to health outcomes. A four-year longitudinal study of employed parents. J Occup Organ Psychol 70: 325–335
Galinsky E, Stein PJ (1990) The impact of human resource policies on employees. Balancing work/family life. J Fam Issues 11(4):368–383
Giffords ED (2009) An examination of organizational commitment and professional commitment and the relationship to work environment, demographic and organizational factors. J Soc Work 9(4):386–404
Greenhaus JH, Collins K, Shaw JD (2003) The relation between work–family balance and quality of life. J Vocat Behav 63:510–531
Greiner LE, Ennsfellner I (2010) Management consultants as professionals, or are they?. Organ Dyn 39:72–83
Halpern DF (2005) How time-flexible work policies can reduce stress, improve health, and save money. Stress Health: J Int Soc Invest Stress 21(3):157–168
Harrington B, Ladge JJ (2009) Work–life integration: present dynamics and future directions for organizations. Organ Dyn 38(2):148–157

Hewlett SA, Luce CB (2006) Extreme jobs. The dangerous allure of the 70-hour workweek. Harv Bus Rev Dec:49–59

Higgins C, Duxbury L, Johnson KL (2000) Part-time work for women: does it really help balance work and family?. Hum Resour Manage 39(1):17–32

Hill EJ, Weiner SP (2003) Work/life balance policies and programs. In Edwards JE, Scott JC, Raju NS (eds) The human resources program-evaluation handbook. Sage, Thousand Oaks, CA, pp 447–468

Hill EJ, Märtinson V, Ferris M (2004) New-concept part-time employment: a work–family adaptive strategy for women professionals with small children. Fam Relat 53:282–292

Hill EJ, Mead NT, Dean LR, Hafen DM, Gadd R, Palmer AA, Ferris MS (2006) Researching the 60-hour dual-earner workweek. An alternative to the 'opt-out revolution'. Am Behav Sci 49(9):1184–1203

Hochschild AR (2000) The time bind: when work becomes home and home becomes work. Holt, New York, NY

Human Resources Development Canada (2002). The business case for work–life balance

Hyland MA, Jackson SE (2006) A multiple stakeholder perspective. Implications for measuring work–family outcomes. In Pitt-Catsouphes M, Kossek EE, Sweet SA (eds) The work and family handbook. Multi-disciplinary perspectives methods, and approaches. Lawrence Erlbaum Associates, Mahwah, NJ, pp 527–549

Ilies R, Schwind Wilson K, Wagner DT (2009) The spillover of daily job satisfaction onto employee's family lives. The facilitating role of work–family integration. Acad Manage J 52(1):87–102

Jackson M (2005) The limits of connectivity: technology and 21st-century life. In Halpern DF (ed) From work–family balance to work–family interaction. Changing the metaphor. Lawrence Erlbaum Associates, Mahwah, NJ , pp 135–150

Jacobshagen N, Amstad FT, Semmer NK, Kuster M (2005) Work–family-Balance im Topmanagement [Work-family balance of top executives]. Zeitschrift für Arbeits- und Organisationspsychologie A&O 49(4):208–219

Johnson MD (2004) Life–Work balance should be the real issue in war for talent. volume 29-06-2004

Judge TA, Boudreau JW, Bretz RD Jr. (1994) Job and life attitudes of male executives. J Appl Psychol 79(5):767–782

Judiesch MK, Lyness KS (1999) Left behind? The impact of leaves of absence on managers' career success. Acad Manage J 42(6):641–651

Kasper H, Meyer M, Schmidt A (2005) Managers dealing with work–family-conflict. An explorative analysis. J Managerial Psychol 20(5):440–461

Kibeom L, Carswell JJ, Allen NJ (2000) A meta-analytic review of occupational commitment: relations with person- and work related variables. J Appl Psychol 85(5):799–811

Kinnunen U, Mauno S (1998) Antecedents and outcomes of work–family conflict among employed women and men in Finland. Hum Relat 51(2):157–177

Kirchmeyer C (1995) Managing the work–nonwork boundary: an assessment of organizational responses. Hum Relat 48(5):515–536

Konrad AM, Mangel R (2000) The impact of work–life programs on firm productivity. Strateg Manage J 21(12):1225–1237

Kossek EE, Ozeki C (1998) Work–family conflict, policies, and the job–life satisfaction relationship. A review and directions for organizational behavior – human resources research. J Appl Psychol 83(2):139–149

Kossek EE, Ozeki C (1999) Bridging the work–family policy and productivity gap. A literature review. Commun Work Fam 2(1):7–32

Kossek EE, Barber AE, Winters D (1999) Using flexible schedules in the managerial world: the power of peers. Hum Resour Manage 38(1):33–46

Kossek EE, Lautsch BA, Eaton SC (2006) Telecommuting, control, and boundary management: correlates of policy use and practice, job control, and work–family effectiveness. J Vocat Behav 68(2):347–367

Lee MD, MacDermid SM, Williams ML, Buck ML, Leiba-O'Sullivan S (2002) Contextual factors in the success of reduced-load work arrangements among managers and professionals. Hum Resour Manage 41(2):209–223

Leiter MP, Durup MJ (1996) Work, home, and in-between: a longitudinal study of spillover. J Appl Behav Sci 32(1):29–47

Litrico J-B, Lee MD (2008) Balancing exploration and exploitation in alternative work arrangements. A multiple case study in the professional and management services industry. J Organ Behav 29(8):995–1020

Lobel SA, Kossek EE (2000) Human resource strategies to support diversity in work and personal lifestyles: beyond the "family friendly" organization. In Kossek EE, Lobel SA, (eds) Managing diversity. Human resource strategies for transforming the workplace. Blackwell Business, Cambridge, MA, pp 221–244

Løwendahl BR (2005) Strategic management of professional service firms. CBS Press, Copenhagen

Lu K, Chang L, Wu H (2007) Relationships between professional commitment, job satisfaction, and work stress in public health nurses in Taiwan. J Prof Nurs 23(2):110–116

Lu J-F, Siu O-L, Spector PE, Shi K (2009) Antecedents and outcomes of a fourfold taxonomy of work–family balance in Chinese employed parents. J Occup Health Psychol 14(2):182–192

Lyness KS, Brumit Kropf M (2005) The relationships of national gender equality and organizational support with work–family balance: a study of European managers. Hum Relat 58(1):33–60

Major DA, Germano LM (2006) The changing nature of work and its impact on the work–home interface. In Jones F, Burke RJ, Westman M (eds) Work and organizational psychology. Work–life balance: a psychological perspective. Psychology Press, Hove, pp 13–38

Mauno S, Kinnunen U (1999) The effects of job stressors on marital satisfaction in Finnish dual-earner couples. J Organ Behav 20(6):879–895

May TY-M, Korczynski M, Frenkel SJ (2002) Organizational and occupational commitment: knowledge workers in large corporations. J Manage Stud 39(6):775–801

Mayerhofer H, Müller B, Schmidt A (2008) Working everytime and everywhere. Herausforderungen für Führungskräfte [Working everytime and everywhere. Challenges for leaders]. In Mühlbacher J, Kasper H (eds) Management development. Wandel der Anforderungen an Führungskräfte. Festschrift für Helmut Kasper zum 60. Geburtstag, Linde, Wien, pp 103–121

Milliken FJ, Dunn-Jensen LM (2005) The changing time demands of managerial and professional work. Implications for managing the work–life boundary. In Kossek EE, Lambert SJ (eds) Work and life integration. Organizational, cultural, and individual perspectives. Lawrence Erlbaum Associates, Mahwah, NJ, pp 43–59

Moore F (2007) Work–life balance: contrasting managers and workers in an MNC. Empl Relat 29(4):385–399

Morrow PC, Wirth RE (1989) Work commitment among salaried professionals. J Vocat Behav 34(1):40–56

Muse L, Harris SG, Giles WE, Feild HS (2008) Work–life benefits and positive organizational behavior: is there a connection?. J Organ Behav 29(2):171–192

Nogueras D (2006) Occupational commitment, education, and experience as a predictor of intent to leave the nursing profession. Nurs Econ 24(2):86–93

Nordenflycht Avon (2010) What is a professional service firm? Toward a theory and taxonomy of knowledge-intensive firms. Acad Manage Rev 35(1):155–174

Peeters MC, Montgomery AJ, Bakker AB, Schaufeli WB (2005) Balancing work and home: how job and home demands are related to burnout. Int J Stress Manage 12(1):43–61

Peper B, van Doorne-Huiskes A, Laura dD (eds) (2005) Flexible working and organisational change. The integration of work and personal life. Edward Elgar Publishing, Cheltenham

Perlow LA, Porter JL (2009) Making time off predictable & required. Harv Bus Rev 87(10): 102–109

Perrewe P, Hochwarter W, Kiewitz C (1999) Value attainment: an explanation for the negative effects of work–family conflict on job and life satisfaction. J Occup Health Psychol 4(4): 318–326

Perrone KM (2009) Traditional and nontraditional work and family roles for women and men. J Career Dev 36(1):3–7

Perry-Smith JE, Blum TC (2000) Work–family human resource bundles and perceived organizational performance. Acad Manage J 43(6):1107–1117

Rapoport R, Rapoport RN (1969) The dual career family: a variant pattern and social change. Hum Relat 22(1):3–30

Robertson M, Swan J (2003) "Control, what control...": culture and ambiguity within a knowledge intensive firm. J Manage Stud 40(4):831–858

Rothbard NP, Dumas TL (2006) Research perspectives: managing the work–home interface. In Jones F, Burke RJ, Westman M (eds) Work and organizational psychology. Work–life balance: a psychological perspective. Psychology Press, Hove, pp 73–89

Sawada T (2009) The relationship of occupational and organizational commitments with professional activities and burnout tendencies in nurses. Jpn J Psychol 80(2):131–137

Scheltwort S (2004) Ein Job fürs Leben [A job for a lifetime]. Junge Karriere 11:18–30

Schneider H, Juncke D, Gerlach I (2007). Betriebliche Familienpolitik in auditierten Unternehmen und Institutionen (Working Paper Nr. 3). Münster.

Schor JB (1998) The overworked American: the unexpected decline of leisure. Basic Books, New York, NY

Smith J, Gardner D (2007) Factors affecting employee use of work–life balance initiatives. N Z J Psychol 36(1):3–12

Smith D, Hall M (2008) An empirical examination of a three-component model of professional commitment among public accountants. Behav Res Account 20(1):75–92

Sullivan SE, Mainiero LA (2007) The changing nature of gender roles, alpha/beta careers and work–life issues. Theory-driven implications for human resource management. Career Dev Int 12(3):238–263

Taylor BL, Delcampo RG, Blancero DM (2009) Work–family conflict/facilitation and the role of workplace supports for U.S. Hispanic professionals. J Organ Behav 30(5): 643–664

Thomas LT, Ganster DC (1995) Impact of family-supportive work variables on work–family conflict and strain: a control perspective. J Appl Psychol 80(1):6–15

Thompson CA, Beauvais LL, Lyness KS (1999) When work–family benefits are not enough. The influence of work–family culture on benefit utilization, organizational attachment, and work–family conflict. J Vocat Behav 54(3):392–415

Tiedje BL (2004) Processes of change in work/home incompatibilities: employed mothers 1986–1999. J Soc Issues 60(4):787–800

Valcour MP, Hunter LW (2005) Technology, organizations, and work–life integration. In Kossek EE, Lambert SJ (eds) Work and life integration. Organizational, cultural, and individual perspectives. Lawrence Erlbaum Associates, Mahwah, NJ, pp 61–84

van Engen ML, Dikkers JSE, Vinkenburg CJ, Rooy EPde (2009) Career success of fathers and mothers: the impact of motherhood ideology, work–home culture, and work–home arrangements. Gedrag Organ 22(2):146–U1

van Steenbergen EF, Ellemers N (2009) Is managing the work–family interface worthwhile? Benefits for employee health and performance. J Organ Behav 30(5):617–642

van Steenbergen EF, Ellemers N, Mooijaart A (2007) How work and family can facilitate each other: distinct types of work–family facilitation and outcomes for women and men. J Occup Health Psychol 12(3):279–300

Wallace JE (1997) It's about time: a study of hours worked and work spillover among law firm lawyers. J Vocat Behav 50(2):227–248

Welch DE, Worm V (2006) International business travelers: a challenge for IHRM. In Stahl GK, Björkman I (eds) Handbook of research in international human resource management. Edgar Elgar Publishing, Cheltenham, pp 279–298

Wharton AS, Blair-Loy M (2006) Long work hours and family life: a cross-national study of employees' concerns. J Fam Issues 27(3):415–436

Wright P, Kehoe R (2009) Organizational-level antecedents and consequences of commitment. In Klein HJ, Becker TE, Meyer JP (eds) Commitment in organizations. Accumulated wisdom and new directions. Routledge, New York, NY, pp 285–307

Yongqing F (2006) Turnover propensity and its causes among Singapore nurses: an empirical study. Int J Hum Resour Manage 29(1):162–164

Chapter 2
Introducing Theoretical Approaches to Work-Life Balance and Testing a New Typology Among Professionals

Johanna Rantanen, Ulla Kinnunen, Saija Mauno, and Kati Tillemann

Clark (2000) defines work-family balance as "satisfaction and good functioning at work and at home, with a minimum of role conflict" (p. 751). In this chapter, we examine how professionals have succeeded in achieving work-life balance in their lives. First, we examine classic and current approaches to multiple roles and then introduce a typology of work-life balance based on the synthesis of the presented theoretical foundation. We propose four types of work-life balance; beneficial, harmful, active, and passive. The employees belonging to each type are expected to differ qualitatively from each other in relation to psychological functioning and role engagement. Second, we empirically investigate (a) how typical these four types of work-life balance are among three samples of professionals (Finnish university professionals, Finnish managers, and Estonian managers), and (b) whether professionals belonging to the different work-life balance types differ from each other in terms of their psychological functioning and work role engagement as expected according to the typology of work-life balance.

2.1 What is Work-Life Balance? A Glance at the Theoretical Background

2.1.1 Role Theories: The Foundation for Work-Life Balance

Work-family research has long been guided by the role stress theory, wherein the negative side of the work-family interaction has been put under the spotlight. Recently, the emphasis has shifted towards the investigation of the positive interaction between work and family roles as well as roles outside work and family lives, and scholars have started to deliberate on the essence of work-life balance (Jones et al., 2006). It should be noted that the term work-*life* is used throughout this chapter from here on as it is more comprehensive than the term work-*family*. However,

J. Rantanen (✉)
University of Jyväskylä, Jyväskylä, Finland
e-mail: k.johanna.rantanen@jyu.fi

S. Kaiser et al. (eds.), *Creating Balance?*, DOI 10.1007/978-3-642-16199-5_2,

when the work of other scholars is referred to, the terms *work-life* and *work-family* are used according to the citations.

It is generally agreed that work-life balance is important for an individual's psychological well-being, and that high self-esteem, satisfaction, and overall sense of harmony in life can be regarded as indicators of a successful balance between work and family roles (Clark, 2000; Clarke et al., 2004; Marks and MacDermid, 1996). However, there is a lack of consensus on how work-life balance should be defined, measured, and researched, and thus, the theorizing of what constitutes work-life balance, how it develops, and what factors enable or hinder it, is still in progress (Grzywacz and Carlson, 2007; Jones et al., 2006; Voydanoff, 2005). Greenhaus et al. (2003) have also questioned the self-evident assumption that work-family balance always leads to favorable outcomes since according to them this is an empirical question which has not yet been firmly answered due to miscellaneous definitions of work-family balance.

The origins of research on work-life balance can be traced back to studies of women having multiple roles. Barnett and Baruch (1985) investigated the psychological distress connected to the balance of rewards and concerns generated by individual women's multiple roles as paid worker, wife and mother. They found that positive role quality – more rewards than concerns experienced in a given role – was related to low levels of role overload, role conflict and anxiety. Based on their research, Barnett and Baruch defined role balance as a "rewards minus concerns" difference score which could range from positive to negative values.

Tiedje and her colleagues (1990) approached the same research question from the perspective of a typology of role perception. They argued that women may perceive their work and family roles in multiple, qualitatively different ways, and thus they based their typology on both the role conflict and enhancement hypotheses. According to the *conflict hypothesis*, multiple roles with infinite demands are likely to cause role strain and conflict for individuals because the resources they have to meet these demands are finite and scarce (Goode, 1960). The core statement of the *enhancement hypothesis*, in turn, is that multiple roles provide benefits in the form of privileges, status security, psychological energy and personal growth which expand individual resources and facilitate role performance (Marks, 1977; Sieber, 1974).

More specifically, Tiedje and colleagues (1990) regarded role conflict and role enhancement as independent dimensions, and therefore they argued that it is possible to experience simultaneously either (a) high conflict and low enhancement, (b) high enhancement and low conflict, (c) low conflict and low enhancement, or (d) high conflict and high enhancement. They found that regardless of the level of enhancement, women who experienced high role conflict were more depressed and less satisfied as parents than women belonging to the low conflict-high enhancement group. On the basis of studies by Barnett and Baruch (1985) and Tiedje and colleagues (1990), it may be concluded that high rewards and enhancement combined with low concerns and conflict experienced across the roles in one's life is beneficial for an individual's well-being, and hence these experiences characterise role balance.

However, Marks and MacDermid (1996) conceptualise balance quite differently. According to them, role balance is not an outcome but rather "both a behavioral pattern of acting across roles in a certain way and a corresponding cognitive-affective pattern of organizing one's inner life of multiple selves" (Marks and MacDermid, 1996, p. 421). Specifically, according to Marks and MacDermid (1996) there are two ways to engage multiple roles; as either positive or negative role balance. *Positive role balance*, in Marks and McDermid's theory (cf. Barnett and Baruch, 1985), refers to the tendency to engage in every role with equally high effort, devotion, attention and care, whereas *negative role balance* refers to the tendency to engage in roles with apathy, cynicism, low effort and low attentiveness.

Due to these behavioral and cognitive-affective tendencies, it is theorised that positive role balance will lead to role ease and that negative role balance will lead to role strain (Marks and MacDermid, 1996) – role ease and strain corresponding with role enhancement and conflict, respectively. In the case of positive role balance, role conflict is either prevented or solved before acute problems of role management become chronic; this is achieved by addressing the demands of each role on time, with effort and attention. For example, avoiding unnecessary breaks, calls and e-mails while working, prioritising job responsibilities, and updating one's professional skills, may substantially facilitate managing job responsibilities more efficiently so that the employee's work time does not cut into his or her allocated family time. In contrast, for individuals of whom a negative role balance is typical, occasional incidents of role conflict are likely to accumulate due to their indifference towards role-related tasks and duties, creating an ongoing state of unfulfilled demands. For example, ignoring one's spouse's emotional concerns and avoiding private life responsibilities, such as taking care of one's children or household chores may, over time, escalate into constant and daily disagreements, which can also negatively affect job performance due to the consequential worsening of mood and concentration.

2.1.2 Overall Appraisal and Components Approach: Contemporary Views on Work-Life Balance

More recent views about work-life balance can be classified into the overall appraisal approach to work-life balance, and the components approach to work-life balance (Grzywacz and Carlson, 2007).

Overall appraisal refers to an individual's general assessment concerning the entirety of his or her life situation. For example, work-family balance has been defined as "satisfaction and good functioning at work and home, with a minimum of role conflict" (Clark, 2000, p. 751), "equilibrium or maintaining overall sense of harmony in life" (Clarke et al., 2004, p. 121), and "global assessment that work and family resources are sufficient to meet work and family demands such that participation is effective in both domains" (Voydanoff, 2005, p. 825). When an overall appraisal approach is applied, work-life balance is typically assessed with general questions (e.g., "All in all, how successful do you feel in balancing your work and personal/family life?": Clarke et al. 2004).

A *components approach* to work-life balance emphasises balance as a direct formative latent construct (Edwards and Bagozzi, 2000), which means that work-family balance consists of multiple facets that precede balance and give meaning to it (Grzywacz and Carlson, 2007). For example, according to Greenhaus et al. (2003), work-family balance consists of time balance, involvement balance, and satisfaction balance. According to Frone (2003), in turn, work-family balance consists of work-family conflict and work-family facilitation (corresponding with role conflict and enhancement, respectively). The advantage of the components approach over the overall appraisals approach to work-life balance is that one can use conceptually based measures of balance that tap into the different aspects of work-life balance. These aspects form the overall evaluation of how well an individual is meeting role-related responsibilities (Grzywacz and Carlson, 2007).

Following the theory of role balance (Marks and MacDermid, 1996), Greenhaus et al. (2003, p. 513) have defined work-family balance as "the extent to which an individual is equally engaged in – and equally satisfied with – his or her work and family role". Furthermore, according to these scholars, work-family balance consists of three dimensions of which *time balance* refers to equal time devoted, *involvement balance* refers to equal psychological effort and presence invested, and *satisfaction balance* refers to equal satisfaction expressed across work and family roles. Greenhaus et al. (2003) regard work-family balance as a continuum where imbalance in favor of the work role lies at one end, and imbalance in favor of the family role lies at the other end, and *balance* lies in the middle favoring neither work nor family role.

In the above conceptualisation, work-life balance and imbalance are not seen as inherently beneficial or detrimental, respectively, for psychological well-being and quality of life. Instead, Greenhaus et al. (2003) state that it should be empirically tested whether equal time, involvement, and satisfaction balance is better for an individual than imbalance in favor of either the work or family role. In their study, it turned out that among individuals with a high level of engagement across roles, those reporting the highest quality of life were those who invested more in the family than the work role, that is, they showed an imbalance in favor of family. In regard to their level of engagement, the equally balanced individuals scored lower in quality of life than those favoring family over work, but higher than those favoring work over family. Thus, those who invested most in work had the lowest quality of life.

Frone (2003), in turn, has presented a *four-fold taxonomy of work-family balance*, in which work-family balance is defined as "low levels of inter-role conflict and high levels of inter-role facilitation" (p. 145). The four-fold taxonomy is based on the notion of bi-directionality between work and family domains, meaning that participation in the work role may interfere with or enhance the performance in the family role, and likewise, participation in the family role may interfere or enhance performance in the work role (Frone et al., 1992; Greenhaus and Beutell, 1985; Grzywacz and Marks, 2000; Kirchmeyer, 1992). Accordingly, work-life balance (low conflict, high facilitation/enhancement) is hypothesised to occur in two directions: from work to nonwork domains and vice versa. Thus, according to Frone (2003), the measurable four components of work-life balance are work-to-family/nonwork conflict, family/nonwork-to-work conflict, work-to-family/ nonwork enhancement, and family/nonwork-to-work enhancement.

2.1.3 Outcomes of Work-Family Balance

Grzywacz and Carlson (2007) studied the variance explained in work and family related outcomes, such as satisfaction and stress, by two means: using a single item of work-family balance (overall appraisal) and by using the four-fold taxonomy of work-family balance (i.e., the components approach presented by Frone (2003)). They found that the components approach produced systematically higher explanation rates than did the overall appraisal: for example, in the case of job stress, the respective explanation rates were 45% as against 18%. Therefore Grzywacz and Carlson (2007) recommend the use of measures of work-family conflict and work-family enrichment (i.e., facilitation, enhancement) for investigating and assessing the experience of work-family balance.

Grzywacz and Carlson (2007) point out one limitation, however, namely that the four-fold taxonomy of work-family balance together with other previously reviewed definitions of work-life balance tends to overemphasise balance as a psychological construct, that is, as the experience of an individual, and thus fails to capture the contextual and social perspective of work-life balance. For example, daily interaction and conciliation of needs and responsibilities between work and nonwork members exemplifies such a contextual and social nature of work-life balance. An extended definition of work-life balance, taking into account this limitation, was therefore developed as follows: work-family balance is the "accomplishment of role-related expectations that are negotiated and shared between an individual and his or her role-related partners in the work and family domains" (Grzywacz and Carlson, 2007, p. 458).

Studies measuring the bidirectional work-nonwork conflict and enhancement have most often relied on the variable-oriented approach to work-life balance with the goal of finding out which of the four work-life balance components is correlated with what outcomes. For example, it has been found that work-to-nonwork conflict is related to various forms of psychological ill-being (fatigue, distress, job exhaustion, and dissatisfaction at work and home), whereas nonwork-to-work conflict has most often been found to be related only to fatigue and low family satisfaction (Geurts et al., 2005; Kinnunen et al., 2006; Wayne et al., 2004). Work-to-nonwork enhancement, in turn, is associated with high job satisfaction, low job exhaustion and low psychological distress, while nonwork-to-work enhancement has shown a positive association with family satisfaction (Aryee et al., 2005; Kinnunen et al., 2006; Wayne et al., 2004).

In conclusion, the different views of work-life balance suggest that the phenomenon has at least three important aspects. First of all, it seems that work-life balance is unlikely to be a unidimensional construct but rather a conglomeration of multiple measurable constructs. Many scholars suggest that work-life balance consists of high rewards, resources and enhancement combined with low concerns, demands and conflict experienced by individuals across their life roles (Barnett and Baruch, 1985; Frone, 2003; Grzywacz and Carlson, 2007; Tiedje et al., 1990; Voydanoff, 2005). Second, in addition to role-related resources and demands over which individuals may not always have control, work-life balance seems to stem also from individuals' own actions and attitudes. It is assumed that acting with equal devotion and being equally satisfied with one's life roles reflects a work-life balance that produces the ability to manage multiple roles successfully (Greenhaus et al., 2003; Marks and MacDermid, 1996). Third, the achieved balance between work and nonwork roles is expected to lead to satisfaction and well-being in life.

2.2 A Typology of Work-Life Balance: A Person-Oriented Approach to Work-Nonwork Interaction

2.2.1 Building a Bridge Between Different Views of Work-Life Balance

The notion that work-life balance consists of multiple constructs, such as work-nonwork conflict and enhancement in both directions, is very important because the components approach captures more of the phenomenon than the overall appraisal approach (Grzywacz and Carlson, 2007). At the same time, however, it may obscure the entirety of an individual's work-life balance experience. For example, it is easy to agree that high role enhancement combined with low conflict constitutes work-life balance and that the opposite – low enhancement, high conflict – constitutes work-life imbalance because enhancement has been shown to be linked with high psychological well-being while conflict has been linked with low psychological well-being (Frone, 2003; Geurts et al., 2005; Kinnunen et al., 2006). But what if one experiences both high enhancement and high conflict or alternatively both low enhancement and low conflict simultaneously – do these experiences reflect work-life balance? Does high work-nonwork enhancement counterbalance the negative effect of high work-nonwork conflict on well-being? Additionally, is high enhancement still needed to produce a positive effect on well-being when work-nonwork conflict is not experienced? To find an answer to these questions, a new four-dimensional typology of work-life balance has been suggested (Rantanen, 2008).

According to the proposed four-dimensional typology illustrated in Fig. 2.1, individuals can belong to beneficial, harmful, active or passive work-life balance types. In line with role conflict theories (Frone, 2003; Greenhaus and Beutell, 1985; Kahn et al., 1964), role enhancement theories (Barnett and Hyde, 2001; Marks, 1977; Sieber, 1974; Wayne et al., 2007) and the demands-resources approach (Bakker and Geurts, 2004; Voydanoff, 2005), the term *beneficial balance* refers to the proposition that the simultaneous experience of work-nonwork enhancement and absence

Work-nonwork enhancement experienced	**Work-nonwork conflict experienced** Yes	No
Yes	**Active balance** - high resources - high demands	**Beneficial balance** - high resources - low demands
No	**Harmful balance** - low resources - high demands	**Passive balance** - low resources - low demands

Psychological functioning differentiating factor

Role engagement differentiating factor

Fig. 2.1 Typology of work-life balance (revised from Rantanen, 2008, p. 25, figure 5)

of work-nonwork conflict facilitates psychological functioning and well-being. This occurs because the resources provided and gains attained from the participation in multiple roles are experienced to exceed the demands of these roles. The term *harmful balance*, in turn, refers to the proposition that simultaneous experience of work-nonwork conflict and absence of work-nonwork enhancement threatens psychological functioning and well-being, because the combined demands of multiple roles are experienced as exceeding the benefits that these roles provide. Thus, the main differentiating factor between beneficial and harmful work-life balance is psychological functioning, due to disparity and imbalance in role-related resources and demands.

Furthermore, active and passive work-life balances are considered to represent the opposite ends of the spectrum of role engagement, based on the theory of role balance by Marks and MacDermid (1996). They suggest that role balance reflects a behavioral and cognitive-affective pattern of acting and feeling across roles with either high or low dedication. Thus, *active balance* in the present typology refers to the proposition that individuals may be highly engaged in their life roles both by choice (will to succeed and achieve happiness in different life spheres) and/or by necessity (due to tough demands from different life spheres). *Passive balance*, on the other hand, refers to the proposition that the simultaneous absence of work-nonwork conflict and enhancement experiences may reflect low engagement across life roles (conflicting role demands are perhaps avoided but also rewards are not gained) or a composition of life roles that are less demanding or challenging.

2.2.2 Formation of the Typology of Work-Life Balance

The typology of work-life balance is intentionally described without reference to the bidirectionality of work-nonwork interaction (i.e., work can affect nonwork and vice versa). This is because the aim of the present typology is to capture the individuals' work-nonwork interaction experience in a holistic manner. Here the bidirectionality is not ignored but rather the work-to-nonwork and nonwork-to-work conflict and enhancement factors are considered as closely connected subdimensions of the higher constructs of work-nonwork conflict and enhancement. This view is supported by the fact that often when work-to-nonwork conflict is experienced, its counterpart nonwork-to-work conflict is also reported; the same applies to work-to-nonwork and nonwork-to-work enhancement (Aryee et al., 2005; Grzywacz and Bass, 2003; Kinnunen et al., 2006; Wayne et al., 2004). Therefore, although both work-nonwork conflict and enhancement are to be assessed according to the principle of bidirectionality, within this typology of work-life balance they are combined into the total experience of conflict and enhancement.

The types of work-life balance (Fig. 2.1) are formed by dichotomising the total of the work-nonwork conflict and enhancement experiences into two groups, using the arithmetic mean of the scale as a cut-off point and then cross-tabulating these dichotomised work-nonwork conflict and enhancement experiences in order to obtain the beneficial, harmful, active, and passive balance types. The arithmetic mean of the work-nonwork interface scale was chosen as a cut-off point because work-family conflict is generally reported according to a proportion of population experiencing conflict at least "sometimes", which is a scale midpoint (Bellavia and Frone, 2005). For example, in a scale from 1 to 5 the mean scores below 2.5 constitute non-experience of the work-nonwork conflict and enhancement (1 = never or 2 = seldom), while a mean score of 2.5 or above signifies experiencing work-nonwork conflict and/or enhancement (3 = sometimes, 4 = often, or 5 = very often). This means that the work-life balance types are anchored in the response scale instead of mean or median splits of the variable mean scores, which are sample-specific and hinder the comparison of results between different samples.

This typology of work-life balance represents a holistic and person-oriented approach to work-life balance and tests Voydanoff's (2005) view that the work-family balance is a result of one's global assessment of the fit between demands and resources within the work and family domains. In addition, Grzywacz and Bass (2003, p. 258), who examined the work-family fit with a variable-oriented approach focusing separately on each dimension and direction of work-nonwork interaction, noted that in the absence of strong theory, the numerous possibilities for how each direction of work-nonwork conflict and enhancement may be combined complicate the precise specification of the work-family balance (i.e., work-family fit). Therefore, we suggest that one alternative is to consider work-nonwork interaction experiences in their entirety because "the totality gets its characteristic features and properties from the interaction among the elements involved, not from the effect of each isolated part on the totality" (Bergman et al., 2003, p. 9).

2.3 Work-Life Balance Among Professionals: The Typology of Work-Life Balance in Practice

2.3.1 Prevalence of the Different Balance Types

Our typology of work-life balance has been studied within a community-based sample of 42-year-old employees with spouse and/or children (n = 213) in Finland (Rantanen, 2008). It was found that 48% of these middle-aged respondents belonged to the beneficial balance type, whereas only 9% belonged to the harmful balance type. Of the respondents, 26% belonged to the passive balance type and 17% belonged to the active balance type. The harmful balance type was represented by a greater number of upper level white-collar workers and by more men than the other types. Those belonging to the harmful balance type had the lowest psychological well-being scores, as well as the least amount of personal time (outside of work and family activities), and they invested less thought and action into their family and their own health compared with the individuals belonging to the other types.

For the present chapter, the typology of work-life balance was examined among three samples of upper level white-collar workers with spouse and/or children: Finnish university professionals (including staff with a minimum of a Master's degree or supervisory position, $N = 1482$), Finnish managers ($N = 1214$), and Estonian managers ($N = 396$). Based on the aforementioned findings, our expectation was that the harmful balance type would be more prevalent among these professionals due to their demanding jobs. We also know from related studies that professional work can be rewarding (Kinnunen et al., 2008), which suggests that these professionals may also be part of the beneficial and active balance types if the rewards gained either exceed or are equal, respectively, with the demands of their roles.

The details for sample characteristics are described in Table 2.1. Characteristic of the university professionals in relation to the other two samples is a higher prevalence of females, higher level of education (in comparison to Finnish managers), higher proportion of persons under age 30, and having more than two children (in comparison to Finnish managers). The sample of Finnish managers, in relation to the other two samples, is distinguished by a higher prevalence of males (in comparison to university professionals), lower level of education, higher proportion of persons over age 49, and a higher proportion of respondents whose children are not living at home with them. The Estonian managers, in relation to the two other samples, were represented by a higher proportion of males (in comparison to university professionals), higher level of education (in comparison to Finnish managers), and a higher proportion of persons between the ages of 30–39.

The work-life balance types among these three samples were formed as described in Sect. 2.2.2. Among Finnish university professionals, work-nonwork conflict was measured with eight items derived from the scale by Carlson et al. (2000): four for work-to-nonwork (e.g., "The time I must devote to my job keeps me from participating equally in household responsibilities and activities") and four for nonwork-to-work (e.g., "Due to stress at home, I am often preoccupied with family matters at work") direction. Work-nonwork enhancement was measured with eight items from the scale by Carlson et al. (2006): four for work-to-nonwork (e.g., "My involvement in my work helps me to acquire skills and this helps me be a better family member") and four for nonwork-to-work (e.g., "My involvement in my family makes me feel happy and this helps me be a better worker") direction. The response scale for all items ranged from "1 = completely disagree" to "7 = completely agree", and the Cronbach's alphas were 0.78 for work-nonwork conflict and 0.88 for work-nonwork enhancement. The cut-off point for dichotomisation of work-nonwork conflict and work-nonwork enhancement experiences was 3.5.

Table 2.1 Sample characteristics (%)

	University professionals ($N = 1482$)	Finnish managers ($N = 1214$)	Estonian managers ($N = 396$)
Gender			
Male	38	72	60
Female	62	28	40
Age			
Under 30	13	1	6
30–39	29	17	44
40–49	28	34	33
Over 49	30	47	18
Education			
Polytechnical school (or less)	4	40	16
University degree[a]	55	57	84
Licentiate/doctor[b]	41	3	
Married or cohabiting			
Yes	91	94	87
No	9	6	13
Number of children			
No children	31	38	28
1	17	21	23
2	33	18	35
3 or more	19	13	14

[a]University degree, which means a minimum of 3 years for a Bachelor's degree among Estonian managers, and a Master's degree among university professionals and Finnish managers
[b]Information not available in the data of Estonian managers

Among Finnish and Estonian managers work-nonwork conflict was measured with the scale by Carlson et al. (2000) as among Finnish university professionals but with twelve items: six for work-to-nonwork and six for nonwork-to-work direction. Work-nonwork enhancement, in turn, was measured with six items from the scale by Grzywacz and Marks (2000): three for work-to-nonwork (e.g., "The things I do at work help me deal with personal and practical issues at home") and three for nonwork-to-work (e.g., "Talking with someone at home helps me deal with problems at work") direction. The response scale for all items ranged from "1 = completely disagree" to "5 = completely agree", and the Cronbach's alphas were 0.85 and 0.88 (among Finnish and Estonian managers, respectively) for work-nonwork conflict and 0.74 and 0.79 for work-nonwork enhancement. The cut-off point for dichotomisation of work-nonwork conflict and enhancement was 2.5.

The prevalence rates for work-life balance types between these professional samples and in relation to the sample of 42-year-old Finnish employees illustrated in Fig. 2.2 were both expected and surprising. Professionals often have both more

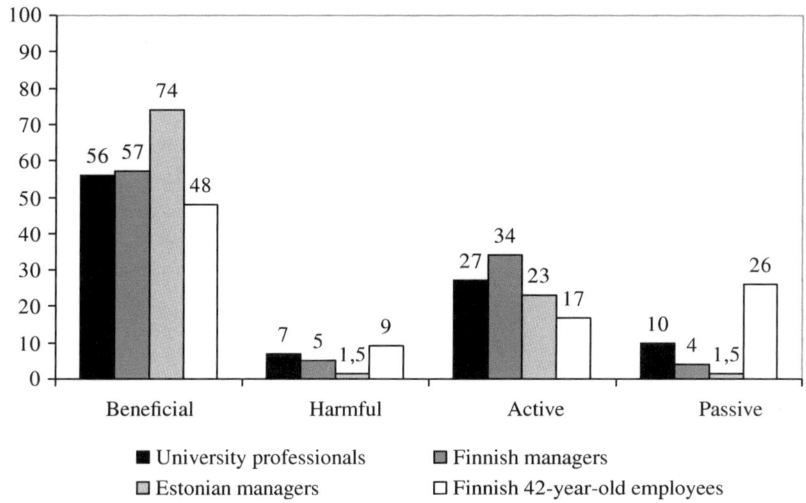

Fig. 2.2 Prevalence (%) of work-life balance types between the samples

work-related demands and resources than blue-collar and lower level white-collar workers (Kinnunen et al., 2008; Lundberg and Frankenhaeuser, 1999), and they are also characterised as being higher in their agency and striving for achievement than the latter two occupational groups (Judge et al., 2002; Tokar et al., 1998). In the light of these aspects, the finding that about one quarter of these professionals belonged to the active work-life balance type (i.e., high levels of both conflict and enhancement) and only 10%, or even much less, belonged to the passive work-life balance type (i.e., low levels of both conflict and enhancement) was not surprising. Thus, the relation between the prevalence of the active and passive balance types was the opposite in these professional samples in comparison to the aforementioned community-based sample (having a lower representation of professionals) where the passive balance was more prevalent than the active balance (Rantanen, 2008).

In contrast, the fairly high prevalence of the beneficial work-life balance and low prevalence of the harmful work-life balance was an unexpected finding. Both of these phenomena were extremely pronounced among the Estonian managers, of whom altogether 74% belonged to the beneficial balance type and only 1.5% (i.e., six persons) to the harmful balance type. Professionals and managers have been found to suffer from higher work-family conflict than other occupational groups, presumably because of high responsibilities, tight deadlines, and long working hours associated with their work (Bellavia and Frone, 2005; Byron, 2005). Therefore, one might expect that the prevalence of the beneficial balance would be lower, whereas the prevalence of the harmful balance would be higher among professionals in comparison with other occupational groups. This point is also suggested by research findings in other studies where higher education and high socio-economic status have been found to be linked to low enhancement from family to work-life (Kinnunen et al., 2006; Wayne et al., 2004). The fact that the ratio between the beneficial and harmful balance is in favor of the beneficial balance is not surprising as such, because enhancement experiences are generally more common than conflict experiences between work and family (Aryee et al., 2005; Grzywacz and Bass, 2003; Kinnunen et al., 2006; Wayne et al., 2004).

The high prevalence of the beneficial balance among the professionals studied may be explained by the typological approach which combines the experiences of enhancement and conflict in both directions, work-to-nonwork and nonwork-to-work. Even though the professionals may show a somewhat higher amount of work-nonwork conflict and lower amount of work-nonwork enhancement at their mean level, when dichotomised as non-experience and experience, the enhancement experience seems clearly to outweigh the conflict experience. Thus, it seems that professional work truly provides important tools for reconciliation of one's work and nonwork roles, or as another option, these professionals have found ways to adapt their nonwork life to support their challenging jobs.

When the three samples were compared with each other, the prevalence of the passive balance was highest among university professionals. This particular group was characterised by a lower level of education and not having a permanent intimate relationship or any children, although they were not especially young and represented equally well different age groups. Thus, these factors seem to confirm the assumption made in the typology of work-life balance that the composition of life roles may be less demanding for persons belonging to the passive balance type when it comes to interplay between work and nonwork domains. The prevalence of an active balance was higher among Finnish managers in comparison with the other two samples; among these managers, having one or two children was characteristic. In addition, the prevalence of the harmful balance was the highest among university professionals, and among these more men than women belonged to the harmful balance type. The prevalence of the active balance was highest among Finnish managers, and the prevalence of beneficial balance was highest among Estonian managers.

Thus, the present findings might further suggest that among professionals, managers in particular are not a disadvantaged group when it comes to their abilities and possibilities in regard to reconciling work and nonwork roles. One reason for this might be that managers, especially those who also face parenting demands, have developed good coping strategies to handle the interplay between work and nonwork. Better coping skills may stem from the kind of job a manager has, which often requires a good deal of organizational and delegating skills – these are useful skills in work-nonwork reconciliation. It might also be that managerial work is so demanding from the viewpoint of work-life balance that those individuals who enter into managerial positions have had to think through what "being a manager" means for work-nonwork reconciliation. Therefore, they are well prepared to meet these demands. Yet another reason for university professionals belonging to the harmful balance type more often than managers may be that Finnish universities have changed a lot during recent years, in which time many of their policies have become more similar to those in private sector companies. These changes in work culture (e.g., long working hours, increased competition, continuous change) may have had a negative effect on university employees' well-being and brought them to reflect negatively on their work-life balance.

Whereas slightly over half of the university professionals and Finnish managers belonged to the beneficial type, altogether 74% of the Estonian managers showed a beneficial work-life balance. The Estonian managers were more often in the range of 30–39 years of age than the other two samples (44% vs. 29% and 17%, respectively). As people in this age group are quite likely to have small children, the high proportion of respondents reporting no perceived work-nonwork conflict, but instead only work-nonwork enhancement, was unexpected. One possible explanation lies in the fact that the Estonian managers may find themselves in a relatively better situation compared to the general population in terms of being able to afford outsourcing chores ranging from childcare to housekeeping and cooking. In addition, the degree of inequality in the distribution of family income in Estonia is higher than in Finland (in 2008 the Gini index for Estonia was 34, and for Finland 29.5; Central Intelligence Agency, 2008), supporting the view that the Estonian managers, in comparing their situation with that of their fellow countrymen, may have the necessary financial means of minimising the possible antecedents to conflict by securing high quality childcare and/or housekeeping. Due to this, they may be more likely to attribute positive feelings and emotions to their jobs, which may in part explain the high proportion of beneficial balance among them.

2.3.2 Differences in Psychological Functioning and Role Engagement Between the Balance Types

Psychological functioning is understood here as a broader phenomenon than merely psychological well-being. Personality characteristics and strategies that enable and enhance well-being and adjustment in life are also part of the definition when referring to good psychological functioning (Caspi et al., 2005; DeNeve and Cooper, 1998). Accordingly, the indicators of psychological functioning studied herein were high vigor and low exhaustion/stress at work, as well as having scored high in positive core-self evaluations (i.e., emotional stability, self-efficacy, self-esteem, and locus of control: Judge et al. (1997)) and regarding oneself as healthy and satisfied with one's life. Role engagement is also defined broadly, including both objective and subjective indicators of role engagement such as actual time spent in a given role and psychological involvement directed to a given role (Greenhaus et al., 2003). In addition, the definition of role engagement also consists of those behavioral and cognitive-affective tendencies that reflect the overall style of engaging in life and its roles (e.g., effort vs. apathy: Marks and MacDermid (1996)). The herein studied indicators of role engagement were high weekly working hours, over-commitment to work, high organizational involvement and low turnover intentions.

According to the typology of work-life balance, the main differentiating factor between the beneficial and harmful balance types is psychological functioning and the main differentiating factor between the active and passive types is role engagement (cf. Fig. 2.1). Our results – based on the General Linear Model of Profile

Table 2.2 Results of profile analyses

Profile	Differences between work-life balance types in each variable[a]: University professionals	Finnish managers
Psychological functioning	***F* (12) = 40.21*****	***F* (12) = 25.06*****
Vigor at work	Ben > Act > Pass > Harm	Ben > Act > Pass, Harm
Low job exhaustion/stress[b]	Ben > Pass > Act > Harm	Ben, Pass > Act, Harm
Core self-evaluations/ Self-Efficacy[c]	Ben > Pass, Act > Harm	Ben, Act > Harm
Self-reported health	Ben > Pass, Act > Harm	Ben > Pass, Act, Harm
Life satisfaction	Ben > Act, Pass > Harm	Ben > Act > Pass > Harm
Work role engagement	***F* (6) = 9.98*****	***F* (6) = 10.65*****
Weekly working hours	Act > Ben	Act, Harm > Ben, Pass
Organisational involvement/ Overcommitment[d]	Ben, Act > Pass, Harm	Harm, Act > Pass, Ben
Turnover intentions	Harm, Act, Pass > Ben	Harm, Pass, Act > Ben

[a]Differences between types ($p < 0.05$) in single variables are based on parameter estimates of the MANCOVA models
***$p < 0.001$
Ben = Beneficial balance, Act = Active balance, Pass = Passive balance, Harm = Harmful balance
[b]In the data of Finnish managers fatigue at work was measured with stress item instead of job exhaustion
[c]In the data of Finnish managers self-efficacy was measured instead of core self-evaluations
[d]In the data of Finnish managers overcommitment was measured instead of organizational involvement

Analyses (i.e., application of MANCOVA) with sample characteristics (Table 2.1) as covariates – were mostly in line with these expectations as summarised in Table 2.2. These profile analyses were only conducted among Finnish university professionals and Finnish managers because there were only six persons in the group of harmful balance type, as well as in the group of passive balance type, among Estonian managers; this was not sufficient to enable valid comparisons between the types.

Professionals belonging to the beneficial type had the highest psychological functioning regarding all studied indicators, while professionals belonging to the harmful type had the lowest, as illustrated in Table 2.2 and Fig. 2.3. These findings are in line with expectations and confirm the most common view on work-life balance, which is that work-life balance consists of minimum conflict and high enhancement between work and nonwork life spheres (e.g., Clark, 2000; Frone, 2003). We would like to point out, however, that low conflict combined with high enhancement could also be interpreted as an *ideal*, or as in this case, *beneficial*, work-life balance. This is because, in general, the active and passive types who fell in between the beneficial and harmful types did not show extremely poor psychological functioning (Table 2.2 and Fig. 2.3). This means that *satisfactory*

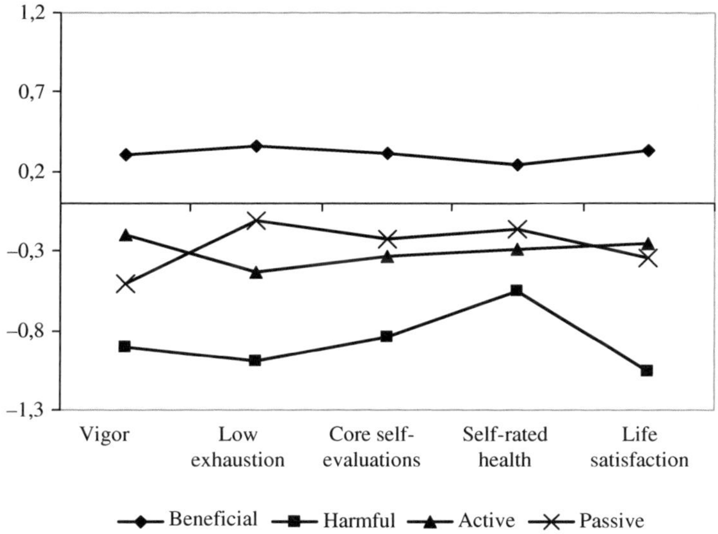

Fig. 2.3 Psychological functioning among Finnish university professionals according to work-life balance types (values are standardized means)

work-life balance in terms of psychological functioning can also be reached when the work-nonwork conflict and enhancement experiences are equivalent (i.e., both are high or low).

Interestingly, professionals belonging to the active type showed more vigor at work than professionals belonging to the passive type, the downside of which was that they also showed more job exhaustion and stress than the passive type. This, however, fits well with the definition of the active and passive work-life balance types. The experiences of devotion and energy combined with exhaustion and stress at work may well reflect the presence of both high work-related resources and demands for the active type, whereas the opposite might be true for the passive type, hence showing a combination of lower vigor and lower exhaustion, respectively.

In general, professionals belonging to the active balance type showed high work role engagement concerning all indicators except (low) turnover intentions (Table 2.2, Fig. 2.4). Professionals of the active type had higher turnover intentions than professionals belonging to the beneficial balance type, whose turnover intentions were the lowest both among university professionals and Finnish managers. Also as expected, professionals belonging to the passive type showed lower work role engagement (lower working hours, lower organizational involvement and lower over-commitment to work) than professionals belonging to the active type. For the beneficial type, in addition to low turnover intentions, higher than average organizational involvement, as well as low over-commitment to work, was typical. For the harmful type, high weekly working hours and high over-commitment to

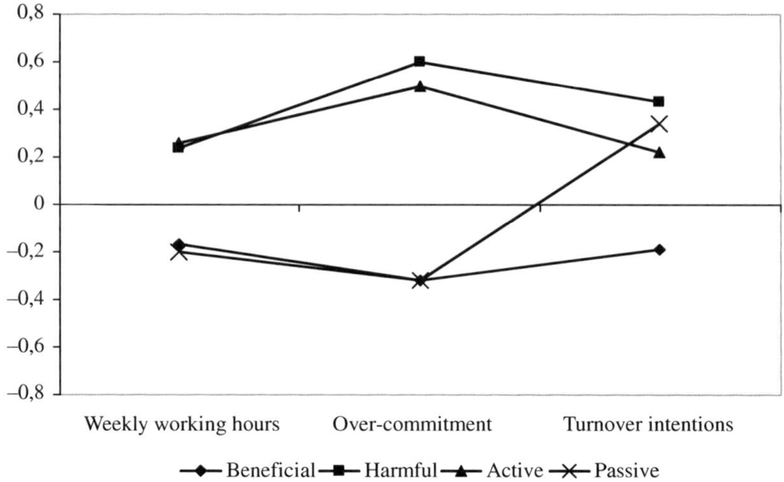

Fig. 2.4 Work role engagement among Finnish managers according to work-life balance types (values are standardized means)

work were combined with low organizational involvement and high turnover intentions. All these findings were especially clear among Finnish managers as illustrated in Fig. 2.4, whereas among university professionals the differences in work-role engagement between the balance types were less pronounced.

In summary, *the typology of work-life balance* seems to function well. As expected, the main differentiating factor between the beneficial and harmful balance types was psychological functioning, and between active and passive balance types it was role engagement. However, the differences between balance types were not as clear in regard to role engagement as they were in relation to psychological functioning. It should be mentioned that our study did not have the scope to use indicators of role engagement in the domain of nonwork (i.e., family and private life) as an analytical factor because the research data at our disposal were designed primarily to study psychological phenomena of working life.

2.4 Theoretical and Practical Conclusions

Our *typology of work-life balance* combines some of the basic premises of theories pertaining to role conflict (Kahn et al., 1964), role enhancement (Marks, 1977; Sieber, 1974), and role balance (Marks and MacDermid, 1996) and also aims to incorporate the main underlying points from recent theories on work-family balance (Bakker and Geurts, 2004; Clark, 2000; Frone, 2003; Greenhaus et al., 2003; Voydanoff, 2005) into one model. The typology of work-life balance thereby corresponds with the overall appraisals approach, as the four different balance types refer to individuals' overall experience of the quality of work-nonwork interaction in their lives, as well as corresponding with the components approach, as the four balance types are based on the multidimensional measurement of work-nonwork interaction (Grzywacz and Carlson, 2007). As such, the typology covers the essential experiences contributing to work-life balance, while at the same time offering a global description of work-life balance and a cost-effective way of modelling work-life balance at an individual level.

Our typology is not without limitations, however. For example, it does not specify whether psychological well-being is an antecedent or an outcome of specific types of work-life balance, the question of which would require longitudinal data. Neither does our typology take into account the social context of an individual such as social interactions at work and in nonwork domains that are expected to contribute to shaping one's experience of work-life balance (cf. Grzywacz and Carlson, 2007). In the future, these two aspects also need to be studied in relation to the typology of work-life balance. Furthermore, more holistic theories on work-family balance, which would consider different life contexts, would also be considered valuable.

From a practical point of view, the present findings draw a relatively positive picture of the work-life balance of professionals: for most professionals, the reconciliation of work and nonwork roles seems to be either ideal (beneficial type 56–74%) or at least satisfactory (active type 23–34%, passive type 1.5–10%) when psychological functioning is held as a criterion. Accordingly, the most problematic group, that of harmful balance types who experience reconciliation of work and nonwork roles solely as burdensome and subsequently show poorest psychological functioning, was rather small (1.5–7%).

Although these prevalence results relate to the way the typology is formed (i.e., using cut-off points under which the presence of the experience is ignored), they show that professionals are in many respects in a good position. This is related to many factors. For example, they have control and autonomy in their work, which helps in reconciling the demands from various life domains, and their financial situation – especially in Estonia – enables them to acquire unburdening domestic help. In addition, it is very likely that professionals, and managers in particular, have consciously thought about and negotiated over the issues related to work-life balance concerning their family. Therefore, they might have better coping abilities in this regard. However, a more careful investigation of those work/nonwork circumstances, organisational practices, and personal strategies that are most relevant to professionals who belong to the beneficial work-life balance type might offer valuable tips on how to support more efficiently those at greatest risk, that is, professionals belonging to the harmful balance type.

The present findings also confirm the self-evident assumption – for which Greenhaus et al., (2003) called for empirical evidence – that the combination of low work-nonwork conflict and high enhancement leads to favorable outcomes of well-being, and that the opposite leads to detrimental outcomes. Nonetheless, the present findings also showed, in accord with others (Clark, 2000; Greenhaus et al., 2003), that the combination of low work-nonwork conflict and high enhancement is not the only strategy for achieving work-life balance and good well-being. Segmentation, that is, the absence of both positive and negative interaction between work and nonwork roles, also seems to be a solution that produces at least satisfactory well-being. In fact, in a highly demanding context, such as combining professional work with nonwork demands, restraining oneself from over-investing in work and nonwork roles may act as a protective shield against loss of personal resources. This may especially be the case if work and nonwork roles do not return the individual investments, that is, there is an imbalance between individual efforts and perceived rewards.

From an organizational perspective, our study offers an important message. A crucial general question for organizations is how to promote role enhancement and prevent work-nonwork conflict among employees. Our findings show that those professionals who belonged to the beneficial balance type reported fewer turnover intentions compared to the other balance types. This finding suggests that it is particularly important *both* to promote work-nonwork enhancement *and* to prevent work-nonwork conflict if organizations want to keep their highly professional employees for a long time. A good starting point to the building of a work environment supporting this beneficial work-life balance is to take the following two things into account. First, the need for policies and, second, supervisors supporting employee needs in order to balance work and nonwork responsibilities (cf. e.g., Kinnunen et al., 2005). We believe that the latter point is more crucial in countries like Finland, where the state is especially active in forming work-family policies (e.g., day care, child care leave system, reduced working hours). Supervisors are in a key role for building a work-nonwork culture consisting of a beneficial work-life balance, that is, a culture which is sensitive to employees' needs. A supportive work-nonwork culture also means increasing employees' entitlement to make use of existing policies within organizations.

References

Aryee S, Srinivas ES, Tan HH (2005) Rhythms of life: antecedents and outcomes of work-family balance in employed parents. J Appl Psychol 90:132–146

Bakker AB, Geurts SAE (2004) Toward a dual-process model of work-home interference. Work Occup 31:345–366

Barnett RC, Baruch GK (1985) Women's involvement in multiple roles and psychological distress. J Pers Soc Psychol 49:135–145

Barnett RC, Hyde JS (2001) Women, men, work, and family. Am Psychol 56:781–796

Bellavia GM, Frone MR (2005) Work-family conflict. In Barling J, Kelloway EK, Frone MR (eds) Handbook of work stress. Sage, Thousand Oaks CA, pp 113–147

Bergman LR, Magnusson D, El-Khouri BM (2003) Studying individual development in an interindividual context: a person-oriented approach. Lawrence Erlbaum, Mahwah NJ

Byron K (2005) A meta-analytic review of work-family conflict and its antecedents. J Vocat Behav 67:169–198

Carlson DS, Kacmar MK, Williams LJ (2000) Construction and initial validation of a multidimensional measure of work-family conflict. J Vocat Behav 56:249–276

Carlson DS, Kacmar MK, Wayne JH, Grzywacz JG (2006) Measuring the positive side of the work-family interface. Development and validation of a work-family enrichment scale. J Vocat Behav 68:131–164

Caspi A, Roberts BW, Shiner RL (2005) Personality development: Stability and change. Ann Rev Psychol 56:453–484

Central Intelligence Agency (2008) The world fact book: distribution of family income – Gini index. https://www.cia.gov/library/publications/the-world-factbook/fields/2172.html. Accessed 24 Nov 2008

Clark SC (2000) Work/family border theory: a new theory of work/family balance. Hum Relat 53:747–770

Clarke MC, Koch LC, Hill EJ (2004) The work-family interface: differentiating balance and fit. Fam Consum Sci Res J 33:121–140

DeNeve KM, Cooper H (1998) The happy personality: a meta-analysis of 137 personality traits and subjective well-being. Psychol Bull 124:197–229

Edwards JR, Bagozzi RP (2000) On the nature and direction of relationships between constructs and measures. Psychol Methods 5:155–174

Frone MR (2003) Work-family balance. In Quick JC, Tetrick LE (eds) Handbook of occupational health psychology. American Psychological Association, Washington, DC, pp 143–162

Frone MR, Russell M, Cooper ML (1992) Antecedents and outcomes of work-family conflict: testing a model of the work-family interface. J Appl Psychol 77:65–78

Geurts SAE, Taris TW, Kompier MAJ, Dikkers JSE, van Hooff MLM, Kinnunen UM (2005) Work-home interaction from a work psychological perspective: development and validation of a new questionnaire, the SWING. Work Stress 19:319–339

Goode WJ (1960) A theory of role strain. Am Sociol Rev 25:483–496

Greenhaus JH, Beutell NJ (1985) Sources and conflict between work and family roles. Acad Manage Rev 10:76–88

Greenhaus JH, Collins KM, Shaw JD (2003) The relation between work-family balance and quality of life. J Vocat Behav 63:510–531

Grzywacz JG, Bass BL (2003) Work, family, and mental health: testing different models of work-family fit. J Marriage Fam 65:248–262

Grzywacz JG, Carlson DS (2007) Conceptualizing work-family balance: implications for practice and research. Adv Dev Hum Resour 9:455–471

Grzywacz JG, Marks NF (2000) Reconceptualizing the work-family interface: an ecological perspective on the correlates of positive and negative spillover between work and family. J Occup Health Psychol 5:111–126

Jones F, Burke RJ, Westman M (2006) Work-life balance: a psychological perspective. Psychology Press, New York, NY

Judge TA, Locke EA, Durham CC (1997) The dispositional causes of job satisfaction: a core evaluations approach. Res Organ Behav 19:151–188

Judge TA, Bono JE, Ilies R, Gerhardt MW (2002) Personality and leadership: A qualitative and quantitative review. J Appl Psychol 87:765–780

Kahn RL, Wolfe DM, Quinn RP, Snoek JD, Rosenthal RA (1964) Organizational stress: studies in role conflict and ambiguity. Wiley, Oxford

Kinnunen U, Feldt T, Mäkikangas A (2008) Testing the effort-reward imbalance model among Finnish managers: the role of perceived organizational support. J Occup Health Psychol 13:114–127

Kinnunen U, Feldt T, Geurts S, Pulkkinen L (2006) Types of work-family interface: well-being correlates of negative and positive spillover between work and family. Scand J Psychol 47: 149–162

Kinnunen U, Mauno S, Geurts S, Dikkers J (2005) Work-family culture in organizations: theoretical and empirical approaches. In Poelmans SAY (ed) Work and family: an international research perspective. Lawrence Erlbaum, Mahwah NJ, pp 87–120

Kirchmeyer C (1992) Perceptions of nonwork-to-work spillover: challenging the common view of conflict-ridden domain relationships. Basic Appl Soc Psychol 13:231–249

Lundberg U, Frankenhaeuser M (1999) Stress and workload of men and women in high-ranking positions. J Occup Health Psychol 4:142–151

Marks SR (1977) Multiple roles and role strain: some notes on human energy, time and commitment. Am Sociol Rev 42:921–936

Marks SR, MacDermid SM (1996) Multiple roles and the self: a theory of role balance. J Marriage Fam 58:417–432

Rantanen J (2008). Work-family interface and psychological well-being: a personality and longitudinal perspective. Jyväskylä Studies in Education, Psychology and Social Research 346. University of Jyväskylä, Jyväskylä

Sieber SD (1974) Toward a theory of role accumulation. Am Sociol Rev 39:567–578

Tiedje LB, Wortman CB, Downey G, Emmons C, Biernat M, Lang E (1990) Women with multiple roles: role-compatibility perceptions, satisfaction, and mental health. J Marriage Fam 52:63–72

Tokar DM, Fischer AR, Subich LM (1998) Personality and vocational behavior: a selective review of the literature, 1993–1997. J Vocat Behav 53:115–153

Voydanoff P (2005) Toward a conceptualization of perceived work-family fit and balance: a demands and resources approach. J Marriage Fam 67:822–836

Wayne JH, Musisca N, Fleeson W (2004) Considering the role of personality in the work-family experience: relationships of the big five to work-family conflict and facilitation. J Vocat Behav 64:108–130

Wayne JH, Grzywacz JG, Carlson DS, Kacmar KM (2007) Work-family facilitation: a theoretical explanation and model of primary antecedents and consequences. Hum Res Manage Rev 17:63–76

Chapter 3
From 'Balancers' to 'Integrators'? Young Professionals' Talk About 'Work' and the Rest of 'Life' in the UK

What Does 'Work-Life Balance' (WLB) Mean to Today's Generation of University Graduates and Future Professionals and What Support Do They Expect?

Uracha Chatrakul Na Ayudhya and Suzan Lewis

3.1 Introduction

The topic of WLB has gained wide academic and public attention in recent times. However, an emergent critical perspective signifies the problematic nature of the WLB discourse (e.g., Fleetwood, 2007; Lewis et al., 2007). One problem is that much of the current work-life debates and research tend to focus on workers who have caring and/or other family responsibilities, thereby confining our understanding of work and personal life experiences to a particular demographic cohort and phase in the life course. Nevertheless, the WLB discourse is part of a wider context in which today's young people, not yet in the workforce, develop impressions of what employment will entail. This raises the question of what WLB means to younger cohorts.

There is some evidence that young British people today emphasize and value the ability to have what they think of as a 'balanced life' once they enter employment (see Lewis et al., 2002; Smithson and Lewis, 2000; Sturges, 2008; Sturges and Guest, 2004). This suggests that the new generation of young workers may no longer subscribe to a more traditional 'relational' psychological contract, implicitly based on trust, loyalty, and a level of job security (Smithson and Lewis, 2000). Instead, many are anticipating 'transactional' psychological contracts (Rousseau and Wade-Benzoni, 1995), where some expect a 'balanced lives contract' that entails them accepting "lack of long term security and less than optimum conditions in exchange for flexibility and reasonable hours, in order to achieve work-life balance" (Smithson and Lewis, 2000, p. 695). Sturges and Guest (2004) report that although the graduates in their study maintain that they highly value WLB they

U. Chatrakul Na Ayudhya (✉)
Middlesex University Business School, London, UK
e-mail: u.chatrakul@mdx.ac.uk

S. Kaiser et al. (eds.), *Creating Balance?*, DOI 10.1007/978-3-642-16199-5_3,

find themselves in a bind of increasingly long hours and unsatisfactory balance between work and personal life. Notably, their study reveals that these graduates are becoming increasingly disillusioned with the extent to which employers' promises and employee expectations of workload and working hours are met. Consequently, Sturges and Guest (2004) argue that organizations need to manage the issue of WLB prudently, particularly in relation to the workloads and working hours that they expect from graduates, or else risk violating young employees' psychological contracts and ultimately their commitment to the organization.

In the UK, graduate surveys indicate that a 'balance' between personal life and career is the one of most important career goals for university students after graduation and imply that organizations need to address young adults' desire for WLB through providing opportunities to work flexibly or risk their ability to attract and retain valuable young employees (Universum, 2003; Unisys, 2004). At present, there are only a handful of studies that have examined work-life issues from the viewpoints of the new generation of young adults. The majority of studies have been conducted in North American (e.g., Families and Work Institute, 2004; Gerson, 2004; Loughlin and Barling, 2001; Smola and Sutton, 2002) and Western European countries (e.g., Brannen et al., 2002; Lewis et al., 1999), including the UK (e.g., Lewis and Smithson, 2001; Smithson and Lewis, 2000; Sturges, 2008; Sturges and Guest, 2004), but with little attention to young workers from other parts of the world.

As a new generation of professionals are entering the workforce at a time of social and economic change, researchers in work and organizational psychology and organizational and management science are turning their attention to examining the impact of generational experiences on the outlooks and expectations of today's young workers. The recruitment and retention rationale is a popular one within existing studies on generational differences in work values and attitudes as depicted by the following quotes:

> . . .there can be little doubt that today's young workers will soon constitute the workforce of tomorrow. Understanding how their future work attitudes and behaviours will be shaped is too important a question to be ignored. (Loughlin and Barling, 2001, p. 555)

> Companies that make modifications to benefits and business practices are more likely to attract (and retain) the best and the brightest of today's and tomorrow's workforce. Millennials and the generations behind them may bring their own set of changing values. Managers may need to be creative in accommodating those needs while still watching the bottom line. (Smola and Sutton, 2002, p. 380)

The topicality of this issue in the UK has given rise to the publication of a guide entitled "The Graduate's Guide to the Best Work-Life Balance Employers in Britain" (The Work-Life Balance Trust and Robertson Cooper Ltd., 2002). Furthermore, the importance of WLB to young British workers, in particular the desire for flexible working, has become a theme that has captured media interest. This is reflected in newspaper article titles in recent years such as "Don't just give me cash and a car – give me a laptop and a free Friday afternoon" (The Times, 2004) and "It's no longer just about the money" (Dight, 2008).

The composition of today's UK workforce is diversifying in a number of ways, including age, gender, disability, ethnicity, and nationality. In the UK, there is increasing recognition that such changes in workforce demographics present new challenges for organizations and government, where "all are struggling to understand and respond to the consequences of far-reaching changes within the population, the labour market, and society as a whole" (Williams and Jones, 2005, p. 2). In light of changing demographics in the labour market, it is then no longer appropriate to assume that workers have similar needs and expectations from work and outside of work or to assume that their needs remain constant over the life course. In particular, with increased geographical mobility, new graduates in the UK come from diverse national backgrounds. So while British-born graduates are likely to have heard a great deal about WLB, not least because of the government's "WLB campaign" in 2000 (see DTI and Scotland Office, 2000; DfEE, 2000), those from other countries may be less familiar with the discourse.

This chapter builds on research on young professionals by drawing on a study that explores the ways in which young people who are in transition from university to work and employment talk about WLB and the support that they expect. It looks at one example of demographic diversity by exploring the views and experiences of British-born and Asian-born graduates in the UK. In particular, it examines variability in the ways in which a group of young people from British and Asian contexts make sense of the WLB discourse in the here-and-now Britain as they and their personal, social, and cultural contexts undergo transformation.

3.2 The Study

This chapter draws on a qualitative study of 30 British-born and Asian-born[1] university undergraduates and postgraduates from British universities about to embark on professional careers. These young future professionals were aged between 21 and 28 and studied a range of disciplines. They participated in interviews and focus groups, in which they discussed what they hoped to gain from their future careers as well as other aspects of their future lives. They talked about what WLB meant to them and their perceptions of how they would manage to combine their career and personal lives at this phase in the life course. They discussed their priorities, considered whether these would change over the life course, and what support (if any) they expected from employers and state policy regulation to enable them to live what they felt would be 'balanced' lives. All interviews and focus groups were recorded, transcribed, and thematically analyzed. The analysis was conducted within a life course perspective (Giele and Elder, 1998) so that young people's accounts can be contextualized within time and place. Some key themes that the young British and Asian people discussed are considered below and these are illustrated in the findings and analysis by quotations. These quotations are not presented as being 'representative' of the sample, but are used to demonstrate the complex and fluid nature young

[1] Of the 30 participants, 15 were British-born and 15 were Asian-born graduates. In the latter group, the nationalities of graduates included (in alphabetical order) Chinese, Indian, Indonesian, Malaysian, Singaporean, and Thai.

people's accounts of WLB before entering their professional careers (see Lewis and Smithson, 2001).

3.3 Findings and Analysis

In the next section, we discuss three major findings. First, we discuss how a notion of 'individual choice' strongly underpins the way these young people talked about WLB. However, even though they all subscribed to the ideals of WLB and individual choice, there were variations in their accounts of these choices. Four emerging patterns depicted how they hoped to implement their individual choices in relation to WLB. Second, we argue that precisely because these young people felt that WLB is a matter of individual choice, they did not feel entitled to expect much employer and workplace WLB support as early career young professionals. Finally, we argue that this does not, however, mean that they did not have WLB needs. We build on this argument by making a distinction between young graduates' WLB wishes and their WLB expectations from their future employers.

3.3.1 Young Future Professionals' Diverse Orientations to WLB

In the study, the young people were asked to consider and discuss what the term 'WLB' meant to them in their own words, instead of being given a working definition. The findings show that although some British graduates were more conversant with the WLB discourse than their Asian counterparts, both groups were able to impart their understandings of WLB in general and personal terms. At the general level, both young British and Asian people perceived WLB as an issue of individual choices, with individuals perceived to be responsible for attaining their ideal points of 'balance'. Choices are always socially embedded and constrained (Crompton et al., 2007; Tomlinson, 2006) and inevitably, the WLB options that young professionals have available to them will be limited by social, professional, and organizational structures, norms, and values. Yet, these future professionals, of both British and Asian origin, seemed only partially aware of such factors and how they impinge on their choices.

The positioning of WLB as an individual responsibility was also reflected in the young people's personal definitions of WLB, which tended to be based on the level of importance or relevance of WLB to them at this point in the life course. The findings showed that there were variations in their perceptions of personal relevance, ranging from WLB being a low priority as they were more concerned with establishing their careers first to WLB being an issue that they identified with and planned to enact upon as they were about to enter their professional careers.

Notably, there were further variations among these young people's ways of talking about WLB. Here, we focus on four patterns of 'WLB orientation preferences' that emerged from the findings. WLB orientation preferences refer to the ways in which these young people talked about the choices they would ideally make for prioritizing their work and personal life in their forthcoming transition into work and employment. The participants can be characterized as 'balancers', 'careerists', 'career-sacrificers', or 'integrators'. Variations between the four pattern groups were based on how the participants talked about how they would like to incorporate 'work' together with the rest of 'life' once they start their professional careers. Although differentiation between patterns was not made on the basis of gender or national and cultural backgrounds, and there were no overall differences in these respects, these important contextual factors were likely to have contributed to the shaping of variations in these young people's preferences. We now discuss each of the four patterns in turn.

3.3.1.1 Balancers

The young people who we characterize as balancers were those who appeared to subscribe to the notion of striking an equal 'balance' between their work and personal life. They talked about ways of achieving this in their early professional careers. They anticipated the attainment of their ideal WLB to be difficult in early career, but WLB was still regarded as an important individual goal at this phase of the life course. For example:

> Well, I want to be able to maintain a balance... I don't want to be (...) on one end of the scale or worried about work or about home. I would like to keep a nice balance. I would like to be able to separate between work and having a life outside of work, but to keep them fairly equal really, so that I did enjoy both – but make sure I had time for both! (...) Um, probably not at first [laughs]. When you first get into something new I think you'll need to spend a lot of time probably on training at first and put a lot more effort into the work. But then hopefully it will balance out and I'll have more time to do things out of work. (Joanne, British woman, age 24)

Similarly to Joanne, a PhD student from the UK, Titho, an MBA student from India, described his preference for having structured working hours. For balancers like Joanne and Titho, this was an important way of maintaining their ideal 'balance' through clear separation of work and non-work boundaries. However, both Joanne and Titho seemed to accept that their preferences for balance might not necessarily translate into actual practice immediately at this phase of their career.

> Yeah, because it [structured working hours] gives you a life for yourself to spend! It gives you a social life. It gives you a family life, which everybody does enjoy, because if you see that's what the whole package is like that. Because you're working fixed time, you've got time with the family for a fixed time, you're trying to see friends for a fixed time. That's what you definitely expect. Whether it happens or not [laughs] is something different. (Titho, Indian man, age 27)

While some balancers preferred to have clear separation of 'work' and 'personal life', others did not. For the latter group of balancers, they were clear about the importance of having flexibility in how they managed their work and personal life boundaries. To them, this was key to their ability to redress instances of perceived imbalance.

3.3.1.2 Careerists

Like the balancers, the young people whom we depict as careerists also subscribed to the notion of 'balance' in the attainment of WLB. However, careerists were clear that WLB was an important individual goal only at a later phase in the life course and not at the start of their professional careers. Despite the understanding that the attainment of WLB will be difficult in early career, this was not seen as detrimental, as the establishment of professional career was considered the top priority at this phase of the life course. For example:

> I said before that I'm afraid that when I start working, the areas of my life would suffer. Because the amount of time I devote to other areas of my life will be channelled into my work. So, yeah, so that's what I'm afraid of, you know? (. . .) But I'm not overly, um, [pause]. If I do not achieve the balance between work and other areas of life, then it's not the end of the world. (. . .) I'm prepared to sacrifice a lot for my work, basically. // Cos I think I come from a perception that it's part and parcel of the working life. (Yong Jian, Singaporean man, age 25)

As with the balancers pattern there were variations within the careerist pattern. For instance, some careerists were leaning towards this pattern on a more permanent basis, especially those who had indicated that they did not plan on having a family or children in the future. Other careerists who talked about putting their personal lives on hold at this stage spoke about how they envisaged this as a temporary measure; a strategy for establishing their careers first before making plans for future commitments.

In many of the interviews, participants referred to the influence of their parents' life experiences in shaping their WLB orientation preferences. For example:

> It's OK to dedicate your life to a job, you know. I think it's all right, because that's how my dad do it, anyway. But there's a limit. And, uh. [Pause] And hard work is not something that, um. . . Hard work is a must if you want to be successful. That's how my dad taught me. (Ita, Indonesia women, age 22)

> To see your parents having to work hard and maybe not getting much out of it makes you appreciate that if you work hard, you know, it could be a lot easier to get more incentive and more family stuff. (. . .) My dad taught me that "don't be stupid with your money. Just work hard so you can have an easy life. Just kill it now and at least later on you won't have to do what I'm doing." (Hritik, British man, age 22)

Both quotes highlight the interconnectedness of lives or the importance of linked lives (Moen and Sweet, 2004) and how such connections must also be taken into account when we examine young professionals' WLB preferences and choices.

3.3.1.3 Career-Sacrificers

In the third pattern, we portray career-sacrificers as young people who subscribed to the view that while work and employment are necessary and important for a comfortable life, personal life takes priority over and above work. These young people were clear that the people in their personal lives come before the establishment of their professional careers.

> ...if it was a case of, leave my partner or leave my work, I'd rather leave my work right now. If there was an issue where my partner had to move somewhere for a job, I would happily move rather than lose him. So really I'm more, 'a job's a job'. (Frank, British man, age 27)

Like the rest of the young people in the study, career-sacrificers did not have children at the time of the interviews. Yet, their accounts indicated that they have what they viewed as family commitments and therefore their WLB orientation preferences were geared towards balancing paid work and family responsibility (to either their partner or parents and siblings). This is in contrast to the accounts of the majority of the participants whose WLB orientation preferences were about balancing work and leisure and/or the rest of life.

3.3.1.4 Integrators

Integrators form the final pattern of WLB orientation preference. In the case of these young people, the notion of 'balance' was being redefined through their talks of bringing together and merging work with personal life. The ability to 'integrate' the two domains was paramount to them.

These young people appeared to identify with some aspects of the WLB discourse, particularly the idea that WLB is an important aspiration for them to achieve at this life course phase. However, they were altering the way 'balance' is defined and constructed. To them, balance was less about sacrificing one domain, either work life or personal life, for the other. Instead, it was more about bringing them together and merging them.

> You know, it is more a case of integration than it is balance. It is more a rush of new experiences that are going to have to be learnt and assimilated and fit into a new pattern. // I think 'integration' seems more appropriate, um, certainly at the stage I'm at, rather than 'balance'. (Ari Mazel, British man, age 27)

> Um, I guess, uh WLB to me means doing something that you enjoy doing, doing something worthwhile. Uh, having enough money to be comfortable. Um, working with friends or making new friends at your place of work. Um, yeah, I guess like trying to transcend the fact that you are at work with the fact that you can have relationships with people who you're working with. (...) I'd really like to do something that would fit into my own life, would fit into the things that I'm interested. (Duncan, British man, age 26)

Notably, there was an explicit preference among integrators to have their work reflecting their personal values and interests and who they are as individuals.

> Probably the first thing is if you enjoy your work. Cos I think the ideal situation would be to be working and to kind of not see it as work, but just as something that you do and that you just enjoy spending time doing. (Ji, British man, age 26)

The four patterns of WLB orientation preferences illustrate the ways in which these young people talk about WLB and show the diverse. Their interpretations of the WLB discourse, especially the 'life' and 'balance' components in the WLB equation, were far from straightforward and standardized. Moreover, they recognized that their definitions and preferences for WLB were situated in the here-and-now and that these would change as they move across their personal life and professional career trajectories through the life course.

3.3.2 Expectations of Employer WLB Support: Low 'Sense of Entitlement'

As indicated in the four WLB orientation preference patterns above, the young British and Asian people in this study regarded WLB as a matter of individual choice. This has direct relevance for understanding their expectations of receiving WLB support from their future employers and workplaces. In this section, we draw on the concept of 'sense of entitlement' to examine their expectations of employer WLB support. It makes a distinction between supports that are perceived by individuals as 'entitlements' or rights (and are, therefore, expected) and those that are perceived as favours that are to be negotiated and/or reciprocated (Lewis and Haas, 2005; Lewis and Smithson, 2001). Social comparison processes are key in individuals' formations of sense of entitlement. If individuals consider a particular form of WLB support to be feasible, normative, and appropriate within their social and cultural contexts, then it is perceived as an entitlement rather than a favour.

Tia: It depends what you can give to the company, I think. (. . .) if you're good at it (or you are not), only the company can be your own judge um how you're doing (. . .) // And if you're good enough, eventually they will give it to you, I think.
Moderator: // (. . .) you think you kind of you have to prove to the company that [overlaps with Tia]
Tia: You're worth it!
Ying: // (. . .) Cos for a company, they can't allow all the employees to leave early on Friday. The pay level, they pay for the working hours. And if anybody wants um to leave, because of their family, the company will lose a lot of money. //
Ying: And I think it's especially important to starters, cos for a starter, the company has to pay a lot of money for their training and probably they're the people that they can contribute to the least to the company. I think it's unfair for a company to pay everything they want.

Findings show that these young British and Asian people had a low sense of entitlement to employer WLB support based on their social comparison processes. In other words, they did not perceive employer WLB support as entitlements or rights and therefore had low expectations of their future employers and workplaces providing support for their individual ideals of 'balance'. Instead of describing forms of support that these young people had low expectations of, we argue that it is more important to examine and understand the underlying processes of their perceptions of limited entitlement. Here, we identify two different processes, both of which are based on their perceptions that employer WLB support is unfeasible, inappropriate, or not normative.

3.3.2.1 Low Sense of Entitlement 1: The 'Bonus' Perception

As the title suggests, the young people who articulated this perception did not perceive employer WLB to be an entitlement, but rather as a 'bonus' or a favour. For example:

Traveller: If they do give, it's a bonus. If they don't, I don't expect it, because there are not that many companies that do that. There are companies that do, but some companies can't have flexible working hours. // (. . .) I have friends that are working with companies

> that deals with US client-base (...) You can't have flexible working hours then, you have to work at the times that the US office are open.
> Interviewer: (...) And do you think that's fair?
> Traveller: Well, they chose the job! No one forced them in to it. If they don't have a problem with it, then that's fine, it's fair. (Traveller, Malaysian man, age 25)

Traveller's account indicated a low sense of entitlement to flexible working hours as a form of man support. This was based on, firstly, his perception that it was not normative for most employers to do so in Malaysia, his national context, where flexible working was not regarded as the norm in comparison to the UK. Secondly, the experiences of his friends in Malaysia further informed his perception of what was normal to expect from future employers.

For many of the young people who perceived employer WLB support as a favour, they perceived the idea of such support to be a positive benefit, but not something that they would expect or take for granted and therefore, not an entitlement. This indicates that while employer WLB support may be deemed feasible among some participants, this perception alone may not necessarily increase their sense of entitlement to such support. The following focus group excerpt extends this point further:

The issue of flexible working hours was raised again among the young people. This time, it was done in a focus group discussion. The two young Asian women above engaged in a complementary interaction (Kitzinger, 1995) to articulate how employer WLB support is not an entitlement for all employees. Tia and Ying's discussion denotes a perception that employers should provide WLB support only to 'good workers' if it is economically feasible for them to do so. They constructed WLB support as a legitimate strategy to reward those who have demonstrated their commitment and high performance in the organization. Therefore, although employer WLB support was seen as feasible, the appropriateness of when it should be implemented depends on whether individual employees have earned it. Tia and Ying did not exclude themselves from this criterion, as they argued that this was particularly relevant in the case of new recruits like themselves. By conceptualizing employer WLB support as something to be 'earned' by individuals, the ideal of individual choice and responsibility was maintained and upheld in these young people's talks of WLB support.

3.3.2.2 Low Sense of Entitlement 2: The Scepticism Perception

In the second perception, the young people who exhibited this were wary of the existence and rationale of WLB support in organizations, as shown in this focus group excerpt:

> Nancy: (...) I don't really expect anything from my employers. I sort of have quite, uh, initially not negative, but I carry quite cynical maybe approach to why they're doing things or what you can get out of them and then it should be there, but you can't necessarily expect it. You can't expect that they'll treat you in a nice way or like benefits, really.
> Moderator: Why is that?
> Nancy: Because they're not all nice [laughs].
> :[Laughter from group]

Moderator: Yeah. Do you agree with that?
Titho: I think it is. That's what the mindset was like. It's changing, I would say. But yeah, most of the places you don't tend to have that.

As this focus group discussion shows, some young people appeared cautious in their expectations of their future employment relationships. Nancy, a young British woman, had negative views of employers based on her existing part-time employment experiences. Titho, a young Indian man, had 6 years' full-time employment experience prior to his current MBA studies in the UK and was less cynical by comparison. This demonstrates how personal experiences from different national and cultural contexts shaped these young people's sense of entitlement in various ways.

The accounts of other young British participants also demonstrate a degree of scepticism towards employer WLB support:

Interviewer: And what do you think about companies giving WLB support to workers?
Duncan: I'd be a little bit dubious about that, to be honest. Um, I, but I guess the word "company" brings up an image of like a large corporation. And I wouldn't particularly trust that they would have my best interest at heart. // I think it's very much down to the individual about what the right level is, rather than being suggested how hard you should work, how you should feel about working, you know. (Duncan, British man, age 26)

Duncan's reluctant view indicated a low appropriateness judgement towards employers having a role in individuals' WLB. To him, such support was an intrusion of individual choice and a way of imposing organizational values onto individual workers. This finding challenges the convention that employer WLB support is a way of providing choices for those with non-work commitment (Lewis, Gambles and Rapoport, 2007). The following quote extends this further:

I think it could be quite worrying, condescending, this kind of prescribing culture of '(...) We need to have meetings about your WLB.' I think, certainly if it's something that someone feels they're experiencing problems with – work-life integration or WLB – then certainly they could go up to their supervisor and say (...) But I don't think it's something that should be imposed on anybody. // It's certainly not something I'd be particularly comfortable discussing with my employers or the company. I think it'd be something I'd think should be my issue. (Ari Mazel, British man, age 27)

Ari Mazel's scepticism towards employer WLB support can be seen through his usage of words such as 'condescending' and 'imposed', which denote that he felt entitled to be able to choose if and when he would utilize such support. Instead of having a sense of entitlement to employer WLB support, there was a stronger sense of entitlement to individual choice among some of the young people. By questioning employers' motives for offering WLB support and arguing it could interfere with individuals' personal choice and responsibility, young people like Nancy, Duncan, and Ari Mazel were contesting the normative assumption that employer WLB support is always a 'win-win' for both employers and their employees.

Both the 'bonus' perception and the 'scepticism' perception expressed by the young people were underpinned by their perceptions that WLB is principally an individual responsibility and a personal choice. In other words, the rhetoric of individual choice was considered the norm for many of the young British and Asian people and therefore their sense of entitlement to employer WLB support were weakened accordingly.

3.3.3 Young People's "Wishes" for Employer WLB Support Are Not the Same as Their "Expectations"

Despite the young people's limited sense of entitlement to employer WLB support, this does not mean some of these young people did not have WLB needs. Here, we argue for the importance of differentiating their WLB expectations from needs or wishes. In the study, the young women and men were asked to talk about what kind of WLB support they would like from their future employers (their 'wishes'). They were then asked to consider their perceptions of employers' role in providing them with WLB support and what they would actually expect from their employers (their 'expectations'). Wishes are based on ideals and therefore do not necessarily reflect individuals' expectations, which tend to be based on perceptions of individual choice, sense of entitlement to WLB support, and social, professional, and organizational contexts.

These young people's 'wish list' brings to light the ways in which they talked about concrete forms of employer WLB support that are specific to their current life course stage. The findings indicate two main types of employer WLB support that were described in their personal and concrete accounts. Both types referred to employer support that would enhance their personal choice and exploration as young people.

The first type of WLB support was concerned with employers recognizing their needs and wishes as young people to develop as individuals and to provide them with the necessary support to do so. In their accounts, the young people talked about the desire to develop personally, including time away from work to pursue non-work interests, such as travelling and learning a new language. In addition, they also referred to professional development opportunities as a form of employer WLB support:

> Leila: (. . .) provide us (with) (. . .) training and, uh, travelling [laughs]. // Yes. And um, at least they give you some chance to learn something.
> Interviewer: Yeah. Why is training very important to your WLB?
> Leila: Uh, cos sometimes (. . .) when you work you just feel you don't know these things. But you know, there are some other experts in your company. But, uh, it maybe time consuming or maybe it's not easy to let others to share knowledge with you. So if you organize internal training, it's a good idea. // (. . .) cos I think job is kind of part of my life, so maybe I think if I'm happy job, I will be happy with life. (Leila, Chinese woman, age 25)

This reinforces the finding that at this stage of the life course, many participants perceived their forthcoming work and employment to be a major part of their lives. Consequently, employer WLB support such as training and development that aim to enhance their confidence and job performance was considered a form of employer WLB support. In the next excerpt, the issue of individual development through training was also raised:

> Ying: Well, I think that WLB ( ) is not only about family-work. // It's something about the chance to get promoted. Like you have the chance to learn while you are at work. Um, I think to me that if your company can pay for your training, your professional training, or they can give you some time to leave during the work for you just intensive education, university or college (. . .) And they give you a real chance to promote yourself. (. . .) That's quite a good way to balance the life. Cos if you work for a company that don't provide anything like that, so you have to study (. . .) during your personal life time, which means you won't have the chance to meet your friends at all. //
> Ji: I think that's, uh, very true. (. . .) your free time is your free time. It shouldn't be involved with the company, but if they can recognize you and the problems that you have and like, yeah, education and things if, if some kind of transition between going into your first job and them helping you learn more skills, uh, I don't know, that would be very useful.

As shown in this focus group discussion, some of the young people were broadening the issue of WLB to be more than fulfilling family and work responsibilities. To them, WLB support involves having the opportunity to realize their goals and ambitions in life, including in their employment career. Both Ying and Ji were in agreement that it was important for employers to support them as individuals with personal lives outside of work, as well as newcomers transitioning into the world of work and employment. One way that employers could provide WLB support that would attend to this was to allow them time at work to continue with training and development. By making studying time a part of working time rather than personal time, these young people believed that this could assist their ability to maintain temporal boundaries between work and personal life.

In addition to wanting employers to meet their development needs as a form of WLB support, young people also talked about their desire of being able to quickly integrate into their organizations. One way to achieve this was for employers to provide appropriate inductions that would reassure these new recruits of their new surroundings, colleagues, and policies and procedures that are in place:

> Interviewer: Now in what ways would you like your school [her employer] to help you personally manage your WLB if you choose to?
> Megan: I think the biggest thing for me is just getting used to where things are. (. . .) I think it sounds silly, but that takes ages to get really into the flow of things. So, I think for my profession, it's just somebody taking the time out to kind of show you where things are. You know, rather than just leave you to it. (. . .) // Or just certain policies the school has on things, cos you can read through tons and tons of policies (. . .) sometimes you just remember things a lot better if someone just sits down and explains it to you. // And just getting to know people, I suppose. Cos you have to work alongside them, so if you haven't really met them, it's not a very good way to start. // You feel quite sort of looked after [laughs] if it's done properly.

In the last part of Megan's quote, she referred to her wish of having the opportunity to get to know the people that she works with. This issue of being able to socially integrate and socialize with people at work was also discussed among the other British and Asian participants:

> Um, I don't think I would look to the company to, to provide it, but if they were going to provide it, I guess, are you asking what sort of format it would take? // Um, I think perhaps, like, one way that they could help would be like to organize like social events and to encourage social interaction. (. . .) I guess that's a slightly different way of looking at it. I

> haven't really thought about it in that way until now, but yeah, encourage social interaction, encourage like, um, social events. (...) You know, ideas like that which will, um, encourage, uh, like how they think it'd all encourage harder working, by like, improving the atmosphere within the office between the people and the work. And I think that will be quite important. // (...) I think before (...) I was thinking about it in terms of someone from within the company telling me how to feel good about working as hard as they want me to work. (Duncan, British man, age 26)

To contextualize Duncan's account, it is important to point out that he was initially doubtful about how employers can provide WLB support. However, as our interview progressed, he had changed his response and became more open to the idea of employer WLB support. This illustrates how during the course of an interview or a focus group, participants can and do engage in self-reflections that may result in modifications of their original answers. In the case of Duncan, it can be seen that the idea of having social events that encourage social interactions among workers emerged to him as a useful form of employer WLB support later on in the interview. As work becomes a new part of these young people's life course trajectories, this finding signifies that the ability to integrate into their new workplace is a key part of their transition into their new life course phase.

What emerged from the 'wish list' findings is that through talking about the various supports that they would like, the participants were in effect broadening and redefining the 'life' component of the WLB equation. Current WLB discourses and policies tend to reflect the assumption that 'life' typically purports to caring responsibilities, mainly directed at women (Lewis et al., 2007; Smithson and Stokoe, 2005). This particular view has been criticized for being too narrow and failing to consider equally pertinent life issues faced by other members of the workforce (Eikhof et al., 2007; Kamenou, 2008). The findings have demonstrated that these young people were approaching 'life' in a way that went beyond how it is usually discussed in existing WLB literature. In essence, they show that life encompasses more than family roles and responsibilities. Moreover, they illustrate how these young people were actively attributing diverse meanings of life in their own terms, based on personal experiences and current life course phase.

3.4 Concluding Discussion

Workplace WLB polices tend to be developed as recruitment, retention and well-being measures in response to what have been identified as the needs of a specific sector of the workforce, namely parents and other carers, particularly women. However, our findings demonstrate that it is not only those with current caring commitments who need some control over workload and working time in order to achieve 'balance' (or indeed 'integration') between work and non-work activities, although balance can be interpreted and experienced in diverse ways. Moreover, we see that notions of WLB are not fixed but dynamic and fluid as people move through the life course.

This suggests that employers and human resource professionals need to examine their work-life policies, the ways in which they are framed, and how they play out in practice. Do they address the diverse needs of the workforce across the life course or are they targeted at certain groups? Do they include or exclude the WLB needs of young professionals? Do WLB policies and practices enable professional across the age and generation spectrum to truly exercise their preferred/ideal WLB? And to what extent do policies and practices recognize the fluid nature of their needs and expectations? Reflecting on these questions could encourage a radical shift towards the development of 'life-course friendly' working practices. We argue that this would necessitate a rethinking of the principles and assumptions on which WLB policies and practices currently tend to be based, moving from the principle of need to the principle of mutual benefit.

Discussions about WLB policies or what were initially called 'family friendly' policies were initially a response to the growing number of women entering workplaces designed by men for those without other commitments. Initiatives therefore tend to be implicitly if not explicitly targeted at narrowly defined groups of workers on the basis of assumed needs. For example, managers usually have some discretion about who can take up opportunities to work flexibly and decisions are often made on the basis of assumed family responsibilities or gender appropriate behaviour. The needs of parents and carers are considered to be different from and more legitimate than, for example, the need for leisure or personal development. Thus, the WLB needs of young professionals without caring responsibilities may be regarded as less pressing or deserving of attention. However, this misses the point.

It is true that care needs are more pressing than other commitments. Yet, if WLB supports are defined in this way, they are constructed as 'favours' or 'accommodations'. Assumptions that ideal workers are those who can give unlimited time to their jobs and who do not need time for other commitments (Bailyn, 2006; Lewis et al., 2007) remain unchallenged, despite changes in behaviours and values across generations. Those who deviate from this norm may be supported, but they are also marginalized as 'non-ideal' employees. Taking up WLB policies is too often career limiting (Holt and Lewis, in press). This is often accepted as a trade-off, by employed mothers in particular (Lewis, 2009; Kelliher and Anderson, in press; Mickel and Dallimore, 2009), but it perpetuates gender inequalities and results in an underdevelopment of an organisation's human resources. Women professionals often perceive the trading of career opportunities for flexibility as a choice, but is clearly a very constrained one (Lewis, 2009). Yet, those who are not carers, and even fathers, do not feel entitled to change the way that they work. It reduces the 'choices' open to other professionals who would like to work in different ways. This approach perpetuates disadvantage among those whom the policies are designed to support, and excludes young professionals and others. It can also create backlash among the wider workforce who may feel that parents are gaining unfair advantages (Casper et al., 2007; Young, 1999).

The association of WLB and flexible working debates and policies with parenting and caring is deeply entrenched. Nevertheless, an alternative approach is to focus arguments for WLB support not on the principle of need, but on the potential benefits of mutually flexible forms of work that can meet the dual agenda of enhancing both employee work-life integration or balance and also workplace effectiveness (Lewis and Cooper, 2005; Rapoport et al., 2002). In this approach, the focus is on enhancing outcomes rather than on inputs such as face time or traditional career patterns. The reason for professionals to request flexible working arrangements or to decide to work in flexible ways thus becomes less important than their effectiveness. If WLB policies and practices are framed in this way, they are more likely to meet the diverse and dynamic needs of the workforce.

We have seen that the future young professionals participating in this study had a strong sense of individual responsibility for their own work and non-work lives and a low sense of entitlement to support from employers. Policies that focus on need and favours within a context in which the ideal worker does not seek work life balance will reduce the real choices that are available and reinforce low sense of entitlement to support. The strong sense of self-responsibility among this group suggests that young professionals would be well placed to work out flexible arrangements that are mutually beneficial, with a focus on rights and responsibilities rather than one-sided entitlements.

Furthermore, these young people recognized that their needs and values were dynamic and shifting across the life course and across careers that no longer fit the traditional lock-step pattern (Moen and Sweet, 2004). This implies that it is important to address the needs of a mutigenerational workforce and manage talent at all phases, with implications for organizations aiming to attract, recruit, and retain 'valuable' workers (Loughlin and Barling, 2001; Pitt-Catsouphes and Smyer, 2007; Smola and Sutton, 2002). For example, some professionals may want to have more time for leisure and/or personal development early in their career, become more intensively involved in their work later, cut back during a period of family involvement and then again become intensively involved at an age when they might be expected to retire, while others will have quite different patterns. Employers, individual professionals, and the wider society can all potentially benefit from enabling people to work in 'life course friendly' ways. The challenge for managing the work-personal life needs of professionals in mutually flexible and beneficial ways is to be able to develop policies and practices that are flexible enough to harness this diversity between workers as well as within individuals across the life course.

References

Bailyn L (2006) Breaking the mold. Cornell University Press, Ithaca, NY

Brannen J, Lewis S, Nilsen A, Smithson J (eds) (2002) Young europeans. Work and family: futures in transition. Routledge, London

Casper WJ, Weltman D, Kwesiga E (2007) Beyond family-friendly: the construct and measurement of singles-friendly work culture. J Vocat Behav 70:478–501

Crompton R, Lewis S, Lyonette C (2007) Continuities, change and transformations. In Crompton R, Lewis S, Lyonette C (eds) Women, men, work, and family in Europe. Palgrave Macmillan, Houndmills, pp 230–244

DfEE (Department for Education and Employment) (2000) Work-life balance 2000: Baseline study of work-life balance practices in Great Britain. Summary Report. DfEE, Sudbury

Dight C (2008) It's no longer just about the money. The Times (2008, 18 January). http://business.timesonline.co.uk/tol/business/career_and_jobs/graduate_management/article3196809.ece. Accessed 17 Dec 2008

DTI (Department of Trade and Industry) and Scotland Office (2000) Work-life balance the business case. Electronic version. http://webarchive.nationalarchives.gov.uk/ + http://www.dti.gov.uk/work-life balance/publications.html. Accessed 8 Nov 2010

Eikhof DR, Warhurst C, Haunschild A (2007) Introduction: what work? What life? What balance?. Empl Relat 29:325–333

Families and Work Institute (2004) Generation and gender in the workplace. Families and Work Institute, New York, NY

Fleetwood S (2007) Re-thinking work-life balance: editor's introduction. Int J Hum Resour Manage 18:351–359

Gerson K (2004) Understanding work and family through a gender lens. Commun Work Fam 7:163–178

Giele JZ, Elder GH (1998) Methods of life course research: qualitative and quantitative approaches. Sage, Thousand Oaks, CA

Holt H, Lewis S (in press) 'You can stand on your head and you still end up with lower pay': gliding segregation and gendered work practices in two Danish workplaces. Gender Work Organ

Kamenou N (2008) Reconsidering work-life balance debates: challenging limited understandings of the 'life' component in the context of ethnic minority women's experiences. Br J Manage 19:99–109

Kelliher C, Anderson D (2009) Doing more with less? Flexible working practices and the intensification of work. Hum Relat 63:83–106

Kitzinger J (1995) Introducing focus groups. Br Med J 311:299–302

Lewis S (2009). Flexible working policies, gender and culture change. In European Commission Directorate-General for Research (ed) Women in science and technology. Creating sustainable careers. European Communities, Luxemburg, pp 40–48.

Lewis S, Cooper CL (2005) Work- life integration. Case studies of organisational change. Wiley, Chichester

Lewis S, Haas L (2005) Work-life integration and social policy: a social justice theory and gender equity approach to work and family. In Kossek EE, Lambert SJ (eds) Work and life integration: organizational, cultural, and individual perspectives. Lawrence Erlbaum, Mahwah, NJ, pp 349–371

Lewis S, Smithson J (2001) Sense of entitlement to support for the reconciliation of employment and family life. Hum Relat 54:1455–1481

Lewis S, Smithson J, Brannen J (1999) Young Europeans' orientations to families and work. Ann Am Acad Pol Soc Sci 562:83–97

Lewis S, Smithson J, Kugelberg C (2002) Into work: Job insecurity and changing psychological contracts. In Brannen J, Lewis S, Nilsen A, Smithson J (eds) Young Europeans, work and family: futures in transition. Routledge, London, pp 69–88

Lewis S, Gambles R, Rapoport R (2007) The constraints of a 'work-life balance' approach: an international perspective. Int J Hum Resour Manage 18:360–373

Loughlin C, Barling J (2001) Young workers' work values, attitudes, and behaviours. J Occup Organ Psychol 74:543–558

Mickel A, Dallimore E (2009) Life-quality decisions: tension management strategies used by individuals when making trade-offs. Hum Relat 62:627–668

Pitt-Catsouphes M, Smyer MA (2007) The 21st century multi-generational workplace. Center Aging Work Issue Brief 09:1–12

Rapoport R, Bailyn L, Fletcher JK, Pruitt BH (2002) Beyond work/family balance: advancing gender equity and workplace performance. Jossey-Bass, San Francisco, CA

Rousseau DM, Wade-Benzoni KA (1995) Changing individual-organization attachments: a two-way street. In Howard A (ed) The changing nature of work. Jossey-Bass, San Francisco, CA, pp 290–321
Smithson J, Lewis S (2000) Is job insecurity changing the psychological contract?. Pers Rev 29:680–702
Smithson J, Stokoe EH (2005) Discourses of work-life balance: negotiating 'genderblind' terms in organizations. Gender Work Organ 12:147–168
Smola KW, Sutton CD (2002) Generational differences: Revisiting generational work values for the new millennium. J Organ Behav 23:363–382
Sturges J (2008) All in a day's work? Career self-management and the management of the boundary between work and non-work. Hum Resour Manage J 18:118–134
Sturges J, Guest D (2004) Working to live or living to work? Work/life balance early in the career. Hum Resour Manage J 14:5–20
The Times (2004). Don't just give me a car – give me a laptop and a free Friday afternoon. The Times. http://www.timesonline.co.uk/tol/classifieds/jobs/article433692.ece?token=null&offset=0&page=1. Accessed on 27 May 2004
The Work-Life Balance Trust & Robertson Cooper Ltd (2002) The graduates' guide to the best work-life balance employers in Britain. The Work-Life Balance Trust & Robertson Cooper Ltd, Manchester. Accessed 17 December 2008
Tomlinson J (2006) Women's work-life balance trajectories in the UK: Reformulating choice and constraint in transitions through part-time work across the life-course. Br J Guid Couns 34: 365–382
Unisys (2004) Young workers demand flexibility as work pressure hits friendships and home life. http://www.unisys.co.uk/about_unisys/news_a_events/05270401uk.htm. Accessed 6 July 2004
Universum (2003) The Universum Graduate Survey 2003 Pan-European Edition. Universum Communications, Stockholm
Williams L, Jones A (2005) Changing demographics. The Work Foundation, London
Young MB (1999) Work-family backlash: Begging the question, what's fair?. Ann Am Acad Pol Soc Sci 562:32–46

Chapter 4
Redefining Success: The Foundation for Creating Work-Life Balance

J. Lee Whittington, Rosemary Maellaro, and Timothy Galpin

4.1 Introduction

The line between work and home has become blurred by our ability to communicate and share information 24/7. The work day, indeed the work week, has expanded beyond the boundaries of place and time that formerly provided some buffer between work and the other arenas of life. While employees at all levels of the organization are affected by this blurring of boundaries, this is especially true for managers, technical people, and individual contributors in the middle and upper ranks of an organization – referred to throughout the chapter as professionals – because they are often expected to do "whatever it takes" to accomplish their assignments.

The lack of perceptible and distinct boundaries exacerbates the lack of balance between work life and personal life that many professionals experience. Consequently, professionals spend the majority of their time, physical effort, and – perhaps most importantly – emotional energy on their work. The result of this is that the personal, family, community, and social arenas of life receive only the leftovers. This lack of balance has both immediate and long-term consequences for individuals, families and organizations.

In this chapter we address the disproportionate amount of time spent in the work and non-work life arenas by suggesting that achieving a more balanced life requires a redefining of personal success. While success is typically defined in the context of work and career, we suggest that a more comprehensive view of success is required: a whole-life view of success. This perspective draws on the pentathlon as a metaphor for success.

As with this Olympic competition, our whole-life view of success requires performance in multiple events (life arenas), not just one. As such, the Whole-Life Model (see Fig. 4.1) provides a basis for a more comprehensive view of well-being. Rather than emphasizing career and work to the exclusion of the family, personal,

J.L. Whittington (✉)
University of Dallas, Dallas, TX, USA
e-mail: jleewhitt@aol.com

S. Kaiser et al. (eds.), *Creating Balance?*, DOI 10.1007/978-3-642-16199-5_4,

Fig. 4.1 The pentathlon: a whole-life model of success

spiritual, and community arenas, this model places career and work in the context of overall life development. The underlying dynamic for the model is a cooperative, win-win strategy as opposed to a competitive, win-lose strategy. We elaborate strategies for developing each of these arenas within our broader view of success.

We believe that a more comprehensive view of success is the foundation for achieving work-life balance. However, mere intellectual agreement will not create the desired balance. Therefore, we explore the barriers that prevent professionals from implementing this perspective and introduce the requirement of strategic cheating in order to live out the view of success we advocate.

A key component of strategic cheating is learning to work and lead from personal strengths. Individuals who are able to sustain their effectiveness have discovered the importance of channeling their energy into those arenas that are closely aligned with their giftedness and passion. This is difficult because many high achievers have a natural desire to prove themselves in everything they do. This is often rooted in a strong need for achievement. While this is certainly an important trait, this orientation may in fact be detrimental if it is not disciplined by a focus on the individual's core competencies. Many professionals burn out because they allow their time and emotional energy to be consumed by tasks that are outside their zone of competence. This drains physical and emotional energy and detracts from their ability to make the greatest contribution possible to the organization. Furthermore, this draining of emotional energy impacts the other arenas of life. Overcoming this tendency to spend time and energy on activities outside the zone of competence requires individuals to identify and work from their strengths.

We begin our discussion with a brief review of recent research on the challenges associated with living a balanced and integrated life. We then introduce the Whole-Life Model as the basis for a new definition of personal success that explicitly acknowledges the complexity of our lives and we discuss each arena in the model. The implementation of this perspective requires an intentional effort that we refer to as strategic cheating. The implications of this are discussed along with personal strategies for achieving a more balanced life. We also discuss the role leaders play in taking the initiative to not only create organizational policies that support work-life balance, but also in modeling the balance they are encouraging employees to implement.

4.2 Literature Review

As the line between work and personal life has become blurred by the ability to communicate and share information 24/7, the work day has expanded and now often intrudes into our personal lives (Mariotti, 1998). Therefore, it is important to develop a balance between our personal and work lives. In fact, work-life balance is imperative for personal and organizational success, as well as for personal health and well-being. On the one hand, a lack of balance between work and personal activities can cause burnout, low productivity, absenteeism, dissatisfaction, and stress-related illnesses at all levels of the organization (Edwards and Rothbard, 2000; Kreiner et al., 2006; Parasuraman and Greenhaus, 2002; Rice et al., 1992; Wiley, 2006). On the other hand, people who are able to strike a balance between their work and personal lives – and thus more effectively allocate their time and energy to the myriad demands made on them – can significantly increase their well-being (Mariotti, 1998). Overall well-being has both physical and mental components and "it's about feeling positive, and having enough energy to be productive at work and to enjoy non-working life" (O'Reilly, 2006).

This call for balance is supported by the emerging theory of work-family enrichment (Greenhaus and Powell, 2006), rather than work-life balance, which implies an either/or mind set. According to this theory, participation in roles at work and at home can have positive additive effects on physical and psychological well-being. Individuals who participate in satisfying work and family roles have been known to experience greater well-being than those who participate in only one of the roles or who are dissatisfied with one or more of their roles. In much the same way that a diverse financial portfolio protects an individual's financial well-being from poor performance in a particular type of investment, participation in multiple social roles can buffer people from distress that may originate from one particular role (Greenhaus and Powell, 2006).

Recent research by Bourne et al. (2009) provides support for the idea of living a more balanced life. Results of their study revealed that the majority of participants identified themselves as "dual-centrics" who value their work and non-work roles equally. These dual-centric individuals experienced more overall satisfaction, greater work-life balance, and less emotional exhaustion. The benefits of this dual-centric focus suggest that organizations can improve and enhance employee well-being by embracing them as whole individuals; that is, by acknowledging and valuing their activities in life arenas other than work.

The benefits of creating such a balance accrue to organizations as well as individuals. Companies that support their employees' efforts to live more balanced lives are featured on Fortune's "100 Best" list. These companies have consistently outperformed the S&P 500 (Levering et al., 2000) and they earn significantly higher customer satisfaction ratings than firms not on the list (Simon and DeVaro, 2006). Most importantly, these firms are more effective in their efforts to recruit and retain top talent (Reed and Clark, 2004).

4.3 Redefining Success

Your definition of success is a fundamental element of any assessment you make about balance in your life. Your own definition of success drives your actions. It is not uncommon for people to discover that they have never identified for themselves just how they define and measure success (Gurvis and Patterson, 2005, p. 6).

Given the evidence that leading a balanced life provides benefits to individuals and the organizations they participate in, it will be helpful for individuals to personally redefine success. Too often professionals define success narrowly by focusing on one arena: Work and career. Yet individuals who define success narrowly have lower levels of overall life satisfaction and more emotional exhaustion than those with a more balanced and comprehensive view (Bourne et al., 2009).

Support for this more complex view of success is also provided by Linville's work on self-complexity (Linville, 1985, 1987). According to Linville's (1987) self-complexity buffering hypothesis, individuals with high levels of self-complexity are buffered from distress and strain when a stressful life event occurs in one life arena. When an individual's self-worth is based on their total self-concept, stressful events in one life arena are not as devastating. Individuals whose self-worth is dominated by only one role or arena are more likely to experience higher levels of distress when a stressful event occurs in that arena.

A key element in developing a more balanced life is to make the decision to redefine success in a way that reflects the importance of the non-work dimensions. Borrowing from the pentathlon model developed by Sherman and Hendricks (1990), we offer a whole-life view of success that, in addition to the work arena, includes the four non-work arenas of life shown in 4.1. Within the framework of this model, which emphasizes each of life's five arenas, success is a function of balance of time, commitment, and the investment of emotional energy in these five areas of life. As with the Olympic event, the model suggests that success requires performance in all arenas (events) and not just one.

4.3.1 Non-work Arenas

The personal arena is the private world of the self (MacDonald, 1985). This private world can be viewed as our essence, and the ordering of this arena in life provides an anchoring stability around which the other life arenas may be ordered. This arena includes personal health, exercise, stress management and leisure time activities. Relationships with mentors and protégés would properly fall in the personal arena.

The complexity of contemporary organizational life creates a great deal of stress on individuals. A plan for personal well-being can help people "stay alive" in their organizational roles (Heifetz and Linsky, 2002). According to Heifetz and Linskey, individuals can fail at work by forgetting to pay attention to themselves. In the midst of workplace challenges, the adrenaline is flowing and people forget that they are vulnerable to the limits of physical and emotional capacity.

In order to stay alive in their work role, individuals must learn to be aware of and manage their own hungers. Recovery programs have long advocated the H-A-L-T principle: Never get too hungry, too angry, too lonely, or too tired because these situations create a heightened vulnerability that can diminish a person's ability to act wisely. Managing these hungers requires self-awareness and personal discipline.

While self-awareness and personal discipline are important, too often people are blind to the toll their work role and its expectations are having on them. Individuals need to cultivate an inner circle of people who are willing to help them maintain balance and proper boundaries. The cultivation of these relationships occurs in the community arena where we can develop healthy interpersonal attachments for the purpose of social support. Secure interpersonal attachments may exist in the work, family, and community arenas, while a secure transcendent attachment to God exists within the spiritual arena (Quick et al., 1995). A person's spiritual and secular communities may or may not overlap.

Heifetz and Linsky (2002) extend the community arena by discussing the importance of confidants. Confidants provide a safe place where a person can say everything that is in his or her heart without the need to script or edit the raw feelings and emotions. These are people who can put a person "back together again" when they have just lived through a Humpty Dumpty-like falling off of the wall (Heifetz and Linsky, 2002). Confidants provide a unique perspective because they care about the person and may or may not be connected to the individual's work role. They can listen intently and objectively. Confidants are people who tell the individual what he or she needs to hear. They provide information and insight that a person may not want to hear and will not be able to hear from anybody else. Reciprocally, confidants can be confided in without concern that the individual's revelations will spill back into the organization. People must cultivate confidants and allow them into their lives if they are to run the race of life with endurance.

The spiritual arena includes the cultivation of a relationship with God, as enhanced by personal efforts in prayer, meditation, and study, as well as involvement in a community of like believers. Internalized, intrinsic and individually active religion has been associated with less illness and better physical health (McIntosh and Spilka, 1990).

Cultivating the spiritual arena may require the creation of a readily available sanctuary (Heifetz and Linsky, 2002). A sanctuary is a designated place where the individual can withdraw for reflection and renewal. It is a place that provides emotional and physical security and allows the person to suspend the stress of the workplace. Creation of sanctuary requires the discipline to structure schedules and routines so that the benefit of the sanctuary is not lost in the busyness of the person's life. The form of the sanctuary may vary. It could be a jogging trail, a garden, or a special room that provides the quiet and solace needed to have uninterrupted time alone for reflection and renewal.

The family arena emphasizes responsibilities to spouse and children, yet also includes obligations to siblings and to parents. Planning and budgeting for household management requirements, spending significant time with children and continuously cultivating one's spousal relationship are all activities associated with the family arena.

The whole-life definition of success suggests that one must perform in all five of life's arenas. This imagery suggests the interdependence, not complete independence, of the five arenas. This advocates a Whole-Life Model of well-being and balanced investment in the five arenas that may buffer a manager from distressing events in life (Linville, 1987).

4.3.2 Strategic Cheating

Developing a more comprehensive definition of success is the point of beginning. However, the implementation of this model requires creating boundaries on the commitments an individual makes in terms of time, physical energy and emotional labor. Enforcing these boundaries requires an intentional effort to limit commitments. We call this intentional effort strategic cheating. This is not an immoral activity but rather a disciplined approach to setting boundaries around time and energy commitments, and making strategic trade-offs that reflect our espoused values and priorities. The essence of strategic cheating is essentially about setting and living out priorities. "Setting priorities is about putting something ahead of something else. A priority is something you say yes to even when it means saying no to other important things" (Stanley, 2003, p. 12).

The fact is we all cheat by expending the bulk of our time and emotional energy in one arena at the expense of the others. This is cheating by default and rarely reflects our espoused values. Strategic cheating is an intentional act. It is a decision to set and keep boundaries so that over-commitment in one arena does not damage the other arenas. The dual-centrics discussed earlier provide an example of people who have made the decision to place as much value on their non-work roles as their work roles.

4.4 Individual Work-Life Balance Tactics

A preponderance of research seeking solutions to the work-life balance conundrum has focused more on organizational solutions rather than the individual solutions. This research places an emphasis on studying human resource policies including flextime, family-friendly benefits, and other macro-level organizational aspects (Stebbins, 2001). Likewise, Perlow and Porter (2009) found that companies requiring professionals to take time off maintained high levels of service and experienced better retention rates of valuable talent. Recently researchers have begun to focus on individual-level solutions including decision process theory (Poelmans, 2005) and flexibility enactment theory (Kossek et al., 2005). However, research on work-life balance at the individual level tends to examine stable and/or hard-to-change variables such as gender. For example, Byron (2005) found that male employees tend to experience more work interference with family and female employees tend to encounter more family interference with work. Furthermore, in their meta-analysis of work and family research, Eby et al. (2005) found that a majority of the literature focused on gender with minimal consideration of individual differences such as personality or motivational traits. The reviewers found that professional women have more difficulty relaxing after work and report greater stress in both their work and non-work lives than men. They also found that women benefit less than men do from work arrangements designed to facilitate work-family balance, and that in dual-earner couples women report greater stress symptoms than men.

A significant weakness of most of the studies of work-life balance is that they do not offer actionable guidance to people wanting to effectively integrate all life arenas or to alleviate stress (Kreiner et al., 2009). We believe that providing actionable knowledge to individuals helps them implement practical solutions to work-life balance problems (Cummings and Jones, 2003). "Clearly, individuals play a crucial role in affecting work-home outcomes; they are not mere automatons reacting helplessly to the pressures around them" (Kreiner et al., 2009, p. 705).

Kreiner et al. (2009) found that individuals utilize a variety of work-life balance tactics, which are the various "work-family decisions" they make to recalibrate the work-home boundary negotiation. Their findings suggest that a reciprocal relationship exists between the challenges of work-life boundaries and the tactics employed, that the challenges cue the need for tactics, and that implementation of the tactics can successfully reduce the challenges. These tactics fall into four broad categories.

First, behavioral tactics include using the skills and availability of other individuals who can help with the work-home boundary, such as having staff members screen calls and using voicemail, caller ID, or e-mail to facilitate work boundaries. Performing regular task triage by prioritizing urgent and important work-life demands such as a work deadline and childcare emergencies is also an example of a behavioral tactic. Second, temporal tactics are characterized by manipulating regular or sporadic plans such as blocking off segments of time to do certain work or family tasks, and removing oneself from work-home demands for significant segments of time through vacations, getaways, or retreats. Third, physical tactics include establishing physical borders between work and home, such as having a different room at home from which to complete work related tasks, increasing or reducing the distance between work and home, and using tangible items such as calendars and photos to blend aspects of work and home. Lastly, communicative tactics involve managing others' expectations in advance of a work-home boundary violation, such as stating preferences to co-workers or family ahead of time and confronting violators of work-home boundaries either during or after a boundary violation.

Beyond these four categories there are specific tactics that people effectively utilize to strike their own work-life balance. For example, Dambrin and Lambert

(2008) found that women auditors utilized delegation and imposed new work practices such as setting a clear time to leave the office. They also built a home partnership with their spouse, employed third parties such as housekeepers and nannies, and shifted careers from operational to staff roles. Likewise, Armenio (2004) presents several work-life balance tactics that include developing work templates to increase efficiency, planning for personal and vacation days well in advance, respectfully declining extraneous meetings, and setting aside a certain amount of time to complete routine daily tasks (for example, answering emails or returning phone calls). Lastly, Zimmerman (2004) suggests that creating an end-of-the-day ritual, such as asking each family member about the high and low points of their day, helps a person to quickly transition from work to family mode.

4.5 Implications for Practice and Leadership Education

Employees flourish when senior leaders help them focus on what matters most not only at work but in all aspects of their lives – at home, in their communities, and in their pursuit of physical, emotional, and spiritual well-being. The happy result: committed people driving hard to achieve superior performance. (Friedman and Lobel, 2003, p. 87)

The notion that all domains of an individual's whole life must be aligned and balanced to realize maximum organizational performance and employee satisfaction requires a more nurturing, collaborative, and holistic approach to leadership (Bilimoria, 1998; Friedman, 2004; Friedman, 2006; Friedman and Lobel, 2003; Marshall, 2000). Professionals do not always realize that they have a bounded capacity to work or they may be reluctant to ask their managers for flexibility in meeting their work-life balance needs (Gurvis and Patterson, 2005). Consequently, professionals may need help clarifying their values to determine a work-life balance that is uniquely meaningful to them and they also need a work environment where they feel comfortable actually striking that balance (Bilimoria, 1998; Kossek et al., 2005). Friedman (2006) suggests that leadership and life are both "pieces of the same puzzle" and recommends a model of "total leadership" wherein leaders recognize and respect employees as whole persons and become their allies to help them clarify what is important. By developing employees in this manner, they can "leverage synergies across work, family, community and self" (Friedman, 2006, p. 1270). A leader's failure to facilitate this type of professional development may be viewed as mismanagement of organizational resources (Marshall, 2000).

Therefore, an updated approach to leadership in organizations requires that leaders live congruently with their core values and encourage their employees to do the same (Friedman, 2006; Glubczynski et al., 2003). Accordingly, business schools must also focus on developing these leadership skills, which traditionally have not been addressed in management education programs. This section provides an overview of the organizational leadership skills needed to effectivley implement the Whole-Life Model.

4.5.1 Self-Awareness

To craft their personal definition of success, leaders must begin by becoming very clear about who they are, what they value, and what is important in all five of their life domains (Friedman and Lobel, 2000; Maellaro and Whittington, 2009; Pfeffer and Sutton, 2007). This can be accomplished through the use of value identification instruments and/or exercises, as well as organizational flexibility that encourages employees to pursue meaningful activities in non-work arenas.

Employees and managers can further enhance their self-awareness by participating in 360° feedback surveys, scheduling specific time to reflect, and being open to feedback about how their behavior impacts others. Organizations can support individual efforts to become more self-aware by providing honest performance feedback and by acknowledging those who demonstrate an understanding of the impact of their behavior and who take proactive steps to ensure a positive impact. Management education programs can help students develop self-awareness by introducing them to personal emotional awareness logs, helping them learn to identify trends or behavioral triggers, and increasing their ability and comfort in asking for, receiving, and acting upon behavioral feedback.

4.5.2 Systems Thinking

A new holistic approach to leadership is needed if organizations want to promote greater work-life integration for their employees. The mechanistic view of a life that consists of separate, independent domains continues to prevail in many organizations. This dis-integrated view must be replaced with a more comprehensive and integrative view. To fully achieve a broader notion of success leaders must model the way by bringing their whole selves to work (Houston, 2004). When leaders "pretend that the moment a human being crosses the magical threshold from nonwork to work, different behavioral and social forces come into play" they are conceptualizing work as a "separate land governed by distinct, narrow rules ..." (Pfeffer and Sutton, 2007, p. 3).

This outdated approach is based on the concept of partial inclusion (Katz and Kahn, 1966) and assumes that professionals bring only a "psychological slice" of themselves into the organization. In reality, employees have multiple identities (Houston, 2004) and bring to work each day an entire person who is also part of many other systems (family, church, community, social organizations, schools, etc.). The significance of this is that identity in each role and behavior in each organization will impact and be impacted by their commitments in other arenas. This calls for explicit recognition of the interdependence of the multiple roles and arenas that make up our lives. When leaders apply systems thinking they will view themselves and others as whole entities. This will help them redefine their view of personal success and will allow them to encourage employees to balance and align their life domains as well.

Organizations and management education programs can train professionals to develop their critical thinking skills and increase their ability to take a strategic, big picture approach to their decisions and actions. Effective techniques for developing a systems thinking mindset include challenging old mental models and creating new ones, mapping organizational processes for the purpose of examining their dynamic interrelationships, and story-telling, which is a conscious effort to discern "a coherent story out of the interrelationships of seemingly random events" (Senge et al., 1994, p. 103).

4.5.3 Sensitivity to Employee Concerns

Leaders must set the tone for creating organizations that reflect this perspective by implementing organizational work-life balance policies and practices. To do so effectively requires that they are flexible and understanding of each person's need to strike a unique balance among the five domains (Gurvis and Patterson, 2005). Leaders must also create a culture that encourages open and honest communication with employess about those needs. To do this, leaders must fully develop their interpersonal skills by learning how to initiate meaningful dialogue, use the appropriate communication medium, listen for understanding, and display empathy (as opposed to the extremes of insensitivity or sympathy) with employees (Friedman and Lobel, 2000; Glubczynski et al., 2003; Pfeffer and Sutton, 2007). Organizations can support leaders by recognizing and rewarding effective interpersonal skills; management education programs can provide opportunities for students to practice and develop these skills.

Being authentic and treating employees with respect will also serve leaders well in this area: "if you treat people as whole people, listen, clarify expectations, and make sure they have what they need to maintain their priority list, you foster loyalty and commitment" (Pfeffer and Sutton, 2007, p. 92).

4.5.4 Adaptability

What is needed to balance the five domains of the Whole Life Model is different for each person and will also vary over time; therefore, each person will have a different configuration of needs at any given time (Friedman and Lobel, 2000; Maellaro and Whittington, 2009; Yost, 2007). Leaders will have to learn to be more comfortable with the ambiguities resulting from the "discontinuities, multiplicities, relativity, cyclicality, and relationality" (Bilimoria, 2000, p. 162) of the Whole-Life Model. Doing so will allow them to be more flexible in the face of the dynamic nature of balancing work and non-work demands.

A concerted effort to increase leaders' emotional intelligence (EQ) can bolster their ability to adapt. Leaders must examine their long-held and often pre-conscious assumptions about people and work, and then consider alternative perspectives. They must develop resiliency and learn to remain focused in the midst of constant change, which requires a keen sense of self and knowledge of one's strengths and limitations. Also needed is a conscious awareness of their emotions and the ability to properly manage them (Goleman et al., 2002). Higher EQ will help them embrace the changes in management practices that are necessary to support a more comprehensive definition of success.

4.6 Conclusion

Overcoming the distress created by an imbalanced life requires an intentional effort on the part of professionals. Restoring balance begins by redefining success in an effort to explicitly recognize the complexity and totality of life. Yet, simply redefining success will not resolve the dilemmas. This redefinition must be enacted through an intentional effort to establish and maintain boundaries that will create margin in terms of time, physical, and emotional energy.

The choice to redefine success and "choosing to cheat" by living within established boundaries that reflect espoused values goes against the stream of modern life. Achieving this balance will be much easier if organizations establish and enforce policies that support work-life balance and if they help professional employees develop the skills necessary to take full advantage of them. The most effective of these efforts will occur in those organizations where leaders model balance themselves and work hard to ensure that job demands do not require professionals to cheat the other arenas of their life.

References

Armenio M (2004) Managing the calendar. PA CPA J 75:1

Bilimoria D (1998) What if we taught management as if "all our relations" mattered?. J Manage Educ 22:449–451

Bilimoria D (2000) Redoing management education's missions and methods. J Manage Educ 24:161–165

Bourne K, Wilson F, Lester S, Kickul J (2009) Embracing the whole individual: advantages of a dual-centric perspective of work and life. Bus Horiz 52:387–398

Byron K (2005) A meta-analytic review of work-family conflict and its antecedents. J Vocat Behav 67:169–198

Cummings TG, Jones Y (2003). Conference theme: Creating actionable knowledge. Academy of Management 2004 Annual Meeting Call for Submissions, http://meeting.aomonline.org/2004/theme.htm

Dambrin C, Lambert C (2008) Mothering or auditing? The case of two Big Four in France. Account Audit Account J 21:474–506

Eby LT, Casper WJ, Lockwood A, Bordeaux C, Brinley A (2005) Work and family research in IO/OB: content analysis and review of the literature. J Vocat Behav 66: 124–197

Edwards JR, Rothbard NP (2000) Mechanisms linking work and family: clarifying the relationship between work and family constructs. Acad Manage Rev 25:178–199

Friedman D (2004). When work works – a project on workplace effectiveness and workplace flexibility. Families and Work Institute. http://familiesandwork.org/3w/tips/index.html. Accessed 2 Oct 2009

Friedman SD (2006) Learning to lead in all domains of life. Am Behav Sci 49:1270–1296
Friedman SD, Lobel S (2003) The happy workaholic: a role model for employees. Acad Manage Exec 17:87–98
Glubczynski J, Kossek EE, Lambert SJ (2003) A question of leadership: what can managers do to promote work-life balance for themselves and others?. Leadersh Action 23:12
Goleman D, Boyatzis R, McKee A (2002) Primal leadership: realizing the power of emotional intelligence. Harvard Business School, Boston, MA
Greenhaus JH, Powell GN (2006) When work and family are allies: a theory of work-family enrichment. Acad Manage Rev 31:72–92
Gurvis J, Patterson G (2005) Balancing act: finding equilibrium between work and life. Leadersh Action 24:4–8
Heifetz R, Linsky M (2002) Leadership on the line: staying alive through the dangers of leading. Harvard Business School, Boston, MA
Houston G (2004) How much of themselves should leaders bring to their work?. Leadersh Action 24:14
Katz D, Kahn RL (1966) The social psychology of organizations. Wiley, New York
Kossek EE, Lautsch BA, Eaton SC (2005) Flexibility enactment theory: implications of flexibility type, control, and boundary management for work family effectiveness. In Kossek EE, Lambert SJ (eds) Work and life integration: organizational, cultural, and individual perspectives. Lawrence Erlbaum, Mahwah, NJ, pp 233–250
Kreiner GE, Hollensbe EC, Sheep ML (2006) On the edge of identity: Boundary dynamics at the interface of individual and organizational identities. Hum Relat 59:1315–1341
Kreiner GE, Hollensbe EC, Sheep ML (2009) Balancing borders and bridges: negotiating the work-home interface via boundary work tactics. Acad Manage J 52:704–730
Levering R, Moskowitz M, Garcia F, Vell-Zarb K (2000) The 100 best companies to work for. Fortune 141:82–110
Linville PA (1985) Self-complexity and affective extreme: don't put all your eggs in one cognitive basket. Soc Cogn 3:94–120
Linville PA (1987) Self-complexity as a cognitive buffer against stress-related illness and depression. J Pers Soc Psychol 52:663–676
MacDonald G (1985) Ordering your private world. Thomas Nelson, New York, NY
Maellaro R, Whittington JL (2009) Management development for well-being and survival: developing the whole person. In Cooper CL, Quick JL, Schabracq M (eds) Work and health psychology: the handbook. Wiley, New York, NY, pp 293–325
Mariotti J (1998) Management's search for meaning is an effort at achieving balance. Industry Week. http://www.industryweek.com/articles/on_management_1841.aspx. Accessed 23 Feb 2010
Marshall J (2000) Leadership now includes nurturing. Financ Exec 16:15–16
McIntosh D, Spilka B (1990) Religion and physical health: the role of personal faith and control beliefs. Soc Sci Relig 2:167–194
O'Reilly S (2006) Are you making your staff ill? Personnel today. http://www.personneltoday.com/articles/2006/06/20/35945/are-you-making-your-staff-ill.html. Accessed 23 Feb 2010
Parasuraman S, Greenhaus JH (2002) Toward reducing some critical gaps in work-family research. Hum Resour Manage Rev 12:299–312
Perlow LA, Porter JL (2009) Making time off predictable & required. Harv Bus Rev 87:102–109
Pfeffer J, Sutton RI (2007) Is work fundamentally different from the rest of life and should it be?. In Pfeffer J, Sutton RI (eds) Hard facts, dangerous half-truths, and total nonsense: profiting from evidence-based management. Harvard Business School, Boston, MA, pp 57–84
Poelmans SAY (2005) The decision process theory of work and family. In Kossek EE, Lambert SJ (eds) Work and life integration: organizational, cultural, and individual perspectives. Lawrence Erlbaum, Mahwah, NJ, pp 251–274
Quick JD, Nelson DL, Matuszek PAC, Whittington JL, Quick JC (1995) Social support, secure attachments and health. In Cooper CL (ed) The handbook of stress medicine. CRC Press, Boca Raton, FL, pp 269–287

Reed PS, Clark SM (2004) Win-win workplace practices: improved organizational results and improved quality of life. Women's Bureau US Department of Labor, Washington, DC

Rice RW, Frone MR, McFarlin DB (1992) Work-nonwork conflict and the perceived quality of life. J Organ Behav 13:155–168

Senge PM, Kleiner A, Roberts C, Ross RB, Smith BJ (1994) The fifth discipline fieldbook: strategies and tools for building a learning organization. Doubleday, New York, NY

Simon DH, DeVaro J (2006) Do the best companies to work for provide better customer satisfaction?. Managerial Decis Econ 27:667–683

Stanley A (2003) Choosing to cheat: who wins when family and work collide?. Multnomah Publishing, Colorado Springs, CO

Stebbins LF (2001) Work and family in America: a reference handbook. ABC-CLIO Inc, Santa Barbara, CA

Sherman D, Hendricks W (1990) Your work matters to god. NavPress, Colorado Springs, CO

Wiley SL (2006). Creating work/life balance within your organization. Catalyst, September/ October, 26

Yost CW (2007) Finding the perfect fit. Incentive 181:42

Zimmerman E (2004) Change your life. Sales Mark Manage 156:65

Part II
Antecedents and Consequences Related to Professionals' Work-Life Integration

Chapter 5
Personality and Work-Life Integration

Jesse S. Michel and Malissa A. Clark

Researchers and lay persons alike have long been intrigued by the behaviors and thoughts of human beings. Likewise, personality has been an area of great interest and research within the Industrial-Organizational Psychology and Organizational Behavior disciplines (Viswesvaran et al., 2007; Zimmerman, 2008). Interestingly, though the field of work-life integration has received enormous attention in recent years, very little of this has incorporated personality (Eby et al., 2005). Accordingly, this chapter aims to fill this gap in the literature by proposing and reviewing relationships between personality and work-life integration of professionals.

This chapter will consist of four primary sections. First, we review relevant work-life constructs and provide a typology for the conceptualization of work-life integration-segmentation and how individuals intentionally and unintentionally approach their work and life roles. Second, we review primary personality constructs that are theoretically most related to work-life integration of professionals. Third, we provide a qualitative review of personality and work-life integration research. Finally, we summarize main points and provide suggestions for future research.

5.1 A Typology of Primary Theories of Work and Life

Much of the work-life integration literature has been dominated by a few popular theories (e.g., conflict theory, enrichment theory). However, most conceptualize these work-life linkages on a continuum where segmentation (when work and life roles are kept separate) and integration (when work and life roles have no clear boundaries) represent opposing approaches (Ashforth et al., 2000; Nippert-Eng, 1996). As an extension of this general framework, we provide a typology that incorporates both the intentional and unintentional influences of the individual. Below we explain the tenets of primary work-life theories within this typology (cf. Table 5.1).

Integration theory suggests that work and life domains are inherently intertwined (Edwards and Rothbard, 2000; Lambert, 1990; Zedeck and Mosier, 1990);

J.S. Michel (✉)
Florida International University, Miami, FL, USA
e-mail: jmichel@fiu.edu

S. Kaiser et al. (eds.), *Creating Balance?*, DOI 10.1007/978-3-642-16199-5_5,

Table 5.1 Summary of primary work-life theories and propositions within an unintentional/intentional – integration/segmentation typology

	Integration Work role = Life role Life role = Work role	Segmentation Work role ≠ Life role Life role ≠ Work role
Unintentional Work and life roles are unintentionally integrated or segmented	*Conflict theory* Increased role performance in X (work) → Decreased role performance in Y (life) Increased role performance in X (life) → Decreased role performance in Y (work) *Enrichment theory* High performance in work role → High Positive affect in work role → High Performance in life role High performance in life role → High Positive affect in life role → High Performance in work role *Resource drain theory* Increase of X in the work role = Decrease of X in the life role Increase of X in the life role = Decrease of X in the work role *Spillover theory* X in work → X in life X in life → X in work	*Congruence theory* C → X in work C → X in life
Intentional Work and life roles are intentionally integrated or segmented	*Compensation theory* Increase (decrease) in work role = Decrease (increase) in life role Increase (decrease) in life role = Decrease (increase) in work role *Instrumental theory* X in work → Y in life X in life → Y in work	*Boundary management* X in work ↔ X in life X in work ↔ Y in life Y in work ↔ Y in life Y in work ↔ X in life

meanwhile, unintentional integration refers to work and life domains that are unintentionally intertwined. One of the most popular unintentional integration theories is conflict theory (Burke, 1986; Evans and Bartolomé, 1984; Greenhaus and Beutell, 1985; Zedeck and Mosier, 1990), which suggests that increased role performance in one domain (e.g., work) results in decreased role performance in the other domain (e.g., life).

A second unintentional integration theory is enrichment theory (Greenhaus and Powell, 2006), which compliments conflict theory in that high role performance in one domain will result in high role performance in the other domain, where performance is mediated by positive affect (e.g., high performance in one's work role results in high positive affect, which in turn leads to high performance in one's life role; Greenhaus and Powell, 2006). A third unintentional integration theory is resource drain theory (Edwards and Rothbard, 2000; Small and Riley, 1990; Staines, 1980; Tenbrunsel et al., 1995), which suggests that an increase of resources (time, attention, and energy) spent in one role results in a decrease of resources available in the other role. A final unintentional integration theory is spillover theory (Edwards and Rothbard, 2000; Zedeck and Mosier, 1990), suggesting that there is a similarity in work and life such that aspects of one domain can spill over and influence the other domain. For example, an increase in job satisfaction would subsequently lead to an increase in life satisfaction.

Intentional integration refers to work-life relationships that are purposefully manipulated by the individual. One form of intentional integration is compensation theory (Edwards and Rothbard, 2000; Zedeck, 1992), which suggests an inverse relationship between work and life such that an increase (decrease) in the work (life) role equates a decrease (increase) in the life (work) role. Thus, from a compensation theory stance, role pressures from one domain (e.g., life) are alleviated by role performance in the other domain (e.g., work). A second form of intentional integration is instrumental theory (Zedeck, 1992), suggesting that work and life supplement one another such that role pressures in one domain are tolerated to fulfill role pressures in the other domain, such as salary from work fulfilling survival needs from life.

Segmentation theory proposes that work- and life-domains do not affect one another and are separate life domains (Edwards and Rothbard, 2000; Lambert, 1990; Zedeck and Mosier, 1990); therefore, one domain (such as work) occurs without influence from the other domain (such as life). Accordingly, segmentation can be defined as the degree to which work- and life-domains are kept separate from one another (Kreiner, 2006). As an extension of this, unintentional segmentation refers to work and life domains that are inherently separate. Within the work-life literature, few theories focus on unintentional segmentation. One exception to this is congruence theory (Edwards and Rothbard, 2000; Morf, 1989; Zedeck, 1992), which suggests that a third variable (i.e., some dispositional chararacteristic of the individual) impacts work and life domains. As such, work and life are not intertwined, but can appear spuriously related as commonalities are actually caused by a constant outside of these domains. For example, individuals high in internal locus of control should feel greater control over the aspects of their work and personal life, resulting in similar perceptions of work and life stressors.

Intentional segmentation refers to the purposeful separation and maintenance of work and life boundaries. Like unintentional segmentation, few work-life theories explore this component of the taxonomy. An exception, however, is the increasingly popular boundary management literature (e.g., Kossek et al., 2006). Boundary management revolves around purposeful decisions pertaining to individual preferences for segmentation. Accordingly, those who prefer to segment work and life demands can incorporate a variety of psychological and behavior strategies to do so.

In summary, the typology provided here reviews the primary tenets of work-life theory. More importantly, it unambiguously identifies the role the individual plays in their processes (i.e., intentional versus unintentional) of work-life situations (i.e., integration-segmentation). We provide this distinction not in contrast or complement to other discussions of work-life integration, but as a clarification for the theoretical underpinnings of personality within a work-life framework. Theoretically, congruence theory would suggest that personality is a possible antecedent of similar work and life domain outcomes. As an extension of this, personality should play a role in how individuals perceived the spillover or integration of work and life roles. The strongest evidence of a personality or dispositional based conception of work-life integration would be equal or near equal influences of personality constructs on work-to-life and life-to-work constructs (e.g., work-to-life conflict, life-to-work conflict), suggesting that personality is a driving force behind perceptions in both domains.

5.2 Personality Variables Relevant to Working Professionals' Work and Life Roles

To date, much of the research investigating the antecedents of work-life integration has focused on the organizational factors (e.g., work time demands, family-friendly policies offered by the organization) or family characteristics (e.g., number of children, spousal support) that cause individuals to perceive or experience conflict or facilitation between their work and life domains. Much less research has investigated how personality may influence work-life outcomes (Eby et al., 2005). Interestingly, although many studies of work-life balance survey working professionals, whether there are certain personality characteristics shared by members of this particular group of employees, and how that may be related to their work and life roles (and integration of work-life roles), has not been addressed. We see this as an important point that deserves further attention, as it is possible working professionals may have higher levels of certain personality traits (e.g., proactive personality, self-efficacy) than blue-collar workers, and this may in turn differentially influence how these individuals perceive and manage their work and life roles (e.g., work-life integration).

Broadly, we see two primary mechanisms through which personality influences work-life outcomes for professionals. First, personality traits impact individuals' perceptions relating to work and life; and second, personality traits impact the behaviors individuals engage in when attempting to manage or cope with their work and personal lives. Below, we propose relationships between personality and work-life integration, focusing first on the role between personality and perceptions, followed by a discussion of the role between personality and behaviors. Here we focus on both primary concepts of dispositions and personality (e.g., negative and positive affect, Five Factor Model), as well as personality constructs that are potentially more critical and beneficial for working professionals (e.g., proactive personality, core-self evaluations).

5.2.1 Personality Traits and Perceptions Relating to Work-Life Integration

Negative and positive affectivity have been defined as the general tendency to experience negative and positive moods and emotions, respectively (Watson and Clark,

1984). Negative affect and positive affect have both been found to directly influence work-family conflict (Carlson, 1999; Michel and Clark, 2009). Recent research and congruency theory suggest that negative affect and positive affect are related to work-life outcomes because they influence how people perceive life stressors. For example, Stoeva et al. (2002) found that negative affect was positively related to job and family stress for a sample of senior civil servants, which in turn were related to increased work-to-family and family-to-work conflict, respectively. Furthermore, Michel and Clark (2009) surveyed individuals from a wide variety of occupations and found that perceptions of work-family enrichment, work-family conflict, and satisfaction with work and family were driven largely by trait negative and positive affect. Much research has found that negative affect is highly related to the trait of neuroticism, while positive affect is highly related to the trait of extraversion (Judge et al., 1999). Thus, it is not surprising that research has found similar relationships between neuroticism, extraversion, and work-life outcomes that have been found for negative and positive affect. For example, Kinnunen et al. (2003) surveyed fathers working in jobs ranging from blue-collar to high-level executives and found that neuroticism moderated the relationship between work-to-family conflict and outcomes such as job exhaustion and depression, with those individuals high in neuroticism having stronger work-to-family conflict and negative outcome links than those low in neuroticism. In sum, it appears that individuals high in negative affect and neuroticism perceive more conflict and less facilitation between work and life domains, and individuals high in positive affect and extraversion perceive less conflict and more facilitation between work and life domains.

5.2.2 Personality Traits and Behaviors Relating to Work-Life Integration

In the previous section we discussed how personality may influence perceptions relating to work-life outcomes. We also propose that other personality traits may be related to behaviors which can help individuals to integrate their work and personal lives. The Five Factor Model personality trait of conscientiousness is characterized by efficiency and organization (Barrick and Mount, 1991); thus, we believe that these individuals will be better equipped to balance the demands of work and life. Indeed, research has found that individuals high in conscientiousness had less work-family conflict (Wayne et al., 2004).

Other researchers have examined the relationship between locus of control and work-life outcomes. Internal locus of control is the extent to which an individual perceives that outcomes are the result of individual (internal) variables, as opposed to external variables, such as change (Rotter, 1966). Because individuals with an internal locus of control believe outcomes rest in their own hands, we would expect that these individuals would engage in a greater amount of proactive coping strategies in order to balance work and life. Generally, research has indeed found that internal locus of control is related to decreased work and life stress as well as decreased work-family conflict (Noor, 2002). A personality trait related to internal locus of control is proactive personality. Cunningham and De La Rosa (2008), using a sample of university professors, found that proactive personality, defined as an individual characteristic that reflects a person's inclination to take personal initiative to ensure positive outcomes in their environment, was related to reduced time-based family-to-work conflict. Thus, personality variables associated with an individual's increased ability to perceive control over their situations and act in proactive ways to better their situation appear to help individuals manage work and family lives. This may be especially the case with working professionals, who may have more autonomy and flexibility in their jobs than traditional blue-collar workers, which enables them to engage in more proactive coping strategies.

5.3 Qualitative Integration and Review of Personality and Work-Life Relationships

To support the above relationships we conducted a literature search of work-life integration constructs and personality variables. We were able to locate approximately 750 studies, of which 64 contained both a key work-life integration and personality variable. Though the majority of the literature on personality and work-life relationships is published in peer-reviewed journal articles ($n = 48$), there are a decent number of unpublished dissertations ($n = 16$). To provide a broader examination of literature we decided to include these. As displayed in Table 5.2, the primary journal outlet for personality and work-life relationships is Journal of Vocational Behavior, followed by Personality and Individual Differences. Secondary outlets consist of journals such as Journal of Applied Psychology, Journal of Managerial Psychology, Journal of Occupational Health Psychology, and Journal of Social Psychology. All other journals only contain a single article on personality and work-life integration.

We also wanted to see what were the primary work-life variables examined. As displayed in Table 5.3, the vast majority of the work-life variables examined within the empirical literature are conflict, interference, and other negative spillover variables. The second most prevalent work-life variables examined were the enrichment, facilitation, and other positive spillover variables. A third portion of studies, which examined work-life variables such as balance and juggling, were much less common. Likewise, the vast majority of all work-life constructs focused on work-family, with only a small number of studies examining work-home, work-nonwork, and work-life variables.

Finally, we also wanted to see what the primary personality variables examined were. As displayed in Table 5.4, there was much more variability on the personality side, with no clear dominant variable of previous research interest like there was within the work-life integration variables. The two most examined personality variables were neuroticism/emotional stability and negative affect. This of course is not surprising considering the vast majority of work-life integration studies that incorporate personality focused on conflict, interference, and negative spillover. Likewise, neuroticism/emotional stability and negative affect have previously been suggested to tap into a similar construct space (Judge et al., 1999). The other four factors of the Five Factor Model of personality (extraversion, conscientiousness,

Table 5.2 Frequency of articles by journal or dissertation

Journal	Frequency	%
Dissertation	16	25
Journal of Vocational Behavior	7	11
Personality and Individual Differences	5	8
Journal of Applied Psychology	4	6
Journal of Managerial Psychology	4	6
Journal of Occupational Health Psychology	3	5
Journal of Social Psychology	2	3
Academy of Management Journal	1	2
Community, Work, & Family	1	2
Development and Psychopathology	1	2
Family Relations	1	2
Hellenic Journal of Psychology	1	2
Human Resource Management	1	2
International Journal of Cross Cultural Management	1	2
International Operations	1	2
Journal of Career Assessment	1	2
Journal of Career Development	1	2
Journal of Family Issues	1	2
Journal of Individual Differences	1	2
Journal of Hospitality Management	1	2
Journal of Marriage and Family	1	2
Journal of Occupational and Organizational Psychology	1	2
New Zealand Journal of Psychology	1	2
Organizational Behavior and Human Decision Processes	1	2
Personnel Psychology	1	2
Psychologia: An International Journal of Psychology in the Orient	1	2
Salud Pública de México	1	2
Stress & Health	1	2
Stress Medicine	1	2
Work & Stress	1	2

$k = 64$; % = percentage of total personality/work-life integration articles/dissertations represented by that outlet

openness, and agreeableness) were all examined to some extent. The other affective factor, positive affect, was also largely examined. Several components of core self-evaluations, such as self-efficacy, self-esteem, and internal locus of control, were all prevalent within the work-life literature. There was much less examination of what we would call perseverance type personalities. However, Type A, proactive personality, optimism, individual initiative, achievement striving, confidence, and perfectionism are all examined, many of which suggests a perseverance aspect of personality. Interestingly, there were a few studies that focused on negative or aversive personality constructs such as aggression and a few other similar variables,

Table 5.3 Summary of work-life integration variable prevalence when examined in conjunction with personality

Work-life constructs	*k*	% Total	% Dissertations
Work-family conflict	48	75	29
Work-family interference	9	14	11
Work-family facilitation	5	8	0
Work-family enrichment	3	2	33
Decision to participate in work or family	1	2	0
General work-family conflict	1	2	0
Home/Work interface	1	2	0
Negative work-family spillover	1	2	0
Positive spillover	1	2	0
Positive work-family spillover	1	2	0
Work-family gratification	1	2	0
Work-family juggling	1	2	0
Work interference w/home	1	2	0
Work-life balance satisfaction	1	2	0
Work-life conflict	1	2	100
Work-life facilitation	1	2	100
Work non-work interference	1	2	0
Work personal conflict	1	2	0

k = number of articles/dissertations; % total = percentage of articles/dissertations that examine the personality variable; % dissertations = percentage of dissertations that makeup the personality variable k

which some have suggested to be facets or components of aggression, such as guilt and impulsivity. Post hoc, this seems logical as the vast majority of work-life integration research has focused on the conflict, interference, and negative spillover aspects.

Next, we present more specific findings relating to personality and work-life outcomes, with a focus on dispositional affect, core self-evaluations, the Five Factor Model, and perseverance type personalities.

5.3.1 Dispositional Affect and Work-Life Constructs

Dispositional affect is one most commonly examined personality constructs within the work-life integration literature. More specifically, we found 17 articles that examine negative affect and nine that examined positive affect. All the articles that examine negative affect (as well as positive affect) also examined some form of work-life conflict, predominantly work-family conflict. Only a few studies examined affect and positive work-life interactions. A few studies examined multiple forms of affect (i.e., negative and positive) and work-life interactions (e.g., positive and negative forms of integration; directional forms such as work-to-life and life-to-work; facet levels such as time, strain, and behavior).

In general, the literature suggests that negative affect has a strong relationship with negative forms of work-life integration, while positive affect has modest

Table 5.4 Summary of personality variable prevalence when examined in conjunction with work-life integration

Personality variable	*k*	% Total	% Dissertations
Neuroticism/emotional stability	23	36	26
Negative affect	17	27	18
Extraversion	13	20	31
Conscientiousness	11	17	36
Self-efficacy	11	17	18
Positive affect	9	14	11
Self-esteem	9	14	22
Agreeableness	8	13	13
Locus of control	8	13	38
Openness/intellect	8	13	25
Type A I & II	4	6	25
Aggression/hostility	3	5	33
Core self-evaluation	2	3	50
Perfectionism	2	3	0
Proactive personality	2	3	0
Adaptive perfectionism	1	2	0
Cognitive reappraisal	1	2	0
Difficult temperament	1	2	0
Guilt	1	2	0
Impulsivity/sensation seeking	1	2	0
Individual initiative	1	2	0
Maladaptive perfectionism	1	2	0
Optimism	1	2	0
Sociability	1	2	0
Trust	1	2	0

k = number of articles/dissertations; % total = percentage of articles/dissertations that examine the personality variable; % dissertations = percentage of dissertations that makeup the personality variable k

relationships with negative forms of work-life integration and modest to strong relationships with positive forms of work-life integration (e.g., Carlson, 1999; Karatepe and Uludag, 2008; Michel and Clark, 2009). For example, with the data from the 1995 National Survey of Midlife Development in the United States, Voydanoff (2005) found that negative affect was significantly related to both work-to-family conflict ($r = 0.34$) and family-to-work conflict ($r = 0.31$). Meanwhile, negative affect was also significantly related to both work-to-family facilitation ($r = -0.10$) and family-to-work facilitation ($r = -0.17$). Another study by Mitchelson (2009) found that negative affect was related to time-based family-to-work conflict ($r = 0.37$), strain-based work-to-family ($r = 0.32$) and family-to-work ($r = 0.46$) conflict, and behavior-based work-to-family ($r = 0.30$) and family-to-work ($r = 0.28$) conflict.

In a study that examined both negative and positive forms of work-life integration with a diverse international working sample, Michel and Clark (2009) found that negative affect is a strong predictor of both work-to-family ($r = 0.50$) and family-to-work ($r = 0.57$) conflict, but not work-to-family and family-to-work enrichment.

Meanwhile, positive affect was a moderate predictor of work-to-family ($r = -0.17$) and family-to-work ($r = -0.18$) conflict, and strong predictor of work-to-family ($r = 0.46$) and family-to-work ($r = 0.48$) enrichment. Beyond mere bivariate relationships, through a series of model testing via structural equation modeling techniques, Michel and Clark found that negative affect was a significant driver of perceptions of work-family conflict, while positive affect was a significant driver of perceptions of work-family enrichment. Further, when tested as a structural model, linkages between conflict and enrichment to satisfaction outcomes became nonsignificant after taking into account dispositional affect. These findings suggest that affective dispositions are a very strong predictor of perceptions of work-life forms of conflict, enrichment, and satisfaction outcomes.

Finally, Ilies and colleagues (2007) provided one of the most sophisticated examinations of affect within the work-family conflict literature. In a longitudinal examination utilizing an experience sampling methodology, they examined the influence of workload and affect at work on work-family conflict, affect at home, and social activity. Bivariate between individual correlations found that work negative affect ($r = 0.20$) and home negative affect ($r = 0.26$), along with home positive affect ($r = -0.36$), were significant predictors of work-family conflict. Interestingly, none of these variables were significantly related to social behaviors at home at the bivariate between individual level. However, adapting a hierarchical linear model approach, work positive affect was significantly related to home positive affect, work negative affect was significantly related to home negative affect, while home positive affect (but not negative affect) was significantly related to spouse rated social behaviors at home. These findings suggest affect is a strong predictor of work-life perceptions, and when modeled within person, affect or mood can have significant spillover effects into other domain affect or mood and subsequent spousal perceptions of social activity and behavior.

5.3.2 Core Self-Evaluations and Work-Life Constructs

Core self-evaluations have recently been a hot topic within the Industrial-Organizational Psychology and Organizational Behavior disciplines (e.g., Harris et al., 2009; Judge, 2009; Kammeyer-Mueller et al., 2009). Further, prior to the "core self-evaluations" terminology, self-efficacy, self-esteem, internal locus of control, and neuroticism/emotional stability have long been studied within the literature. Similarly, our literature review suggests that these individual facets are often studied in isolation within the work-life literature. As displayed in Table 5.4, neuroticism/emotional stability is the most often examined personality variable within the work-life literature. Self-efficacy, self-esteem, and internal locus of control are all modestly examined. Meanwhile, core self-evaluations as conceptualized by Judge and colleagues (Judge et al., 1997) have received significantly less attention. Again, the majority of the literature on core-self evaluations examined negative forms of work-life integration, and predominantly work-family conflict.

In general, the literature suggests that these forms of core self-evaluations are negatively related to negative forms of work-life integration, and positively related to positive forms of work-life integration (e.g.,, Beauregard, 2006; Boyar and Mosley, 2007; Nikandrou et al., 2008). Starting with emotional stability, Horwitz et al. (2008) found that emotional stability was negatively related to negative work-to-family (r = –0.32) and family-to-work (r = –0.29) spillover, and positively related to positive work-to-family (r = 0.14) and family-to-work (r = 0.22) spillover. A similar pattern of results was also found by Rotondo and Kincaid (2008; work-to-family conflict, r = 0.34; family-to-work conflict, r = 0.29; work-to-family facilitation, r = –0.10; family-to-work facilitation, r = –0.10). Meanwhile, in a study based on private-sector workers, Witt and Carlson (2006) found that emotional stability was negatively related to work-to-family conflict (r = –0.28) and family-to-work conflict (r = –0.29), but not work-to-family or family-to-work enrichment. Lastly, Mitchelson (2009) found that emotional stability was related to time-based, strain-based, and behavior-based facets of work-to-family and family-to-work conflict. These relationships were stronger for strain-based (r = 0.30–0.36) then time- or behavior-based (r = 0.19–0.22) forms of conflict.

In regards to self-esteem and self-efficacy, there is a general trend. For example, with a female Greek sample of multiple management levels, Nikandrouet al. (2008) found that self-esteem was negatively related to work-family conflict (r = –0.23). In a study on UK public sector employees, Beauregard (2006) comments self-efficacy was negatively related to both work-to-home (r = –0.18) and home-to-work (r = –0.31) conflict, as was self-esteem (r = –0.14; r = –0.15). An interesting study by Erdwins et al. (2001) found that domain or role self-efficacy was related to work-family conflict. Specifically, both parent self-efficacy (r = –0.27) and job self-efficacy (r = –0.33) were significantly related to perceptions of work-family conflict. However, after controlling for age, education, income, and children, job self-efficacy, but not parent self-efficacy, was incrementally predictive of work-family conflict. Similarly, Cinamon et al. (2007) found that parental self-efficacy was negatively related to both work-to-family (r = –0.29) and family-to-work (r = –0.27) conflicts. This suggests role or state self-efficacy and trait or general self-efficacy operate differently.

In regards to internal locus of control, the general finding is that greater internal locus of control results in a reduction of negative work-life interactions. For example, Noor (2002) found that internal locus of control is negatively related to work-family conflict (r = –0.20). Similar findings have been found for overall and facet level forms of work-to-family and family-to-work conflict (Andreassi and Thompson, 2007).

Finally, in one of the few studies that examined Judge's construct of core self-evaluations, Boyar and Mosley (2007) found that core self-evaluations were negatively related to both work-to-family (r = –0.32) and family-to-work (r = –0.45) conflict, but not work-to-family and family-to-work facilitation. These relationships held when examined concurrently via structural equation modeling. Further, facet level examination showed that self-esteem and internal locus of control related to both directions of work-family conflict, while emotional stability was also related to family-to-work conflict. Meanwhile, only internal locus of control was related to either forms of work-family facilitation; $r = 0.25$ for both.

5.3.3 Five Factor Model and Work-Life Constructs

As the Five Factor Model of personality has given much consistency to personality research in general (Costa and McCrae, 1992), it has also given rise to much of the personality research within the work-life integration literature. All five factors have been examined to some extent, with neuroticism/emotional stability being the most frequently included personality variable. As we have already covered neuroticism/emotional stability, we will now turn to the remaining four factors. In addition, we focus only on studies that included the full five factors to determine the relative importance, so to speak, of each variable on work-life integration.

Of the studies that examined the full Five Factor Model, it seems that agreeableness and conscientiousness are the most predictive of work-life integration constructs. However, almost all of the studies currently published only examine work-family conflict. For example, agreeableness was negatively related to work-to-family conflict, with effect sizes ranging from $r = -0.14$ to $r = -0.20$ (Bruck and Allen, 2003; Kinnunen et al., 2003; Rantanen et al., 2005). Agreeableness was also related to family-to-work conflict ($r = -0.25$; Bruck and Allen, 2003). A closer facet level examination by Mitchelson (2009) revealed significant relationships for time-, strain-, and behavior-based family-to-work conflict, and a significant relationship for behavior based work-to-family conflict, suggesting that relationships occur at multiple levels of work and life (e.g., time aspects, strain aspects). Meanwhile, several studies found significant effects for conscientiousness and family-to-work conflict with effect sizes ranging from $r = -0.14$ to $r = -0.22$ (Bruck and Allen, 2003; Kinnunen et al., 2003). There was also some support for the relationship between conscientiousness and family-to-work conflict ($r = -0.16$; Rantanen et al., 2005). However, closer examination by Mitchelson (2009) again revealed facet level relationships, this time indicating significant relationships for strain- and behavior-based work-to-family conflict, and time-, strain-, and behavior-based family-to-work conflict. Meanwhile, there was much less support for extraversion and openness; nonetheless, some studies found some modest yet significant relationships (Kinnunen et al., 2003; Mitchelson, 2009; Wierda-Boer et al., 2009).

One particular insightful study was conducted by Wayne et al. (2004), in which a large national random sample examined the relationships between the Five Factor Model of personality and both work-family conflict and facilitation. After controlling for gender, marital status, parental status, education, and hours worked, Wayne and colleagues found significant personality influences on each form of conflict and facilitation. For work-to-family conflict, agreeableness and conscientiousness were both unique negative predictors, while neuroticism was a positive predictor, collectively accounting for 15% of variance in work to family conflict. For family-to-work conflict, conscientiousness was a unique negative predictor, while neuroticism was a unique positive predictor, collectively accounting for 13% of variance in family-to-work conflict. For work-to-family facilitation, extraversion and openness were unique positive predictors, while neuroticism was a unique negative predictor, collectively accounting for 7% of variance in work-to-family facilitation. Finally, for family-to-work facilitation, extraversion, agreeableness, and conscientiousness were all unique positive predictors, collectively accounting for 8% of the variance in family-to-work facilitation. Overall, these results suggest that each of the Five Factor Model traits are unique and significant predictors of some form of positive or negative work-life integration.

5.3.4 Perseverance Type Personalities and Work-Life Constructs

Of the personality variables examined within the work-life integration literature, perseverance type personalities have received little attention. However, the personality variables of productivity, perfectionism, and individual initiative seem particularly fruitful in regards to work-life integration of professionals.

In regards to proactive personality, Aryee et al. (2005) found significant negative relationships for proactive personality and work-to-family ($r = -0.19$) and family-to-work ($r = -0.21$) conflict, along with significant positive relationships with work-to-family ($r = 0.21$) and family-to-work ($r = 0.17$) facilitation. Nearly identical results were also found for optimism. However, only proactive personality and work-to-family facilitation remained significant when entered into a multiple regression. Meanwhile, other research suggests that proactive personality is related only to time-based family-to-work conflict ($r = -0.20$; Cunningham and De La Rosa, 2008).

In an examination of perfectionism, Beauregard (2006) found that both adaptive ($r = 0.14$) and maladaptive ($r = 0.30$) perfectionism were positively related to work-to-home conflict, while maladaptive perfectionism was positively related to home-to-work conflict ($r = 0.24$), but adaptive perfectionism was negatively related to home-to-work conflict ($r = -0.24$). In a series of multiple regressions, maladaptive perfectionism was a unique predictor of work-to-home conflict, accounting for an additional 3% of the variance; however, both maladaptive and adaptive perfectionism were unique predictors of home-to-work conflict, each accounting for an additional 5% of the variance, again with a negative relationship between adaptive perfectionism and home-to-work conflict.

Another example of a perseverance type personality is individual initiative. We identified one study that examined this personality variable. Specifically, Bolino and Turnley (2005) found a very strong relationship between individual initiative and work-family conflict ($r = 0.42$). Further, after controlling for age, gender, marital status, number of children, organizational tenure, salary, and even negative affect, individual initiative remained significant and accounted for an additional 9% of the variance in work-family conflict. Interestingly, this relationship was also moderated by gender, with low initiative men experiencing higher work-family conflict than low initiative women, while high initiative men and women experienced similar work-family conflict.

5.4 Summary and Suggestions for Future Research

We started this chapter with a typology of work-life integration theories. Within this typology, congruence theory was specified as an unintentional segmentation theory that proposes a constant or third variable impacts both work and life domains similarly. Accordingly, congruence theory provides the foundation for personality within a work-life context. However, this chapter has shown that personality plays not only a role in work and life roles (Sect. 5.2), but also in the perceptions of work-life interaction constructs (Sect. 5.3). Our review strongly supports the role of personality in the perceptions of work-life integration. Proven variables of importance include, but are not limited to, dispositional affect and mood, core self-evaluations, the Five Factor Model, and perseverance type personalities.

5.4.1 Suggestions for Future Research

Upon examination of Table 5.3, it seems clear that additional research needs to be conducted on work-life integration constructs outside of work-family and negative spillover or conflict areas. For example, the vast majority of the research conducted to date has focused on work-family conflict or similar constructs. To better understand the work-life interface, we clearly need to move beyond work-family and conflict. From the examination of Table 5.4, it seems that aggression and similar constructs may be fruitful in the examination of negative spillover and conflict; additional aversive personality constructs could prove fruitful. Likewise, other prosocial constructs may play a role in the perceptions of positive spillover and enrichment/facilitation and subsequent behaviors.

Much more research is needed in order to understand the influence of personality on work-life integration. However, simply just examining the relationship between personality traits and work-life conflict, work-life facilitation, etc., is not enough. In order to tease apart how personality influences perceptions and behaviors relating to work-life outcomes, it is imperative that measures of perceptions (e.g., perceptions of work and family stressors) as well as measures of behaviors (e.g., objective or subjective assessments of the specific coping behaviors used and the frequency of their use) be included along with measures of work-life integration.

In addition, future research could further utilize intraindividual methodologies (cf. Ilies et al., 2007) as neither work nor life demands are static. For example, people's job demands can fluctuate based on the time of year (e.g., accountants become increasingly busy as April 15th approaches), or even from day to day. People's family demands can fluctuate as well, with more childcare demands put on parents during traditional school breaks, for example. Therefore, research should begin to take into account the role that personality plays in managing dynamic work and life stressors, for example. Such methodology can also be used to determine the role that transient moods versus personality traits have on perceptions and behaviors of work and life.

A final suggestion could be that of a more ideographic and social-cognitive conception of the individual within the work-life context.

5.4.2 Towards a Cognitive-Affective Conception of Work and Life

To provide a work-life meta-theory, we could turn towards the increasingly popular paradigm of social-cognition (e.g., Cervone and Shoda, 1999a; Dweck and Leggett, 1988; Greenwald and Banaji, 1995). That is, a primary goal of social-cognition in general and social-cognitive theories in particular, is to develop a "common language for understanding both consistency and variability in social behavior" (Cervone and Shoda, 1999b, p. 4). As such, Mischel and Shoda's Cognitive-Affective Processing System (CAPS) is a particularly apt starting point for generating such psychological explanations (Mischel, 1999; Mischel and Shoda, 1995, 1998, 1999; Shoda and Mischel, 1998). However, it is important to point out that CAPS theory is merely one viable option to explain the underlying structure of work-life dynamics, and that alternative meta-theories could be developed.

As a general overview, CAPS theory extends two fundamental assumptions. First, CAPS assumes that there are individual differences in "chronic accessibility." That is, individuals differ in the activation levels or ease with which particular mental representations become activated. Further, behavior can be explained through mental representations or cognitive-affective units that refer to the various cognitions and affects that are available. These cognitive-affective units consist of encodings (constructs for the self, other, events or situations), expectations and beliefs (expectations and beliefs towards outcomes), affects (feelings, emotions, and affective responses), goals and values (desirable outcomes), and competencies and self-regulatory plans (potential behaviors and scripts that one can do). The second assumption of the CAPS model states that there are individual differences in the stable organization of relations among units. Stated alternatively, there are distinct interconnections among specific cognitive-affective units unique to the individual, in addition to the chronic accessibility of individual units. Further, these interconnections can consist of various positive or inverse relationships, which in turn activate or constrain various cognitions, affects, and behaviors. However, if some individuals have the same cognitive-affective units, but differences in the stable organization of relations among units, this would result in different behaviors. As such, it's not the units or connections themselves that generates a behavioral proclivity, but the combination of cognitive-affective units and how they are connected to one another than engenders such behavior.

The CAPS model views cognitive-affective units as mediating components between the features of the work-life situation and subsequent behavior. These cognitive-affective units interact dynamically and influence each other reciprocally as the individual selects and interprets the situation and behaves accordingly.

Further, it is the relative importance of each mediating unit to the individual, along with their various interactions, that makeup the core of the cognitive-affective processing system. As a result, individuals who share a common organization of cognitive-affective units (e.g., encodings, expectations and beliefs, goals and values) and interconnections should possess similar behavioral proclivities. Thus, if our goal is to better understand the complex nature of work-life integration, future research should examine the underlying nature or cause of behavior from a social-cognitive framework. Accordingly, work-life researchers could move towards a work-life taxonomy that takes into account both (1) the intentional/unintentional and integration/segmentation of work-life integration, and (2) the psychological features of the situation, and the underlying cognitive-affective system that drives behavioral expression. Work-life research could explore the various types of cognitions and affects (e.g., encodings, expectations and beliefs, goals and values) that engender different behavioral proclivities relating to work and life roles. For example, what encodings influence one's reaction to flextime, how does the cognitive accessibility of these encodings influence flextime, and how do the individual differences in the stable organization of relations among these encodings among other cognitive-affective units influence proper use versus abuse of flextime? Though this research, a comprehensive explanation of work-life integration could be provided.

Acknowledgements We would like to thank David Jaramillo for his assistance with the literature review, retrieval of articles, and creation of summary tables.

References

Andreassi JK, Thompson CA (2007) Dispositional and situational sources of control: relative impact on work-family conflict and positive spillover. J Managerial Psychol 22: 722–740

Aryee S, Srinivas ES, Tan HH (2005) Rhythms of life: antecedents and outcomes of work-family balance in employed parents. J Appl Psychol 90:132–146

Ashforth BE, Kreiner GE, Fugate M (2000) All in a day's work: boundaries and micro role transitions. Acad Manage Rev 25:472–491

Barrick MR, Mount MK (1991) The big five personality dimensions and job performance: a meta-analysis. Pers Psychol 44:1–26

Beauregard TA (2006) Predicting interference between work and home: a comparison of dispositional and situational antecedents. J Managerial Psychol 21:244–264

Bolino MC, Turnley WH (2005) The personal costs of citizenship behavior: the relationship between individual initiative and role overload, job stress, and work-family conflict. J Appl Psychol 90:740–748

Boyar SL, Mosley DC (2007) The relationship between core self-evaluations and work and family satisfaction: the mediating role of work-family conflict and facilitation. J Vocat Behav 71: 265–281

Bruck CS, Allen TD (2003) The relationship between big five personality traits, negative affectivity, type A behavior, and work-family conflict. J Vocat Behav 63:457–472

Burke RJ (1986) Occupational and life stress and the family: conceptual frameworks and research findings. Int Rev Appl Psychol 35:347–369

Carlson DS (1999) Personality and role variables as predictors of three forms of work-family conflict. J Vocat Behav 55:236–253

Cervone D, Shoda Y (1999a) The coherence of personality: social-cognitive bases of consistency, variability, and organization. Guilford, New York, NY

Cervone D, Shoda Y (1999b) Social-cognitive theories and the coherence of personality. In Cervone D, Shoda Y (eds) The coherence of personality: social-cognitive bases of consistency, variability, and organization. Guilford, New York, NY, pp 3–33

Cinamon RG, Weisel A, Tzuk K (2007) Work-family conflict within the family: crossover effects, perceived parent-child interaction quality, parental self-efficacy, and life role attributions. J Career Dev 34:79–100

Costa PT, McCrae RR (1992) Four ways five factors are basic. Pers Individ Differ 13:653–665

Cunningham CJL, De La Rosa GM (2008) The interactive effects of proactive personality and work-family interference on well-being. J Occup Health Psychol 13:271–282

Dweck CS, Leggett EL (1988) A social-cognitive approach to personality and motivation. Psychol Rev 95:256–273

Eby LT, Casper WJ, Lockwood A, Bordeaux C, Brinley A (2005) Work and family research in IO/OB: content analysis and review of the literature (1980–2002). J Vocat Behav 66:124–197

Edwards JR, Rothbard NP (2000) Mechanisms linking work and family: clarifying the relationship between work and family constructs. Acad Manage Rev 25:178–199

Erdwins CJ, Buffardi LC, Casper WJ, O'Brien AS (2001) The relationship of women's role strain to social support, role satisfaction, and self-efficacy. Fam Relat 50:230–238

Evans P, Bartolomé F (1984) The changing pictures of the relationship between career and family. J Occup Behav 5:9–21

Greenhaus JH, Beutell NJ (1985) Sources of conflict between work and family roles. Acad Manage Rev 10:76–88

Greenhaus JH, Powell GN (2006) When work and family are allies: a theory of work-family enrichment. Acad Manage Rev 31:72–92

Greenwald SB, Banaji MR (1995) Implicit social cognition: attitudes, self-esteem, and stereotypes. Psychol Rev 102:4–27

Harris KJ, Harvey P, Kacmar KM (2009) To social stressors impact everyone equally? An examination of the moderating impact core self-evaluations. J Bus Psychol 24:153–164

Horwitz BN, Luong G, Charles ST (2008) Neuroticism and extraversion share genetic and environmental effects with negative and positive mood spillover in a nationally representative sample. Pers Individ Differ 45:636–642

Ilies R, Schwind KM, Wagner DT, Johnson MD, DeRue DS, Ilgen DR (2007) When can employees have a family life? The effects of daily workload and affect on work-family conflict and social behaviors at home. J Appl Psychol 92:1368–1379

Judge TA (2009) Core self-evaluations and work success. Curr Dir Psychol Sci 18:58–62

Judge TA, Higgins CA, Thoresen CJ, Barrick MR (1999) The big five personality traits, general mental ability, and career success across the life span. Pers Psychol 52:621–652

Judge TA, Locke EA, Durham CC (1997) The dispositional causes of job satisfaction: a core evaluations approach. Res Organ Behav 19:151–188

Kammeyer-Mueller JD, Judge TA, Scott BA (2009) The role of core self-evaluations in the coping process. J Appl Psychol 94:177–195

Karatepe OM, Uludag O (2008) Affectivity, conflicts in the work-family interface, and hotel employee outcomes. Int J Hospital Manage 27:30–41

Kinnunen U, Vermulst A, Gerris J, Mäkikangas A (2003) Work-family conflict and its relations to well-being: the role of personality as a moderating factor. Pers Individ Differ 39:1669–1683

Kossek EE, Lautsch BA, Eaton SC (2006) Telecommuting, control, and boundary management: correlates of policy use and practice, job control, and work-family effectiveness. J Vocat Behav 68:347–367

Kreiner GE (2006) Consequences of work-home segmentation or integration: a person-environment fit perspective. J Organ Behav 27:485–507

Lambert SJ (1990) Processes linking work and family: a critical review and research agenda. Hum Relat 43:239–257

Michel JS, Clark MA (2009) Has it been affect all along? A test of work-to-family and family-to-work models of conflict, enrichment, and satisfaction. Pers Individ Differ 47:163–168

Mischel W (1999) Personality coherence and dispositions in a cognitive-affective personality system (CAPS) approach. In Cervone D, Shoda Y (eds) The coherence of personality: social-cognitive bases of consistency, variability, and organization. Guilford, New York, NY, PP. 37–60

Mischel W, Shoda Y (1995) A cognitive-affective system theory of personality: reconceptualizing situations, dispositions, dynamics, and invariance in personality structure. Psychol Rev 102:246–268

Mischel W, Shoda Y (1998) Reconciling processing dynamics and personality dispositions. Ann Rev Psychol 49:229–258

Mischel W, Shoda Y (1999) Integrating dispositions and processing dynamics within a unified theory of personality. In Pervin LA, John OP (eds) Handbook of personality: theory and research. Guilford, New York, NY, pp 197–218

Mitchelson JK (2009) Seeking that perfect balance: perfectionism and work-family conflict. J Occup Organ Psychol 82:349–367

Morf M (1989) The work/life dichotomy. Quorum, Westport, CT

Nikandrou I, Panayotopoulou L, Apospori E (2008) The impact of individual and organizational characteristics on work-family conflict and career outcomes. J Managerial Psychol 23:576–598

Nippert-Eng EC (1996) Home and work: negotiating boundaries through everyday life. University of Chicago Press, Chicago, IL

Noor NM (2002) Work-family conflict, locus of control, and women's well-being: Tests of alternative pathways. J Soc Psychol 142:645–662

Rantanen J, Pulkkinen L, Kinnunen U (2005) The big five personality dimensions, work family conflict, and psychological distress: a longitudinal view. J Individ Differ 26:155–166

Rotondo DM, Kincaid JF (2008) Conflict, facilitation, and individual coping styles across the work and family domains. J Managerial Psychol 23:484–506

Rotter JB (1966) Generalized expectancies for internal versus external control reinforcement. Psychol Monogr Gen Appl 80:1–28

Shoda Y, Mischel W (1998) Personality as a stable cognitive-affective activation network: characteristic patterns of behavior variation emerge from a stable personality structure. In Read S, Miller LC (eds) Connectionist models of social reasoning and social behavior. Lawrence Erlbaum, Mahwah, NJ, pp 175–208

Small SA, Riley D (1990) Toward a multidimensional assessment of work spillover. J Marriage Fam 52:51–61

Staines G (1980) Spillover versus compensation: a review of the literature of the relationships between work and nonwork. Hum Relat 33:111–129

Stoeva AZ, Chiu RK, Greenhaus JH (2002) Negative affectivity, role stress, and work-family conflict. J Vocat Behav 60:1–16

Tenbrunsel AE, Brett JM, Maoz E, Stroh LK, Reilly AH (1995) Dynamic and static work-family relationships. Organ Behav Hum Decis Process 63:233–246

Viswesvaran C, Deller J, Ones DS (2007) Personality measures in personnel selection: some new contributions. Int J Sclcct Assess 15:354–358

Voydanoff P (2005) The differential salience of family and community demands and resources for family-to-work conflict and facilitation. J Fam Econ Issues 26:395–417

Watson D, Clark LA (1984) Negative affectivity: the disposition to experience negative emotional states. Psychol Bull 96:465–490

Wayne JH, Musisca N, Fleeson W (2004) Considering the role of personality in the work-family experience: relationships of the big five to work-family conflict and facilitation. J Vocat Behav 64:108–130

Wierda-Boer HH, Gerris JRM, Vermulst AA (2009) Managing multiple roles: personality, stress, and work-family interference in dual-earner couples. J Individ Differ 30:6–19

Witt LA, Carlson DS (2006) The work-family interface and job performance: moderating effects of conscientiousness and perceived organizational support. J Occup Health Psychol 11:343–357

Zedeck S (1992) Introduction: exploring the domain of work and family concerns. In Zedeck S (ed) Work, families and organizations. Jossey-Bass, San Francisco, CA

Zedeck S, Mosier KL (1990) Work in the family and employing organization. Am Psychol 45: 240–251

Zimmerman RD (2008) Understanding the impact of personality traits on individuals' turnover decisions: a meta-analytic path model. Pers Psychol 61:309–348

Chapter 6
Potential Antecedents and Consequences of Work-Family Conflict: A Three Country Study

Ronald J. Burke, Lisa Fiksenbaum, Ghada El-Kot, Mustafa Koyuncu, and Weng Jing

Work-family issues have received increasing attention by both organizations interested in effective human resource management and by researchers (cf. Brough and Kalliath, 2009). Work-family concerns were initially included in the occupational stress field but have now emerged as an important research focus on its own. Work-family issues, whether seen in terms of conflict, balance or synergy, have important consequences for individuals, families and organizations. Work-family conflict (WFC) has been found to be associated with lower levels of work satisfaction, organizational commitment, absenteeism, and psychological well-being (Allen et al., 2000; Byron, 2005; Kossek and Ozeki, 1998; Parasuraman and Greenhaus, 1997). WFC was also found to be associated with lower levels of family satisfaction (Eby et al., 2005; Rice et al., 1992). Finally, employee dissatisfaction and heightened turnover resulting from work-family issues also have implications for organizational performance (Goff et al., 1990; Jones et al., 2005). Organizations providing their employees with opportunities for work-family balance are better managed, more productive, and have higher external reputations (Bloom et al., 2009). Work-life balance is an important element in talent management and employee retention. The challenge for organizations and their management is to determine how they can be productive and competitive while encouraging and supporting employees in their efforts to achieve work-life balance. Work-life issues have become a worldwide concern.

Work-family conflict, or work-family integration, seems to be a particularly important concern for managers and professionals, not forgetting that all employees have concerns about this. Managers and professionals face some unique circumstances. These include: long work hours, being in intense jobs, technology making it possible to work anywhere at any time, holding jobs not constrained to a specified number of hours, and increasing job performance pressures resulting from globalization, cost restraints and the current worldwide economic recession. In addition, managers and professionals are likely to find themselves in dual-career situations with both individuals holding demanding jobs. But managers and professionals

---

R.J. Burke (✉)
Schulich School of Business, York University, Toronto, ON, Canada
e-mail: rburke@schulich.yorku.ca

S. Kaiser et al. (eds.), *Creating Balance?*, DOI 10.1007/978-3-642-16199-5_6,

also enjoy some benefits which potentially address work-family concerns. These include high pay, job satisfaction, some influence over the content of their work, some flexibility and income to purchase tangible support (e.g., child-care).

Research on the antecedents and consequences of work-family conflict have sometimes produced inconsistent results however, influenced in part by the use of different outcomes and different measures of work-family conflict. Carlson et al. (2000), based on previous theorizing and research findings, developed a measure of work-family conflict. Their measures examined both work-family and family-work relationships (bi-directional conflict) and different forms of conflict (time-, strain-, and behavior-based).

- Time-based conflict – The time devoted to one role makes it more difficult to participate successfully in the other role.
- Strain-based conflict – Strains in one role make it more difficult to participate or be successful in the other role.
- Behavior-based conflict – Behaviors expected or required in one role are incompatible with behaviors required or expected in the other role.

Thus there are six dimensions of WFC: work-family and family-work in each of the three forms. They found that these six measures were all positively and significantly correlated (average inter-correlation of 0.64). They also found that females scored significantly higher than males on four of the six measures. In addition, the strain-based measures were more strongly and consistently correlated with both potential antecedents and outcomes than were the time-based measures.

The present studies replicate and extend findings from the original Carlson et al. (2000) study. Data were collected in three countries (Turkey, China, Egypt) since countries likely vary in their cultural values of the role of both work and family and in their policies supporting work-life balance. Their measures of only work-family conflict were used in this research since other research has shown that work-family conflict is generally more strongly and consistently related to work and well-being outcomes (Byron, 2005; Jones et al., 2005). The initial Carlson et al. (2000) study was also extended by including additional potential antecedents and consequences. The former included personal and demographic characteristics, indicators of job demands, and measures of personality associated with work investments; the latter included a wider array of work and well-being consequences.

Carlson (1999), using the same US sample as in her later 2000 study, reported correlations between Type A behavior and negative affectivity, both personality factors, and the three forms of work-family conflict. As predicted, negative affectivity was positively related to levels of each of the three forms of work-family conflict. Contrary to predictions, however, Type A behavior was negatively related to each of the three forms of work-family conflict. She suggests that Type As may be more adaptive, involved in a wider array of activities, be able to compartmentalize work and family, or be so involved in their work that they are oblivious to family matters.

It is important to undertake work-family research in a cross-cultural contest. Managers and professionals work more hours in some countries, the amount of vacation time differs, the centrality of work varies across countries, the priority of family varies across countries, the presence of dual-career couples varies across countries, and organizations, and governments, in different countries differ in their support for work-family integration Thus organizations in the Western world are more likely to offer flexible work hours and government support for child-care.

The following general hypotheses, building primarily on the Carlson et al. findings, were considered in each of the three studies:

1. The measures of time-, strain, and behavior – based work-family conflict would be significantly and positively correlated.
2. The three measures of work-family conflict would have different predictors.
3. Personal demographic and work situation factors would be modest predictors of forms of WFC.
4. Stable individual difference personal characteristics associated with work investment (need for achievement, workaholic behaviors) would be strong predictors of forms of WFC
5. The three measures of work-family conflict would have different relationships with work and well-being outcomes. Strain-based conflict would have more relationships with work and well-being outcomes than would time-based conflict.

6.1 Study I – Manufacturing Managers in Turkey

The Carlson et al. (2000) measures were used in a large sample of employed men and women working in management and professional jobs in the manufacturing sector in Turkey. Turkey is a large developing country having a secular but almost exclusively Muslim population. In addition, the values of Turkey are different from those of Western countries (Hofstede, 1980; House et al., 2004), though these are slowly changing. Turkey places a high value on the family (Aycan, 2004). In addition, human resource management practices in Turkey tend to be patriarchal (Aycan, 2001). Finally, the work ethic among employees in turkey is different from that found in developed Western countries (Ali and Al-Owaihan, 2008).

6.1.1 Procedure

Data were collected from organizations in 16 Turkish cities (e.g., Denizli, Mersin, Nevsehir). Members of the research team contacted organizations in the manufacturing sector in these cities requesting their participation in the research. Cooperating organizations then provided a list of their managers and professionals. Approximately 1500 managers and professionals were contacted; 945 returned questionnaires to the research team of which 877 provided reasonably complete data, a 58% response rate. Questionnaires were completed anonymously. Measures originally develop in English were translated into Turkish using the back translation method.

Organizations fell into a variety of industries including agricultural machinery, textiles, health products, construction, food processing, furniture, metal, carpet production and electrical products. The respondents are best described as a large sample of Turkish managers and professionals in the manufacturing sector.

6.1.2 Respondents

Most respondents were male (77%), married (72%), were between 26 and 35 years of age (47%), had children (74%) (44% had two children), were college/university graduates (63%), held supervisory jobs (82%), worked between 41 and 50 h per week (49%), earned between US$10,000 and US$14,000 income (24%), had 5 years or less of organizational tenure (45%) and 5 years of less of job tenure (62%), worked in organizations having 250 or fewer employees (77%), and worked in production or accounting and finance (30% and 28%. respectively). There was a tendency for a higher proportion of males to work in production and management and a lower percentage of males to work in marketing or accounting and finance.

6.1.3 Measures

6.1.3.1 Work-Family Conflict

Three aspects of work-family conflict were assessed using scales developed by Carlson et al. (2000). Each had three items and assessed time-, strain-, and behavior-based conflict. The respective reliabilities for these were 0.85, 0.83, and 0.75. One item was “The stress from my job often makes me irritable when I get home.”

6.1.3.2 Predictors of WFC

Personal Demographic, Work Situation Characteristics, and Job Demands

A number of personal demographics (e.g., age, gender, level of education, marital and parental status) and work situation characteristics (e.g., organizational level, job and organizational tenure) were measured by single items.

*Job Demands.*Two job demands were included. Work hours were assessed by a single item. Respondents indicated the number of hours they worked in a typical week. Work intensity was assessed by a 15 item scale ($\alpha = 0.85$). Some items were taken from Hewlett and Luce (2006) while others were developed by the researchers. Items included: “an unpredictable flow of work”, “availability to clients 24/7”, and a “large scope of responsibility that amounts to more than one job”.

Personality Characteristics

Four stable individual difference personality characteristics were included.

Need for Achievement (NAch) was measured by a five item scale ($\alpha = 0.66$) developed by Steers and Braunstein (1976). One item was “I try very hard to improve on my past performance at work.”

Workaholic Behaviors. Two workaholic behavior scales developed by Mudrack (2006) were included. Non-required work had four items ($\alpha = 0.85$) (e.g., "Thinking of ways to improve the quality of work provided to customers and/or co-workers."). *Control of others* also had four items ($\alpha = 0.76$) (e.g., "fixing problems created by other people"). A third workaholic component, *Feeling driven to work because of inner pressures*, developed by Spence and Robbins (1992) contained seven items ($\alpha = 0.71$). One item was "I often feel that there is something inside me that drives me to work hard".

6.1.3.3 Work and Well-Being Outcomes

A wide range of outcome variables were included in this study covering both work and extra-work domains. These variables were consistent with those typically used in studies of work and well-being more generally (e.g., Barling et al., 2005; Schabracq et al., 2003). Four work outcomes and three aspects of psychological well-being were considered.

Job satisfaction was measured by a seven item scale ($\alpha = 0.81$) developed by Kofodimos (1993) (e.g., "I feel challenged by my work"). *Career satisfaction* was assessed by a five item scale ($\alpha = 0.86$) created by Greenhaus et al. (1990) (e.g., "I feel satisfied with the progress I have made in my career to date."). *Job stress* was measured by a nine item scale ($\alpha = 0.58$) developed by Spence and Robbins (1992). One item was "sometimes I feel like my work is going to overwhelm me." *Intent to quit* was measured by two items ($\alpha = 0.66$) used previously by Burke (1991). One item was "Are you currently looking for a different job in a different organization? (yes/no).

Exhaustion was measured by a nine item scale ($\alpha = 0.86$), part of the Maslach Burnout Inventory, developed by Maslach et al. (1996). A sample item was "I feel emotionally drained from my work." *Psychosomatic symptoms* was measured by a nineteen item scale ($\alpha = 0.88$) developed by Quinn and Shepard (1974). Respondents indicated how frequently they had experienced each physical symptom (e.g., headaches, difficulty sleeping) in the past year. *Life satisfaction* was assessed by a five item scale ($\alpha = 0.84$) created by Diener et al. (1985). A sample item was "I am satisfied with my life."

6.1.4 Results

6.1.4.1 Descriptive Statistics

The three measures of WFC had acceptable levels of internal consistency reliability and were significantly and positively correlated ($p < 0.001$): time and strain, 0.62; time and behavior, 0.33; and strain and behavior, 0.37; the average correlation being 0.44. These values were similar to those reported by Carlson et al. (2000) in their US study: 0.58, 0.31, 0.54, and an average inter-correlation of 0.48.

The mean values of the three WFC measures were: time, 3.1; strain, 2.9; and behavior, 3.1; the average being 3.1. Carlson et al. (2000) reported mean values of 2.8, 2.6, 2.5, and 2.6, respectively. Respondents in the Turkish sample reported significantly higher levels of WFC than did respondents in the Carlson et al. (2000) study. Carlson and her colleagues reported significant gender differences; females reporting significantly higher levels of strain-based WFC than did males. Males tended to report more WFC on all three measures in the present Turkish study but none of these differences were statistically significant.

6.1.4.2 Analysis Strategy

In order to better understand the sources of the three forms of WFC, correlations were computed between the experience of each form and personal and work situation characteristics, job demands, and the personality factors. Then to examine the potential consequences of the forms of WFC, hierarchical regression analyses were undertaken in which the measures of the four work outcomes and the three indicators of psychological well-being were separately regressed on three blocks of predictors entered in a specified order. The first block of predictors ($N = 5$) consisted of personal demographics (e.g., age, gender, level of education). The second block of predictors ($N = 6$) included work situation characteristics (e.g., organizational level, job and organizational tenure). The third and final block of predictors ($N = 3$) were the forms of WFC. When a block of predictors accounted for a significant amount or increment in explained variance on a given criterion variable ($p < 0.05$), individual items or measures within such blocks having significant and independent relationships with these criterion variables were then identified ($p < 0.05$). These analyses control for the relationships of both personal demographics and work situation characteristics before examining the relationship of forms of WFC and the work and well-being variables of interest.

6.1.4.3 Predictors of Forms of WFC

Personal and Work Situation Characteristics

Correlations were computed between twelve personal demographic characteristics and the three forms of WFC:-time-, strain-, and behavior-based. Only seven of the resulting 36 correlations (19%) were significantly different from zero ($p <$ 0.05). None exceeded 0.10. Older respondents indicated less strain-based WFC; respondents having more children reported more behavior-based WFC; respondents earning higher incomes indicated less strain-based WFC but more behavior-based WFC; those at higher organizational levels indicated less strain-based conflict, respondents having longer organizational tenure also indicated less time-based WFC and less strain-based WFC, and respondents having longer job tenure also indicated less strain-based WFC. These data indicate weak and inconsistent relationships between personal demographic and work situation characteristics and levels of WFC and provide support for the first hypothesis.

Job Demands

Correlations of two job demands (hours worked, work intensity) and the three forms of WFC were computed. Five of the six correlations were significantly different from zero ($p < 0.05$), and all were positive, as predicted in hypothesis 2. Not surprisingly, respondents working more hours per week indicated higher levels of time- and strain-based WFC: respondents reporting higher work intensity reported higher levels of time-, strain-, and behavior-based WFC.

Personality Characteristics

Correlations between four personality characteristics (NAch, three workaholic components) and the three forms of WFC were computed. Five of the resulting twelve correlations (42%) were significantly different from zero ($p < 0.05$) but only three of the five positive in sign as predicted in hypothesis 3. Respondents scoring higher on NAch indicated less strain-based WFC; respondents scoring higher on non-required work also indicated lower levels of strain-based WFC. However respondents scoring higher on feeling driven to work because of inner pressures also reported higher levels of time-, strain-, and behavior-based WFC. These findings provide partial support for the third hypothesis. They also suggest a potentially important role for emotions in the WFC relationship in light of the link between the addictive elements in feeling driven to work (cf. Burke, 2009) and all three measures of WFC.

6.1.4.4 Consequences of Work-Family Conflict

Work-Family Conflict and Work Outcomes

Hierarchical regression analyses in which four work outcomes were separately regressed on the three blocks of predictors were undertaken. All three blocks of predictors accounted for a significant amount or increment in explained variance on job satisfaction. Respondents with children, respondents at higher organizational levels, and respondents indicating lower levels of strain-based WFC reported more job satisfaction ($B = -0.12$, 0.23 and –0.14, respectively). Further, all three blocks of predictors accounted for a significant amount or increment in explained variance on career satisfaction. Respondents at higher organizational levels, those with longer organizational tenure, and respondents indicating lower levels of strain-based WFC indicated higher levels of career satisfaction ($B = 0.21$, 0.13 and –0.10, respectively). Third, two of the three blocks of predictors accounted for significant increments in explained variance on job stress (not personal demographics). Respondents with longer organizational tenure, respondents at lower organizational levels, respondents indicating higher levels of strain-based WFC and those indicating higher levels of time-based WFC indicated higher levels of job stress ($B = 0.16$, –0.10, 0.35, and 0.13, respectively). Finally, all three blocks of predictors accounted for a significant amount or increment in explained variance on intent to quit. Single respondents, respondents having shorter organizational tenure, those having longer job tenure, and those reporting higher levels of strain-based WFC were more likely to intend to quit (B=–0.14, –0.15, 0.14 and 0.18, respectively). These results provide support for the fourth hypothesis.

Work-Family Conflict and Psychological Well-Being

Hierarchical regression analyses in which three indicators of psychological well-being were regressed on the three blocks of predictors were carried out. All, all three blocks of predictors accounted for a significant amount or increment in explained variance on exhaustion. Respondents with higher education, at lower organizational levels and respondents reporting more strain-based WFC also indicated higher levels of exhaustion (B = 0.11, –0.12, and 0.39 , respectively). Furthermore, all three blocks of predictors also accounted for a significant amount or increment in explained variance on psychosomatic symptoms. Women, respondents at lower organizational levels and respondents indicating higher levels of strain-based WFC also reported more psychosomatic symptoms (B = 0.10, –0.10, and 0.16, respectively). Finally, all three blocks of predictors accounted for a significant amount or increment in explained variance on life satisfaction. Respondents at higher organizational levels, those having longer organizational tenure and those indicating less strain-based WFC also reported greater live satisfaction (B = 0.26, 0.16 and 0.13, respectively). These findings provide support for the fifth hypothesis.

Two other observations are worth noting. Strain-based WFC emerged as the strongest and most consistent predictor of both work outcomes and indicators of well-being. Additionally, the WFC measures accounted for a larger amount of explained variance in about half of the analyses carried out.

6.2 Study II – Hotel Managers in Beijing, China

China is a huge country that has exhibited significant economic progress over the past 20 years. Its values are based on Confucianism with the population being atheist. While the hotel and hospitality sectors in China have grown, they still face challenges such as levels of cleanliness and quality of service (Lew et al., 2002).

6.2.1 Procedure

A member of the research team and a vice director of an international training center for hotel managers in Beijing called senior managers of 25 hotels in Beijing asking them to have their properties take part in the research. The directors of human resources of these properties met a research team member who explained the purposes of the study and asked them to nominate respondents in middle management and above to complete anonymous questionnaires. The survey was translated from English into Mandarin and then back-translated. Data were collected from 19 hotels (a 76% participation rate), four 5-star hotels, five 4-star hotels, and ten 3-star hotels. About 330 surveys were distributed and 309 surveys were returned, a considerable response rate of about 90%. Almost 60% of the respondents worked in 3-star hotels with about 30% in 5-star hotels and about 15% in 4-star hotels.

The hotels varied in size, age, location in Beijing, ownership structure and country of management. This sample of hotels was diverse and is best described as a convenience sample.

6.2.2 Respondents

About one-third of the sample ($N = 309$) was between 36 and 40 years of age, 59% were male, most were married (78%), with children (52%), most had high school education (42%), were in middle management (74%), had supervisory duties (95%), had worked in hotels and hospitality before their current jobs (94%), had considerable hospital/hotel sector tenure (about 60% having 10 or more years), most worked in mid-sized hotels having 300–400 employees (30%), and a majority worked 41–50 h per week (41%).

6.2.3 Measures

The measures used in this study were identical to those used in the study of manufacturing managers in Turkey with one exception. The measures of stable personality characteristics excluded one measure of workaholic behaviors (Feeling driven to work because of inner pressures) and included one new measure: Proactive personality. This was measured by a 17 item scale ($\alpha = 0.76$) developed and validated by Bateman and Crant (1993). One item was "I am great in turning problems into opportunities". The majority of the measures used in the hotel manager study had acceptable levels of reliability.

6.2.4 Results

6.2.4.1 Descriptive Statistics

The three measures of WFC had acceptable levels of internal consistency reliability and were significantly and positively correlated ($p < 0.001$): time and strain, 0.63; time and behavior, 0.31; and strain and behavior, 0.39; the average correlation being 0.44.[1]

The mean values of the three WFC measures were: time, 3.3; strain, 2.9; and behavior, 2.9; the average being 3.0.[2] Respondents in the Chinese sample reported significantly higher levels of WFC than did respondents in the Carlson et al. (2000)

---

[1] These values were similar to those reported by Carlson et al. (2000) in their US study: 0.58, 0.31, 0.54, and an average inter-correlation of 0.48.

[2] Carlson et al. (2000) reported mean values of 2.8, 2.6, 2.5, and 2.6, respectively.

US study. Regarding gender differences males reported significantly more time-based and behavior-based WFC in the present Chinese study.

6.2.4.2 Analysis Strategy

Hierarchical regression analyses were undertaken in which the three forms of WFC were first regressed on three blocks of predictors: personal demographics ($N = 5$), work situation characteristics ($N = 8$) (hours worked, work intensity), and personality factors ($N = 3$) (proactive personality, two workaholic job behaviors). Then to examine the potential consequences of the forms of WFC, hierarchical regression analyses were undertaken in which the measures of the four work outcomes and the three indicators of psychological well-being were separately regressed on three blocks of predictors entered in a specified order. The first block of predictors ($N = 5$) consisted of personal demographics (e.g., age, gender, level of education). The second block of predictors ($N = 8$) included work situation characteristics (e.g., organizational level, job and organizational tenure). The third and final block of predictors ($N = 3$) were the forms of WFC. When a block of predictors accounted for a significant amount or increment in explained variance on a given criterion variable ($p < 0.05$), individual items or measures within such blocks having significant and independent relationships with these criterion variables were then identified ($p < 0.05$). These analyses control for the relationships of both personal demographics and work situation characteristics before examining the relationship of forms of WFC and the work and well-being variables of interest. The regression analyses reported below involved sample sizes of 245 or larger.

6.2.4.3 Predictors of Forms of WFC

Job Demands and Personality Characteristics

Two out of three blocks of predictors (personal demographics, work situation characteristics and stable personality characteristics) accounted for a significant amount or increment in explained variance on time-based WFC (personal demographics, work situation characteristics). Women, managers having more education, managers working in more intense jobs, and managers working more hours reported higher levels of time-based WFC ($B = 0.19$, 0.19, 0.19 and 0.12, respectively). Two blocks of predictors accounted for a significant amount or increment in explained variance on strain-based WFC (personal demographics, personality). Managers having higher levels of education, women, and managers scoring lower on proactive personality reported higher levels of strain-based WFC ($B = 0.19$, –0.14, and –0.17, respectively). None of the blocks of predictors accounted for significant amount or increment in explained variance on behavior-based WFC.

6.2.4.4 Consequences of Work-Family Conflict

WFC and Work Outcomes

Hierarchical regression analyses in which four work outcomes were separately regressed on the three blocks of predictors were undertaken. None of the three blocks of predictors accounted for a significant amount or increment in explained variance on job satisfaction. Second, none of the three blocks of predictors accounted for a significant amount or increment in explained variance on career satisfaction. Third, two of the three blocks of predictors accounted for significant increments in explained variance on job stress (work situation characteristics, WFC). Respondents with longer organizational tenure, working in more intense jobs, and respondents indicating higher levels of strain-based WFC indicated higher levels of job stress ($B = -0.22$, -0.15, and 0.33, respectively). Finally, none of the three blocks of predictors accounted for a significant amount or increment in explained variance on intent to quit. Contrary to predictions, WFC seemed to have no effect on the work outcomes included here in this sample of Chinese hotel managers.

WFC and Psychological Well-Being

Hierarchical regression analyses were conducted in which three indicators of psychological well-being were regressed on the three blocks of predictors. First, all three blocks of predictors accounted for a significant amount or increment in explained variance on exhaustion. Respondents without children, respondents working in more intense jobs, respondents having longer organizational tenure, and respondents reporting more strain-based WFC also indicated higher levels of exhaustion ($B = 0.17$, 0.12, and 0.37, respectively) Second, two of the three blocks of predictors (work situation characteristics, WFC) also accounted for a significant amount or increment in explained variance on psychosomatic symptoms. Respondents in more intense jobs, those having longer organizational tenure, and respondents indicating higher levels of strain-based WFC also reported more psychosomatic symptoms ($B = 0.21$, 0.14 and 0.35, respectively). Finally, two of the three blocks of predictors (personal demographics, WFC) accounted for a significant amount or increment in explained variance on life satisfaction. Respondents with more education, and those indicating less strain-based WFC also reported greater live satisfaction ($B = 0.17$ and -0.14, respectively).These findings provide only partial support for the fifth hypothesis.

Three more general observations are worth noting. First, the WFC measures had little effect on work outcomes. Second, strain-based WFC emerged as the strongest and most consistent predictor of indicators of well-being. Third, the WFC measures accounted for a larger amount of explained variance on indicators of psychological well-being than on work outcomes.

6.3 Study III – Managers and Professionals in Egypt

Egypt is a relatively large Muslim country that is becoming increasingly industrialized. It faces considerable challenges however in terms of the quality of its education system in preparing graduates for the modern workplace, autocratic management, work values not supportive of efficiency and limited attention to human resource management (Budhwar and Mellahi, 2007; El-Kot and Leat, 2008; Tyler and Holmes, 2008). Their management practices are now being influenced by those of the West however.

6.3.1 Procedure

Data were collected between October 2008 and January 2009 from service and manufacturing organizations in two Egyptian cities (Alexandria and Cairo). About 50 organizations were contacted in these cities requesting their participation in the research. The 24 cooperating organizations then provided a list of managers and professionals to the researchers. Service organizations included telecommunications, banks, educational institutions, and a maritime service provider. Manufacturing organizations included pharmaceutical, petroleum companies, and production companies focusing upon production of milk, juice and food. Approximately 500 managers and professionals were contacted of which 242 provided completed questionnaires, a response rate of 48%. Questionnaires were completed anonymously in English. The respondents are best described as a large convenience sample of Egyptian managers and professionals in a variety of industries.

6.3.2 Respondents

Over half were male (60%), almost all worked full-time (93%), over half were 30 years of age or younger (61%), most were single (62%), without children (64%), were university graduates (95%), worked 40 h a week or less (42%), were in middle management (33%), supervised others (66%), earned over LE 25,000 a year of income (46%), had relatively short job and organizational tenures (over half having 2 years or less job tenure (60% and over one third having 2 years of less of organizational tenure (37%), and worked in organizations of varying sizes, the average being about 1,000 employees. Respondents fell into several functions: 16 % IT and logistics, 14% marketing and sales, 13% finance, 11% production, 11 and 9% customer service.

6.3.3 Measures

In this study three stable individual difference personality characteristics were included: *Need for Achievement* (NAch) was measured by a five item scale (α = 0.62) developed by Steers and Braunstein (1976) (e.g., "I try very hard to improve on my past performance at work"). Additonally the two workaholic behavior scales developed by Mudrack (2006) were included (see above).

6.3.4 Results

6.3.4.1 Descriptive Statistics

The three measures of forms of WFC had acceptable (but in two cases, only minimally so) levels of internal consistency reliability and were significantly and positively intercorrelated ($p < 0.001$): time- and strain-, 0.61; time- and behavior-, 0.25; and strain- and behavior-, 0.47; the average correlation being 0.44.[1]

The mean values of the three WFC measures were: time-, 2.9, strain, 3.1, and behavior-, 2.9; the average being 3.0.[2] Respondents in the Egyptian sample reported significantly higher levels of WFC than did respondents in the Carlson et al. (2000) US study. Contrary to the US study males and females reported similar levels of time-, strain-, and behavior-based WFC in the present Egyptian study.

6.3.4.2 Analysis Strategy

Hierarchical regression analyses were first undertaken in which the three forms of WFC were separately regressed on two blocks of predictors: personal demographics ($N = 5$) such as age, gender, and level of education, and work situation characteristics ($N = 4$) such as organizational level, job and organizational tenure. Then the three forms of WFC were separately regressed on three blocks of predictors: personal demographics, work situation characteristics, and, in one analysis, stable personality factors ($N = 3$) including Need for achievement, Non-required work and Control of others, and in a second analysis, the third block of predictors contained two job demands (work hours, work intensity), and in a third analysis, the third block of predictors included the use of four recovery experiences (psychological detachment, relaxation, mastery, control).

Then to examine the potential consequences of the three forms of WFC, hierarchical regression analyses were undertaken in which the various outcome measures (work engagement, work outcomes, recovery experiences, and indicators of psychological well-being) were regressed on three blocks of predictors: personal demographics, work situation characteristics, and the three forms of WFC. When a block of predictors accounted for a significant amount or increment in explained variance on a given criterion variable ($p < 0.05$), individual items or measures within such blocks having significant and independent relationships with these criterion variables were then identified ($p < 0.05$). These analyses control for the relationship of both personal demographics and work situation characteristics before examining the relationship of various forms of WFC and the work and well-being variables of interest. The sample size for all regressions reported below was 242.

6.3.4.3 Predictors of Work-Family Conflict

Personal Demographic and Work Situation Characteristics

Hierarchical regression analyses in which the three forms of WFC (time-, strain-, and behavior-based) were regressed on two blocks of predictors: personal demographics and work situation characteristics were carried out. The following comments are offered in summary. First, both blocks of predictors accounted for a

[1]These values were similar to those reported by Carlson et al. (2000) in their US study: 0.58, 0.31, 0.54, and an average inter-correlation of 0.48.

[2]Carlson et al. (2000) reported mean values of 2.8, 2.6, 2.5, and 2.6, respectively.

significant amount or increment in explained variance on time-based WFC. Less educated managers, and managers having long job tenure reported higher levels of time-based work family conflict (B = –0.16 and 0.32, respectively). Second, both blocks of predictors accounted for a significant amount or increment in explained variance on strain-based WFC. Managers who were married and managers at lower organizational levels indicated higher levels of strain-based WFC (B = –0.28 and –0.20, respectively). Third, one of the two blocks of predictors accounted for a significant amount or increment in explained variance on behavior-based WFC (work situation characteristics). Managers working in larger organizations, and managers at lower organizational levels reported higher levels of behavior-based work family conflict (B = 0.33 and –0.19, respectively).

Stable Personality Factors

Stable personality factors accounted for a significant increment in explained variance on all three forms of WFC. Managers scoring higher on NAch reported higher levels of all three forms of WFC (B = 0.31, 0.21 and 0.21, for time-, strain- and behaviorally-based WFC respectively). Manager scoring higher on Control of others reported higher levels of both strain- and behavior-based WFC (B = 0.26 and 0.22, respectively). Managers scoring higher on Non-required work, however, indicated lower levels of behavior-based WFC (B = –0.22).

Job Demands

Job demands accounted for significant increment in explained variance on all three forms of WFC. Managers in more intense jobs indicated higher levels of time- and strain-based work family conflict (B = 0.29 and 0.26, respectively) and managers working more hours per week indicated higher levels of behavior-based work family conflict (B = 0.22).

Work Outcomes

Hierarchical regression analyses in which four work outcomes were separately regressed on the three blocks of predictors were undertaken. The three forms of WFC accounted for a significant increment in explained variance on three of the four work outcomes: job satisfaction, job stress, and intent to quit. Managers indicating higher levels of time-based conflict and lower levels of behavior based WFC reported higher levels of job satisfaction (B = 0.26 and –0.24, respectively), reporting more strain-based work family conflict indicated higher levels of job stress (B = 0.29), and managers indicating more strain-based conflict reported greater intentions to quit (B = 0.24).

Psychological Well-Being

Three indicators of psychological well-being (exhaustion, psychosomatic symptoms, life satisfaction) were separately regressed on the three blocks of predictors. The forms of WFC accounted for a significant increment in explained variance in all three analyses Managers reporting more strain-based conflict also indicated higher levels of exhaustion ($B = 0.24$), more psychosomatic symptoms ($B = 0.27$).and less life satisfaction ($B = -0.17$).

6.4 Discussion

This research attempted to replicate and extend the use of a measure of work-family conflict developed and validated in the US by Carlson et al. (2000). Three forms of WFC were considered: time-, strain, and behavior-based work-family conflict. Data were collected from large samples of female and male managers and professionals employed in various manufacturing organizations in Turkey, various hotels in Beijing, China and in various organizations and industries in Egypt using anonymously completed questionnaires. In addition a greater number of potential predictors and consequences of work family conflict were included here as well.

6.4.1 Replicating the Carlson Et al. Study

Our findings were consistent with those based on the early US study in several areas.

The five hypotheses that guided the research received considerable empirical support. First, the three forms of WFC had generally acceptable levels of internal consistency reliability, low however on both strain- and behavior-based WFC, and were intercorrelated to the same extent in the Turkish, Chinese and Egyptians sample as in the US sample. In addition, forms of WFC were related to outcomes in generally the same way; that is, higher levels of work-family conflict tended to be associated with less favorable work and well-being outcomes. Interesting, as in the US study, strain-based WFC was more strongly related to these outcomes than were either time- or behavior-based WFC. Finally, our findings and those of Carlson (1999) showed that stable personality factors added additional understanding of forms of WFC, and not surprisingly more work-oriented managers indicated higher levels of WFC.

There are several reasons why the findings in the three studies were similar. All respondents were managers and professionals working in global industries. Many respondents and their employing organizations would have contacts with mangers and professionals from the West. Some of the respondents worked in organizations owned by Western organizations. These factors would result in increasingly similar work and managerial experiences.

6.4.2 Extending the Carlson, Kacmar, and Williams Study

The present studies also extended the Carlson et al. (2000) research in several important ways. First, our understanding of these three forms of work-family conflict has been increased; personal and work situation characteristics had weak and inconsistent relationships with the three forms of work family conflict, job demands (work hours, work intensity) had strong and consistent relationships with the three forms of WFC, stable personality characteristics in the form of both need for achievement and workaholic job behaviors increased levels of WFC, and use of particular recovery experiences after work both lessened some forms of WFC while increasing other forms. Thus both individual difference characteristics (e.g., need for achievement) and job demands (level of perceived work intensity) increased levels of WFC. These findings also reflect the fact that work experiences are more likely to affect family well-being than family experiences affect work well-being and functioning (Byron, 2005).

The use of the Carlson et al. (2000) US-developed measures in different countries (Turkey, China, and Egypt) has indicated generally consistent findings with some exceptions. The measures of the three forms of conflict were generally found to be reliable and intercorrelate to the same degree in the four countries studied. There were some differences in level of forms of WFC reported, and the presence or absence of gender differences on forms of WFC, but it was not possible to determine whether this was due to the specific samples, the occupation, or country differences in culture and values. Other consistent findings involved the role of stable personality factors, the role of job demands, and the negative effects of WFC, particularly strain-based conflict on work and psychological well-being. Recent studies have documented the transmission of both positive and negative emotions from work to family, and among family members (Bakker et al., 2008; Westman, 2001, 2005), suggesting that further work on the role of emotions in the workplace and the home is warranted.

Our results also fit with the recent emphasis being given to the role of emotions, both positive and negative, in the workplace (Ashkanasay et al., 2000; Cameron et al., 2003; Fineman, 1996). Emotions have traditionally been ignored in organizational scholarship (Fineman, 1996). In addition, recent studies have documented the transmission of both positive and negative emotions from work to family, and among family members (Bakker et al., 2008; Westman, 2001, 2005).

6.4.3 Country Differences

It is also important to note country differences. Managers in Turkey, China, and Egypt reported higher levels of all forms of WFC than did US respondents. In addition, male and female managers in both Turkey and Egypt reported similar levels of all three forms of WFC whereas female managers in the US reported higher levels of strain-based WFC whereas female managers in China reported higher levels of time based-WFC and lower levels of behavior-based WFC than did their male colleagues. These differences may reflect differences in occupations. There were also country or occupational differences in the levels of both predictors and outcome variables (e.g., work hours, work intensity). Clearly more cross-cultural research involving respondents in the same occupations is needed to sort this out. Other researchers (e.g., Aycan, 2001; Aycan et al., 2000; Wasti, 1998) have cautioned against assuming North American findings and interventions will directly "fit" other cultures.

We collected data from respondents in three countries using scales developed in a fourth country. Although this approach permitted an examination of similarities and differences, country or national culture and values was never measured. The field as a whole has neglected cultural influences in the work family interface. Powell et al. (2009) identify three types of studies here.

- Culture as nation – Comparisons are made across countries though culture is never measured,
- Culture as referent – Comparisons are made between two countries based on ideas of what their culture looks like based on other evidence though again culture is never measured,
- Culture as dimensions – Comparisons are made between countries that include assessments of relevant cultural dimensions. Cultural data may be collected from individuals or based on aggregate data reported by others.

Powell and his colleagues (2009) suggest that four cultural dimensions identified by others would be fruitful: individualism-collectivism (Hofstede, 1980), humane orientation and gender egalitarianism (House et al., 2004) and diffusion-specificity (Trompenaars and Hampden-Turner, 2005). Burke (2009) made the same observation when reviewing the influence of cultural values in women in management research. Further work-family/life integration research conflict should consider the cultural aspect in more detail.

6.4.4 Practical Implications

Some preliminary practical implications follow from these findings. First, individuals and organizations would benefit by offering training sessions that address the consequences of excessive work for family and personal health. Second, organizations might consider ways of reducing levels of work hours and work intensity (Friede et al., 2008). Munck (2001) outlines what the Marriot hotel chain did to reduce both work hours and work intensity of supervisor and managerial staff. Deery (2008) also offers several suggestions in this regard. Work practices matter. Ngo et al. (2009) found that these practices directly influenced firm performance in a sample of multi-national corporations in Hong Kong. The empirical results presented in this chapter undermine the necessity to train employees in dealing with work-life integration as well as the need for reduced work hours and intensity.

Acknowledgements Preparation of this manuscript and conduct of the research was supported in part by York University, Nevsehir University, Beijing Union University and the College of Management and Technology, Arab Academy for Science and Technology and Maritime Transport. Ufuk Durna, Halil Demirer, Mehmet Tekinkus, Cetin Bektas and Fuus Tekin Acar assisted with data collection in the Turkish study. We thank the organizations that cooperated and our respondents for their participation.

References

Ali AJ, Al-Owaihan A (2008) Islamic work ethic: a critical review. Cross Cult Manage 15:5–19
Allen. TD, Herst DEL, Bruck CS, Sutton M (2000) Consequences associated with work to family conflict: a review and agenda for future research. J Occup Health Psychol 5:278–308
Ashkanasay NM, Hartel CEJ, Zerbe WJ (2000) Emotions in the workplace: research theory and practice. Quorum, Westport, CT
Aycan Z (2001) Human resource management in Turkey: current issues and future challenges. Int J Manpower 22:252–260
Aycan Z (2004) Key success factors for women in management in Turkey. Appl Psychol An Int Rev 53:453–477
Aycan Z, Kanungo RN, Mendonca M, Yu K, Deller J, Stahl G, Kurshid A (2000) Impact of culture on human resource management practices: a 10-country comparison. Appl Psychol:Int Rev 49:192–231
Bakker AB, Demerouti E, Dollard MF (2008) How job demands affect partner's experience of exhaustion: integrating work-family conflict and crossover theory. J Appl Psychol 93:901–911
Barling J, Kelloway EK, Frone MR (2005) Handbook of work stress. Sage, Thousand Oaks, CA
Bateman TS, Crant JM (1993) The proactive component of organizational behavior: a measure and correlates. J Organ Behav 14:103–118
Bloom N, Kretschmer T, Van Reenen J (2009) Exploring productivity growth in Europe, America and Asia. London School of Economics, Center for Economic Progress, London
Brough P, Kalliath T (2009) Work-family balance: theoretical and empirical advancements. J Organ Behav 30:581–585
Budhwar P, Mellahi KI (2007) Introduction: human resource management in the Middle East. Int J Hum Resour Manage 18:2–10
Burke RJ (1991) Early work and career experiences of female and male managers and professionals: reasons for optimism?. Can J Admin Sci 8:224–230
Burke RJ (2009) Cultural values and women's work and career experiences. In Steers R, Bhagat R (eds) Handbook of cross cultural research. Cambridge University Press, Cambridge, pp 442–461
Byron K (2005) A meta-analytic review of work-family conflict and its antecedents. J Vocat Behav 67:169–198
Carlson DS (1999) Personality and role variables as predictors of three forms of work-family conflict. J Vocat Behav 55:236–253
Carlson DS, Kacmar KM, Williams LJ (2000) Construction and initial validation of a multidimensional measure of work-family conflict. J Vocat Behav 56:249–278
Cameron KS, Dutton JE, Quinn RE (2003) Positive organizational scholarship: foundations of a new discipline. Berrett-Koehler, San Francisco, CA
Deery M (2008) Talent management, work-life balance and retention strategies. Int J Contemp Hospital Manage 20:792–806
Diener E, Emmons RA, Larsen RJ, Griffin S (1985) The satisfaction with life scale. J Pers Assess 49:71–75
Eby LT, Casper WJ, Lockwood A, Bordeaux C, Brinley A (2005) Work and family research in IO/OB: content analysis and review of the literature (1980–2002). J Vocat Behav 66: 124–197
El-Kot G, Leat M (2008) A survey of recruitment and selection practices in Egypt. Educ Bus Soc: Contemp Middle East Issues 1:200–212
Fineman S (1996) Emotions in organizations. Sage, Thousand Oaks, CA
Friede A, Kossek EE, Lee MD, MacDermid S (2008) Human resource manager insights on creating and sustaining successful reduced-load work arrangements. Hum Resour Manage 47:702–727
Goff SJ, Mount MK, Jamison RL (1990) Employer supported child care, work-family conflict, and absenteeism: a field study. Pers Psychol 43:793–809

Greenhaus JH, Parasuraman S, Wormley W (1990) Organizational experiences and career success of black and white managers. Acad Manage J 33:64–86

Hewlett SA, Luce CB (2006). Extreme jobs: the dangerous allure of the 70-h work week. Harv Bus Rev December:49–59.

Hofstede G (1980) Culture's consequences: international differences in work-related values. Sage, Beverly Hills, CA

House RJ, Hanges PJ, Javidan M, Dorfman PW, Gupta V (2004) Culture, leadership, and organizations: the GLOBE study of 62 societies. Sage, Thousand Oaks, CA

Jones F, Burke RJ, Westman M (2005) Work-life balance: a psychological perspective. Psychology Press, East Sussex

Kofodimos J (1993) Balancing act. Jossey-Bass, San Francisco, CA

Kossek EE, Ozeki C (1998) Work-family conflict, policies and the job-life satisfaction relationship: a review and directions for organizational behavior-human resources research. J Appl Psychol 83:139–149

Lew AA, Yu L, Ap J, Guangrui Z (2002) Tourism in China. Haworth Press, Binghampton, NY

Maslach C, Jackson SE, Leiter MP (1996) Maslach burnout inventory. Consulting Psychologists Press, Palo Alto, CA

Mudrack PE (2006) Understanding workaholism: the case for behavioral tendencies. In Burke RJ (ed) Research companion to working time and work addiction. Edward Elgar, Chichester, pp 108–128

Munck B (2001). Changing a culture of face time. Harv Bus Rev November:3–8

Ngo HY, Foley S, Loi R (2009) Family friendly work practices, organizational climate and firm performance: a study of multinational corporations in Hong Kong. J Organ Behav 30:665–680

Parasuraman S, Greenhaus JH (1997) Integrating work and family: challenges and choices for a changing world. Quorum, Westport, CT

Powell GS, Francesco AM, Ling Y (2009) Toward culture-sensitive theories of the work-family interface. J Organ Behav 30:597–616

Quinn RP, Shepard LJ (1974) The 1972–1973 quality of employment survey. Institute for Social Research, University of Michigan, Ann Arbor, MI

Rice RW, Frone MR, McFarlin DB (1992) Work-nonwork conflict and the perceived quality of life. J Organ Behav 13:155–168

Schabracq MJ, Winnubst JAM, Cooper CL (2003) The handbook of work and health psychology. Wiley, West Sussex

Spence JT, Robbins AS (1992) Workaholism: definition, measurement, and preliminary results. J Pers Assess 58:160–178

Steers RM, Braunstein DN (1976) Manifest needs questionnaire. In Cook JD, Hepworth SJ, Wall TD, Warr PB (eds) Experience of work: a compendium and review of 249 measures and their use. Academic, New York, NY, pp 159–160

Trompenaars F, Hampden-Turner N (2005) Riding the waves of culture: cultural diversity in business. Nicholas Brealey, London

Tyler M, Holmes E (2008) Higher education reforms in Egypt: preparing graduates for Egypt's changing political economy. Educ Bus Soc Contemp Middle East Issues 1:175–185

Wasti SA (1998) Cultural barriers in the transferability of Japanese and American human resources practices to developing countries: the Turkish case. Int J Hum Resour Manage 9:608–631

Westman M (2001) Stress and strain crossover. Hum Relat 54:717–753

Westman M (2005) Crossover of stress and strain in the work-family context. In Jones F, Burke RJ, Westman M (eds) Work-life balance: a psychological perspective. Psychology Press, East Sussex, pp 163–184

Chapter 7
Work-Family Balance, Role Integration and Employee Well-Being

Ann Chunyan Peng, Remus Ilies, and Nikos Dimotakis

Researchers have long been interested in understanding the antecedents of subjective well-being, which has been generally defined as "how people evaluate their lives" (Diener et al., 2003, p. 404). A number of studies have adopted a top-down approach to predicting subjective well-being, whereby a person's dispositional characteristics influence the level of life satisfaction that the person experiences (see Heller et al., 2004). On the other hand, there is also evidence for a bottom-up process whereby a person's levels of satisfaction with specific life domains (e.g., work, marriage, etc) are combined to form an overall level of satisfaction with one's life in general (Heller et al., 2004).

Furthermore, it is understood that individuals are called to perform different roles in each of their specific life domains (e.g., employee, supervisor, husband, parent, etc.), and there is evidence suggesting that the ways in which individuals manage the interface or transition between these roles has implications for their well-being, as well as how successful they are in performing these varied roles (Ashforth et al., 2000). This chapter is concerned with individuals' balance (or imbalance) between work and family roles and with the specific strategies that can be used to best fulfill work and family demands across the work-family role boundary. Perhaps most importantly, the chapter is specifically concerned with the implications of work-family balance and of work-family work boundary management strategies for the well-being of the professionals, who oftentimes face increased workplace challenges stemming from demanding occupations.

Increasing demands at work and at home create challenges for employees in fulfilling multiple roles, and thus the ways in which one strives to manage the interface between one's work and personal life (e.g., achieving work-life balance) has implications for his/her work and non-work feelings and satisfaction, and ultimately for one's psychological well-being. In this chapter, we will not only consider the additive contributions of specific role-related affective experiences and satisfactions (e.g., job affect and satisfaction, marital satisfaction) to psychological well-being, but will also consider the interactive effects of role demands, feelings, and behaviors

A.C. Peng (✉)
Michigan State University, East Lansing, MI, USA
e-mail: peng@bus.msu.edu

S. Kaiser et al. (eds.), *Creating Balance?*, DOI 10.1007/978-3-642-16199-5_7,

on well-being, as well as issues of balance between work and non-work life roles.

Managing the interface between one's work role and other life roles (e.g., family) is especially important for professionals, whose role boundaries can be seen as potentially more blurred or permeable compared to blue-collar, hourly, or part-time employees. Unlike typical 9–5 workers, professionals are required to work intensively and under "non-typical" work schedules (e.g., meeting with clients during the weekend). Highly demanding jobs push professionals to spend more hours at work, which may negatively influence their health and personal life. Meanwhile, technological advancements (e.g., ubiquitous cellular or wireless Internet access, smartphones, etc.) enable professionals to continue to perform work related tasks (checking work email, attending teleconferences, etc.) during times traditionally reserved for non-work roles, potentially leading to growing permeability of the boundaries between work and non-work domains for these employees. Then, if the role boundaries are potentially more permeable for the professional employee, how one actively manages processes that extend across these boundaries can be seen as a specific and conscious decision (i.e., deciding not to read work email even if the technology allows one to do so), which can be in turn linked to important outcomes.

This chapter will begin with a brief review of important issues in work-life balance research (e.g., work-family conflict, enhancement, spillover, etc.). We will then focus on the key concept of work-family role integration, providing (1) a multilevel perspective on the antecedents of work-family integration, (2) a comprehensive consideration of various consequences of work-family integration, and (3) an in-depth discussion about the effects of work-family integration on employee well-being. We will conclude with a discussion of the practical implications of these issues for professional employees and their employers.

7.1 Work-Family Balance and Role Integration

During the last few decades, a great number of studies have investigated the interface between work and family life. This type of research has led to two different perspectives regarding the implications of participating in both work and family roles, namely work-family conflict and work-family enrichment. Rooted in role theory (Merton, 1957), the work-family conflict perspective emphasizes that the expectations or demands from one life role often interfere with one's ability to meet the expectations or demands from another role. Greenhaus and Beutell (1985) specified three categories of work-family conflict, introducing a conceptualization of time-based, strain-based, and behavior-based conflict. In general, this stream of research assumes that the finite nature of one's personal resources (e.g., time, energy, attention) creates the potential for tension when allocating these resources to meet competing demands in both work and family domains, which can in turn lead to the experience of psychological strain (e.g., Glass and Fujimoto, 1994; Moen and Yu, 1998).

To the contrary, the work-family enrichment perspective posits that one's participation in a work role can potentially facilitate participating in family roles and vice versa. This perspective thus proposes that holding multiple roles will ultimately benefit individuals, through such mechanisms as role accumulation and role congruence (Crosby, 1991; Gove and Zeiss, 1987). It suggests that, by occupying multiple social roles, individuals can obtain alternative resources, and this expansion in resources will ultimately outweigh the potential strain that multiple role engagement can create (Crosby, 1991; Rothbard, 2001). Supporting this perspective, studies have shown that a strong engagement in both family and work roles has a positive effect on health and well-being (Baruch and Barnett, 1986; Nordenmark, 2002; Verbrugge, 1986). In one specific instance, for example, employed, married mothers were found to have higher levels of physical and psychological well-being compared to unemployed, married mothers (Thoits, 1983; Waldron et al., 1998).

Given that both work-family conflict and enrichment perspectives have been supported by empirical studies (e.g., Rothbard, 2001), there seems to be value in the position that it is how people manage different life roles, instead of the number of roles in and of itself, that ultimately determines the specific outcomes experienced by the individual. Achieving a balance between work and non-work personal life (e.g., family life, leisure, personal time) is critical to one's well-being (Frone, 2003). According to Kirchmeyer (2000, p. 81), a balanced life consists of "achieving satisfying experiences in all life domains, and to do so requires personal resources such as energy, time, and commitment to be well distributed across domains". In the same vein, work-family balance is defined as "satisfaction and good functioning at work and at home with a minimum of role conflict" (Clark, 2000, p. 349).

7.1.1 Integration and Segmentation

The introduction of the work-family balance concept has inspired a number of studies that examined the outcomes of pursuing or attaining work-family balance for employees and employers. The concept itself, however, is often not rigorously defined and at times hard to distinguish from other related constructs (e.g., life satisfaction). The measurement of work-family balance also suffers from a rather vague approach, in which participants are generally asked to report "How successful do you feel in balancing your paid work and family life?" (Milkie and Peltola, 1999, p. 481), or questions about affective reactions are used to infer balance levels (e.g., "I am satisfied with the balance I have achieved between my work and life", Saltzstein et al., 2001, p. 467). Moreover, surface examinations of work-family balance do not provide insight into the mechanisms or processes underlying the interface between work and family (Greenhaus et al., 2003).

An important concept that is relevant to achieving work-family balance concerns the extent to which people integrate or segment their work and family life roles. The concept of work-family integration originates in boundary theory, which states that each role (e.g., employee, parent) is structured by a boundary or a "mental fence" (Zerubavel, 1991, p. 2) that distinguishes it from other roles in terms of time and space (Ashforth et al., 2000). The degree to which these boundaries are permeable varies across different jobs and individuals, thus keeping the two roles relatively segmented or integrated (Nippert-Eng, 1996). In other words, each individual can be placed on a continuum, ranging from high levels of segmentation on one end to high levels of integration on the other, depending on how they actively manage the boundaries among these two life roles.

A high level of integration between work and family roles indicates a "flexible" or "permeable" boundary that allows cross-role interruptions between work and family. Individuals with a high level of work-family integration might perform activities typically associated with work roles while at home (e.g., replying to work emails during family time), or the other way round (e.g., shopping online for house wares and supplies while at work). On the contrary, a high level of segmentation denotes a boundary management style that keeps each role relatively separate and independent. For example, individuals with high levels of work-family segmentation would avoid undertaking work tasks while at home, or chatting with family members on the phone during their time at work. The way in which individuals integrate or segment their work-family roles can have implications for their role performance and well-being (e.g., Kreiner, 2006), an issue which we will discuss in detail in the next section.

7.1.2 Spillover and Crossover

Another concept that has been frequently employed in understanding the interface between work and family is spillover. Work-family spillover is defined as "the effects of work and family on one another that generate similarities between the two roles" (Edwards and Rothbard, 2000, p. 180). Four types of "similarities" have been distinguished; these relate to spillovers of affect (i.e., mood and satisfaction), values (i.e., beliefs that ascribed to work and family pursuits), skills, and behaviors. Although findings of "negative spillover" have been common, whereby unpleasant work (or family) characteristics are carried on into the family (or work) domain, there is also evidence of positive work-family spillover processes that underlie the existence of synergies between work and family (Ilies et al., 2009; Song et al., 2008). Compared to the dominant work-family conflict approach that has tended to overemphasize the competition of resources between work and family roles, spillover research examines the shared characteristics between the two roles instead. The strength of this approach is that it has the potential to accommodate the potential positive syntheses (e.g., common valuable skills) between roles alongside the extant processes of negative spillover (e.g., strain or negative mood spillover).

While spillover approaches describe the similarities between work and family roles within the same individual, crossover approaches adopt an interpersonal perspective instead. Crossover has been conventionally defined as a transmission of strain from one person to another person in the same social environment (Westman, 2001). Unlike spillover, which has been recognized to have both positive and negative aspects, extant studies on crossover have been exclusively focused on how the strain experienced by an individual can influence the well-being of their spouse (Hammer et al., 1997; Takeuchi et al., 2002; Westman and Etzion, 2005), with one exception being a recent study conducted by Song et al. (2008). Such research, however, should not be constrained to only negative processes. Future research efforts can stand to benefit by examining potential instances of positive crossover between family members (for example, crossover of positive moods or domain satisfaction) as well as by extending this research to the context of the workplace (e.g., in examinations of positive crossover among colleagues or between supervisor and subordinates).

7.2 Predictors of Work-Family Integration

Despite the increasing attention researchers have focused toward issues of work-family integration, only a limited number of studies have empirically examined its antecedents. In order to facilitate future efforts, we suggest a multi-level framework to help conceptualize and understand the factors that can affect an individual's positioning on the work-family integration-segmentation continuum. In the framework presented below, we seek to integrate the various factors of interest from a number of suggested levels of analysis, ranging from social, to occupational and organizational, and finally to individual factors.

7.2.1 Social Factors

One of the most widely recognized factors that can contribute to increasing levels of integration between work and family is the changing nature of the workforce. During the last few decades, there has been a marked increase in women's participation in the labor force in industrialized societies. Women now constitute the majority of the workforce in the United States, as well as almost half the workforce in the United Kingdom and other developed countries (Toossi, 2002). Despite their increased investment in work, women are still frequently undertaking the majority of family responsibilities such as child-caring. As a result, female employees, those holding professional jobs in particular, frequently experience difficulties in coordinating family and work responsibilities, and report higher levels of work-family conflict (Eby et al., 2005; Friedman and Greenhaus, 2000).

At the same time, women's increased entry into the workforce has affected men's life as well. With their spouses also working, men are nowadays no longer the sole source of family income (or "bread winner") in the majority of families. Instead, they are increasingly expected to share the family responsibilities alongside their spouses. Consequently, the increasing coexistence of both work and family roles for each individual can be seen as leading to a more blurred boundary between work and family domains. As noted by Ashforth et al. (2000), work-family segmentation, which used to be a naturally occurring process, now has to be consciously and carefully implemented for individuals that seek to maintain a non-permeable boundary between roles.

Moreover, advances in information and communication technologies (e.g., widely available internet access, smartphones) make work-family integration easier. In conjunction with increasing levels of demands from a service-oriented economy and changes in work norms (i.e., 24/7 availability), it is increasingly common for employees to pursue work endeavors during family or personal time. New technologies also make various flexible work arrangements (telecommuting, working at home) available and feasible. These work arrangements can lead to more integration between work and family roles, and potentially allow employees to capitalize on higher levels of discretion and flexibility in order to accomplish both family and work responsibilities successfully. The downside, however, is that by integrating work-family lives, employees might experience increased levels of strain, as a result of increased level of conflict between work and family demands and increased role ambiguity (Ashforth et al., 2000).

The specific cultural norms of a country, ethnic group, or community can also be seen as an important influence on individuals' specific levels of role integration. In one way, cultural norms affect how individuals choose to manage work-family boundaries by defining the generally accepted territory for each role and by shaping individuals' held values relevant to each role (Hill et al., 2004). It is possible that in some national or ethnic cultures work roles are considered more important than family roles whereas in others family roles are given more priority, and that general judgements might differ for each gender. This difference in relative importance can, in turn, affect how individuals choose to prioritize between roles and ultimately the way in which work and family roles are integrated or segmented.

This positioning on the integration-segmentation continuum can also be shaped by other culture-level factors such as uncertainty avoidance or masculinity/femininity (Hofstede, 2001). We suspect that people in a culture less concerned about minimizing uncertainty might be more likely to integrate different roles whereas a high level of uncertainty avoidance might contribute to a preference towards role segmentation. Similarly, cultures high in masculinity could influence individuals towards high levels of segmentation because of decreased (and potentially reluctant) participation in specific roles. In all, although a thorough examination of culture-level factors, and generation of propositions predicting their effects is outside the scope of this chapter, these factors can be seen as having interesting implications for work-family research, examinations of which we believe could provide a valuable contribution.

7.2.2 Occupational and Organizational Factors

Work-family integration is more common for professionals and managers, occupational groups that are frequently expected to expend a substantial amount of time and effort in order to meet increasing levels of job demands and to advance their careers (Jacobs and Gerson, 2004). This increased integration, however, is a complex issue that can both facilitate and hinder achieving work-life balance. On the one hand, the long work hours culture promoted and practiced by many organizations induces professionals to spend more time at work, and creates difficulties in achieving work-life balance (Burke, 2008). Compared to blue-collar or hourly workers, professionals who operate under increased organizational expectations in terms of accepted levels of work commitment, are more likely to carry job-related tasks home, and experience more psychological difficulty in maintaining the two life roles separate; the potential of these issues to result in work-family conflict can result in these individuals frequently experiencing higher levels of strain.

On the other hand, professionals generally have more control over their jobs, and can thus feel more comfortable attending to family tasks and responsibilities while at work, when they need to do so. They are also more likely to have access to flexible work arrangements that can enable them to better accomplish family obligations. These flexible work arrangements are a part of various work-family friendly policies that employers have been adopting in larger numbers, as a result of the increased importance both employees and organizations have placed on work-family balance issues. These policies are designed to promote integration of work and family lives for their employees with the aim of reducing or eliminating their experienced levels of work-family conflict (Berg et al., 2003; Osterman, 1995).

However, such work-family policies are not available in all organizations, and the actual way in which these policies are implemented can also vary. Studies have found that the ambiguity associated with participating in these policies in some organizations can actually discourage employees from using them (Blair-Loy, 2001; Lobel et al., 1999). Employees might also hesitate to use these policies because of concerns about their career advancement being delayed or hindered as a result (Blair-Loy and Wharton, 2002). Therefore, companies with more formally established family-friendly policies and supportive systems (e.g., supervisors that are supportive of the policies, guarantees of job and career security relevant to policy use) are more likely to encourage their employees' actual use of these provided benefits, and to ultimately be more sincere and successful in assisting their employees to achieve work-family balance.

At this point it should be noted that research so far has not addressed the directionality of work-family integration. That is, there has not been a conceptual separation of integrating work into the family domain, and vice versa. We suggest that separating work integrated into family (WIIF, integrating work-related activities within family life) from family integrated into work conceptualizations (FIIW, integrating family-related activities into work life) can provide a more meaningful and accurate description of these concepts, and better capture their occurrence in applied contexts. For example, organizational norms might be more relevant to FIIW; an organization with strong norms that prohibit personal activities at work can create a strong situational constraint whereby employees are prevented from integrating family roles into work. In contrast, an organization that values family-friendly policies might be more lenient toward employees performing family duties at work. In a similar fashion, a person with a more "work-friendly" family environment is more likely to be able to integrate work into his or her family life. For example, a supportive spouse can make it easier for an employee to work efficiently at home. A recent interview study showed that couples who are working in the same organization are more likely to integrate work into family but less likely to integrate family into work compared to couples who do not share the workplace (Janning, 2009). Future studies could investigate how job-related or organizational factors (e.g., workplace demands, organizational policies) and family characteristics (e.g., family responsibility, spouse support) can jointly affect one's preference on segmentation or integration between work and family roles. These examinations might also determine to what extent an individual's work-family integration strategies are effective in terms of performing effectively in both roles and the psychological experiences associated with these strategies.

7.2.3 Individual Factors

Although the various occupational demands and organizational norms that individuals are facing at work might induce them to integrate work into their family life, the actual decision regarding how to manage work and family boundaries is, to some extent (and within the extant situational constraints), the individual's own choice (Kossek and Lautsch, 2008). This statement is even more applicable to professionals who generally have more decision latitude or job control in performing their work-related responsibilities (Jacobs and Gerson, 2004). Therefore, the effects of individual differences on work-family integration decisions can be seen as an important factor to consider and examine.

To our knowledge, little research has examined how individual characteristics affect work-family integration. It has been generally assumed that an individual's selected level of work-family integration will be influenced by organizational practices (e.g., flexible work time, family supportive policies) and individual preferences (e.g., Rothbard et al., 2005). However, actual individual factors that influence individual preferences (and thus, how individuals develop a personal style of work-family integration) have yet to be identified and confirmed empirically.

We suggest two important individual characteristics that can impact individual's preferred level of work-family integration. The first is role priority, that is, the relatively importance that an individual ascribes to each of their life roles. Role priority is expected to affect the amount of work-family integration as well its resulting directions (WIIF, FIIW, or both). There is evidence that a high priority role is more likely to be integrated into other relative less important roles (Olson-Buchanan and Boswell, 2006). For example, employees who place higher importance on their work roles might be more motivated to perform work-related tasks at home, whereas employees who place higher importance on their family role might be more likely to perform family-related activities while at work. In cases of roughly equal role importance, potential role integration might occur on an ad hoc basis, with time or energy slack during one role being used as needed to address increased demands in the other role.

Another individual characteristic that might be relevant to work and family boundary management refers to individuals' multitasking capabilities. Individuals differ in how effectively they can perform different tasks simultaneously. Suggested by role boundary theory (Ashforth et al., 2000; Nippert-Eng, 1996), the primary benefit of role integration is to minimize the cost in switching roles and to effectively utilize resources such as time. However, the risk of highly integrated roles is role ambiguity, dealing with which might require more cognitive resources in order to meet dissimilar role requirements. Enacting different roles at the same time can be considered as a multitasking process that requires an effective allocation of different cognitive resources in order to simultaneously pursue diverse goals. Therefore, individuals who are more capable in enacting multiple roles at the same time might prefer a high integration level between work and family roles, whereas those who find dealing with blurring role boundaries more challenging might be more likely to pursue a strategy of segmentation.

Overall then, driven by changes at the societal, organizational, and occupational levels, the boundaries between work and family domains are becoming increasingly blurred. Nevertheless, individuals vary significantly in their preferred and actual style of managing work and family boundaries, with individuals being found at every point of the integration-segmentation continuum. An individual's placement on this continuum, therefore, can be viewed as a function of joint influences from all the levels discussed above, and examinations of each of these levels has the potential to inform theory and practice, especially investigations that seek to integrate these effects, and investigate their potential interplay.

7.3 Consequences of Work-Family Integration

Until relatively recently, it has been generally assumed that higher levels of integration between work and family lives would be beneficial overall for employees and their employers. A main reason for this assumption rests on the implied equivalence of the constructs of work-family integration and work-family balance that has been identified in both academic and policy discussions, in which the two terms have been often used interchangeably (Jones et al., 2006). While it is entirely possible that as an individual increasingly integrates his or her family and work lives, he/she manages to achieve a balance between these two life roles, this benefit is neither automatic nor guaranteed. Similarly, although organizations adopt family-friendly practices more frequently, i.e. these practices are increasingly more common in some form in many modern organizations (Berg et al., 2003) those work-family policies often do not substantively help employees to achieve work-family balance. Instead, they become more of an instrumental tool that help the employers recruit talent and establish a favorable social reputation (Blair-Loy and Wharton, 2002), without necessarily helping employees manage their work and family obligations more effectively. Carefully conceptualizing the terms work-family integration and work-life balance can be of help in informing research as to the factors that influence the outcomes of integration for individuals and organizations.

Indeed, based on role boundary theory, studies have refined the definition of integration or segmentation, and begun to examine the specific consequences of work-family integration (Ashforth et al., 2000; Nippert-Eng, 1996; Williams and

Alliger, 1994). From this perspective, work-family integration is viewed as a style that reflects how people manage the boundaries between work and family roles. Naturally, integration has benefits and costs for the individual, and both of these outcomes can be examined by using the framework of role boundary theory (Ashforth et al., 2000). In general, the primary benefit of role integration is to minimize the cost in switching roles and to effectively utilize the resources such as time, while the primary cost of this approach is higher levels of role ambiguity and role blurring that can confuse and hinder individuals in successfully addressing diverse role requirements.

In this section, we will summarize how work-family integration can impact employees' health and psychological well-being, and discuss the mechanisms through which these effects occur. We aim to examine processes such as spillover to understand when work-family integration can have positive or negative effects on an individual's well-being, and also discuss crossover processes to describe how work-family integration influences individuals' family members as well as how it affects employers.

7.3.1 Outcomes for the Individual

People integrate their work and family lives in various ways. By having a phone meeting with their boss during a family vacation, checking their email on a smartphone while waiting for dinner, and reading a work report in bed, people integrate work into the family domain. On the other hand, by having a conversation with their spouse while at work, leaving work earlier to attend a Parent-Teacher Association (PTA) meeting, or using flexible work arrangements to fulfill family obligations, they also integrate their family activities into the work domain. Although, as mentioned above, the literature has tended not to distinguish the direction of work-family integration (WIIF or FIIW), we believe that making such a distinction can be valuable in predicting the outcomes of such integration for the individual.

WIIF can be seen as more likely to have negative outcomes on one's family role performance but be potentially beneficial for one's performance in the work role. For example, working extra hours at home in order to complete an urgent project report can deal with the temporal increases in job demands, and therefore reduce perceived stress at workplace and improve well-being. However, WIIF might also lead to potential work-family conflict (Kossek et al., 2006; Olson-Buchanan and Boswell, 2006) and poor psychological or physical health. Spending time on a work project at home might increase emotional exhaustion and fatigue, which subsequently hinder an individual's quality interactions with family members and result in low family satisfaction. In a similar fashion, dealing with a family issue while at work (FIIW) might facilitate fulfilling family obligations, but does so at the expense of decreased effectiveness at work. In addition, bringing family issues into work might be perceived as a lack of commitment to work, and therefore adversely impact one's career advancement (Lee et al., 2002).

So far, few studies have examined the various consequences of work-family integration on individual's role performance and satisfaction. Research effort should also address how different types of work-family integration (e.g., WIIF, FIIW) influence individual's physical and psychological health, and identify important contextual factors that promote favorable consequences of work-family integration.

7.3.2 Outcomes for the Individual's Family Members

It seems intuitively apparent that the ways in which individuals manage the boundary between their work and family lives should have an effect on their family members; nevertheless, clear empirical evidence of this is not yet available. For example, bringing work home might interfere with family life such that it becomes difficult, if not impossible, to spend quality time with one's family members, to perform parental responsibilities, or to share housework with one's spouse. This "cross-utilization" of resources (e.g., family time) can then interfere with one's role performance at the family domain, which in turn has the potential to negatively affect the experiences and feelings of their family members (for example, by increasing their levels of family demands or creating perceptions of injustice). A spouse might feel being ignored when their partner is occupied by work duties and is unable or unwilling to discuss a family or personal issue, and children might be disappointed and feel neglected by having a parent unavailable to play with or otherwise spend time with them.

Moreover, the experience of performing work duties at home, in circumstances when doing so results in feelings of stress or anxiety for the focal individual, might also produce negative emotion contagion that negatively affects the well-being of their family members. For example, an employee who tries to meet an important work deadline might convey various negative emotions (e.g., anxiety, irritation) when working at home. These negative displays of emotion can then crossover, through emotional contagion processes, and result in experiences of strain for the other family members. In fact, a growing body of research has examined crossover between one spouse's stress and strains and the stress and strains experienced by the other, and these studies have been generally supportive of the existence of such processes (Hammer et al., 1997; Westman, 2001; Westman et al., 2004).

However, in terms of other family members apart from the spouse (such as children or cohabiting parents), there is virtually no research on how an individual's work-family boundary management styles can affect them, although some indirect conclusions could perhaps be drawn from studies on non-standardized work arrangements (e.g., working evenings, nights or weekends). For example, using data from a representative sample of 4,433 dual-earner Canadian families, a study found children's well-being is adversely impacted by their parent(s)'s undertaking non-standardized jobs (Strazdins et al., 2004). This negative relationship held even after controlling for potentially confounding factors such as socio-economic status, parent part-time or full-time work, and childcare use. Such indirect conclusions underline the role that spillover can potentially play, and the importance of pursuing further investigations of the same.

7.3.3 Outcomes for Employers

Studies have shown that providing family-responsive benefits can have various positive outcomes for organizations, such as increased organizational commitment (Roehling et al., 2001), reduced turnover intentions (Berg et al., 2003), and lower levels of absenteeism (Dalton and Mesch, 1990). Moreover, employers with more family-supportive policies are more favorably rated by job applicants and therefore more likely to recruit and retain talented employees (Berg et al., 2003; Osterman, 1995). These findings would then seem to suggest that allowing employees to integrate their family activities into work can potentially be of benefit towards employers.

On the other hand, policies that encourage increased levels of work-family integration can also have undesirable effects for organizations. For instance, trying to fulfill family demands can restrict employee's work investment and constrain worker productivity (e.g., Moen and Sweet, 2003). As employees bring more family activities into the workplace, the ensuing distraction might result in reduced work effectiveness and productivity. It is also possible that such work policies discussed above might lead to employees consciously or unconsciously prioritizing family over work tasks, resulting in decreased productivity as well.

Despite work-family integration having the potential for both negative and positive consequences on outcomes of interest to employers, it does not seem doubtful that, in general, a balanced work-family life on the part of the employee would contribute to a healthy organizational environment. Companies that pay little concern on their employees' family needs risk being unable to retain talented employees, and the long-term productivity of individual employees that are unable to attain work-life balance will eventually suffer, leading to increased levels of strain.

7.4 Influences on Employee Work-Family Balance and Well-Being

7.4.1 Happiness and Health Outcomes

Due to the societal level changes that we previously discussed, dual career households are becoming increasingly common, but occupations are generally not changing to reflect the increased family obligations that dual career households entail to a similar extent. Since increased role demands can lead to higher difficulty in coordinating work and family roles, employees today are more likely to integrate the two roles than ever before. Segmentation, which used to be a natural process as a result of the boundary between work and life, is now a process that needs to be consciously chosen and carefully implemented. This increase in work-family integration creates a strong need to understand the implications of this process for employee health and well-being.

Some researchers have suggested that greater integration between work and family roles is a way in which work and family can be balanced, and such integration can even facilitate a positive synthesis between the two roles (Friedman et al., 1998). This perspective is consistent with work-family enrichment perspective, in which the skills or resources acquired by holding one role can be of benefit in performing another role effectively. These perspectives thus suggest that relatively permeable boundaries between work and family facilitate transferring resources between roles and capitalizing on this transfer to perform both roles better. Moreover, they also reduce the various tangible and intangible costs associated with role transition. For example, by working at home employees save time and monetary resources by avoiding the daily commute. The psychological cost associated with role switching might also be reduced when the two roles are more integrated, since role transitions become relatively easier.

Yet with increased integration of work and family roles, individuals have greater potential exposure to role ambiguity, which in turn might lead to increased interference or conflict between roles (Ashforth et al., 2000; Desrochers et al., 2005). For example, the literature on alternative work arrangements (such as telecommuting) indicates that the blurred line between work and family domains is a key issue of concern in such arrangements. Employees who work at home might have difficulty working efficiently when being frequently interrupted by family events. In addition, two recent studies have shown that higher levels of work-family role integration are associated with increased work family conflict (Kossek et al., 2006; Olson-Buchanan and Boswell, 2006).

As we discussed before, increasing role demands, job demands in particular, make people integrate their work and family roles more readily, resulting more work-family conflict and spillover between work and family (Ilies et al., 2009). As shown in a recent meta-analysis on the consequences of work-to-family conflict (Mesmer-Magnus and Viswesvaran, 2005), employees with high work-to-family conflict typically experience lower levels of family and job satisfactions. They also report a variety of negative physiological and psychological symptoms. Although the majority of the evidences concerning the consequences of job stressors have been observed via between-person differences, recent studies using experiencing-sampling method demonstrated the within-individual relationships between work stressors and work-family conflict and individual well-being (Ilies et al., 2007; Ilies et al., in press; Judge et al., 2006). For example, Ilies and colleagues (2007) surveyed a sample of 106 employees over a period of 2 weeks and found that workload is associated with work-family conflict, and it reduces positive affect both at work and at home within individuals over time. In a more recent study, they examined within-individual effects of workload on distress at work, blood pressure and daily well-being, and their results showed adverse influence of workload on both physical (blood pressure) and psychological well-being (Ilies, Dimotakis, and De Pater, 2010).

Overall, research has indicated that work-family integration might have implications on individual health and well-being via mechanisms such as affect spillover (Ilies et al., 2007, 2009), and work-family conflict (Ilies et al., 2007). Further empirical investigations are nevertheless needed to further clarify the effects of work-family integration on employee health and psychological well-being.

7.4.2 Positive and Negative Spillover as Mediating Mechanisms

Despite a growing number of studies on the interface between work and family, the mechanisms that underscore the linkages between work-family integration and various individual and organizational outcomes have yet to be determined. Only a few studies have specified and examined the mechanisms that link one's boundary management strategy (e.g., high integration vs. high segmentation) with well-being outcomes (Michel and Hargis, 2008). In the sections above, we discussed that integrating work and family might entail some tradeoffs, with the primary benefit being the minimization of process costs associated with role transitions, but this benefit can come at the potential expense of increasing role ambiguity and cognitive complexity. Similarly, highly integrated roles facilitate the positive syntheses between the roles but can also lead to an increase in negative spillover processes. We suggest that studying these (positive and negative) spillover processes can be of value in understanding the interface between work and family roles by linking the integration/segmentation boundary management style with well-being and other outcomes.

As with the integration of work and family roles, spillover can occur either from family to work or from work to family domains (Grzywacz and Marks, 2000). By surveying around two thousand employed adults, Grzywacz and Marks (2000) found evidence for four potential types of spillover by combining the potential direction (from family to work, or from work to family) and nature (positive or negative) of spillover. Their results indicated that work and family factors that facilitated development (e.g., decision latitude, family support) were associated with less negative and more positive spillover between work and family, while work and family barriers (e.g., job pressure, family disagreements) were associated with more negative and less positive spillover between work and family. This research effort provided important insights on how the characteristics associated with work and family domains can affect the direction and nature of spillover between work and family. Future research might examine how these characteristics across domains interact in affecting spillover between work and family. For example, the association between job pressure and negative spillover from work to family might be stronger when family barriers are present.

Moreover, in a dynamic, within-people investigation, a recent study by Ilies et al. (2009) investigated the role of work-family integration in employees' spillover between their work and family roles. Results indicated that daily levels of job satisfaction were associated with increased levels of marital satisfaction and positive affect at home, supporting a spillover approach. Moreover, these spillover effects were moderated by the level of integration between work and family, with higher levels of work-family integration facilitating both positive and negative spillover from work to family. This approach also provides a contribution by demonstrating strategies that can capitalize on work-family integration by striving to capture positive while minimizing negative spillovers. Future studies might further examine how the two different types of work-family integration (WIIF and FIIW) can affect spillover processes, and how these ultimately relate to individual levels of well-being.

7.4.3 Interdependence Among Domain-Specific Satisfaction and General Life Satisfaction

By occupying different roles in their daily lives, individuals find themselves nested in social structures. The roles they occupy are further associated with a set of expectations which guide individuals' behavior. To achieve a balance in fulfilling the different sets of expectations associated with dissimilar or diverse roles is critical for one's well-being. In examinations of well-being, an important index frequently examined is life satisfaction, which represents a global evaluation of one's satisfaction with his or her life (Diener, 1984). To understand the relationship between work-family integration and life satisfaction can therefore be of both theoretical and practical interest in examinations of the relationship of work-family issues and well-being.

Two basic theoretical models have been proposed to explain individual differences in well-being: the top-down and bottom-up approaches (David et al., 1997; Diener, 1984). The top-down approach is a dispositional perspective, in which personality components (positive or negative affectivity, extroversion, neuroticism, etc.) are considered strong predictors of an individual's levels of life satisfaction, and then these top-level dispositions similarly color one's satisfaction with different aspects of one's life (e.g., job, marriage, etc.). The relative stability of one's reported subjective well-being provides some support to this dispositional approach (for a review, see Pavot and Diener, 1993). In addition, twin studies have also indicated that well-being levels have a significant genetic component (Lykken and Tellegen, 1996; Tellegen et al., 1988).

On the other hand, the bottom-up approach focuses on how situations, events, and context can add up to influence life satisfaction. The frequently revealed positive correlations between specific domain (e.g., job, marriage etc.) and general life satisfaction provide some empirical support for this model. Nevertheless, meta-analytic path analysis efforts (Heller et al., 2004) have suggested that both approaches can be useful and valid in understanding the determinants of life satisfaction, in that it seems that a mixed model received best empirical support from the studies included in the Heller et al. review.

Based on the bottom-up model, an individual's satisfaction with each of his life domains (e.g., work domain) is expected to be associated with life satisfaction. Relevant to work-family issues, poor role performance in one life domain (e.g., being unable to meet parental duties) could have a negative impact on one's general life satisfaction, through changes in one's domain specific levels of satisfaction. This would accordingly suggest that achieving balance with respect to time involvement and performance across work and family domains is important for individual well-being.

On the other hand, domain-specific satisfaction levels can covary, such that a person reporting satisfaction in their family life could also be more likely to be satisfied with their work life. The top-down approach to life satisfaction would explain this relationship by proposing that certain traits such like optimism influence the experience of satisfaction with each role domain. Alternatively, as suggested by the bottom-up approach and work-family spillover research, positive and negative experiences and the resulting satisfaction levels in each domain have the potential to influence individuals experiences in, and evaluations of, other domains, resulting in similar patterns of results. We suggest that the work-family literature is an ideal place to simultaneously test bottom-up and top-down approaches, since the interrelations of the various roles can provide an ideal setting in which both processes can be examined together with their mediating mechanisms.

7.5 General Suggestions and Implications

Managing the boundaries between work and home domains has become an increasingly challenging task for many employees, and this is particularly true of professionals. An interviewee in a recent study stated: "I do not really have big walls around either of them (work and family). If something big is going on, one tends to bleed over into the other..."(Kossek and Lautsch, 2008, p. 17) This difficulty in separating work and family roles, especially in environments that involve increased demands in both domains creates a difficult situation for professional workers, one that does not seem likely to disappear in the near future. To successfully balance work and family lives, and to integrate the two life roles in a more constructive way will require effort on the part of both professional workers and their organizations.

7.5.1 Organizational Practices

Given that work–family conflict has been found to adversely affect organizational outcomes such as employee attendance and performance (Eby et al., 2005; Hammer et al., 2003), employers are understandably interested in potential ways to reduce this phenomenon. This interest is reflected in the fact that employers are increasingly providing various family-friendly practices for their employees. The most commonly used organizational family-supportive practices include telecommuting, flexible work hours, reduced work time, and paid family leave, and it is generally found that having programs such as these available can provide benefits to organizations by providing advantages in recruiting and retaining talent, and improving employee commitment (e.g., Berg et al., 2003).

Unfortunately, these policies are not always implemented effectively, and might sometimes create unforeseen negative effects due to weakened role boundaries and increased role ambiguity, as we have previously argued. For these family-supportive practices to be effective, employers must consider individual differences present in their workforce, and design these practices in a way that can fit individual needs (or at least, the needs of the majority of their employees). For example, offering these benefits to every employee as part of an inflexible benefits package might be an issue of concern to employees who are not currently married or raising children. Ideally, employers should ensure workers have their own voice in shaping work-family programs and benefits. In addition, to encourage employee to use these practices, organizations need to establish formal systems that assuage employee's potential career concerns, and establish norms that respect their usage.

A sincere family-friendly environment, spearheaded by organizational leadership, is also critical for the successfully implementation of these family policies. For example, research has indicated that CEOs with high commitment to promoting work-life balance are associated with reduced levels of workload within the organization (Kossek and Lee, 2005). An organizational environment with supportive supervisors and friendly coworkers can also positively affect the utilization of family-friendly policies.

7.5.2 Personal Strategies

Perhaps the most feasible approach for professional workers to achieve work-life balance is to develop personal boundary management strategies that are optimal for themselves as well as their specific work and family environment. There does not seem to be a single best answer regarding how one should integrate or segment their work-family roles. Instead, in order to find a way that works best for each individual, it requires a careful personal reflection toward one's current boundary management style and the courage to experiment with alternative approaches.

For example, approaches that involve increased segmentation seem to be appropriate in cases in which permeable boundaries between work and family have resulted in increased levels of experienced strain, or reduced role performance (as perceived by either the individual or the people around them within each domain, such as family members). If increased work-family integration has been shown to be problematic, then increased segmentation could be more successful in achieving balance. On the other hand, increased integration could enable professionals to better capture positive spillover effects which are less likely to occur under high segmentation, or help transfer personal and social resources more efficiently from one domain to the other. In the end, the choice of integration or segmentation strategies should depend on each professional's specific circumstances, the resources they have and the demands they are expected to meet.

In conclusion, to meet the dual agenda of improving work performance and fulfilling family responsibilities, professionals and their employers should work together to design processes and practices that facilitate work-family balance, and capitalize on processes such as positive spillover while avoiding the harmful outcomes associated with work-family interference and negative spillover. Professional workers should actively monitor and adjust their approach to managing their personal work-family boundaries such that the segmentation/integration strategies that they use improve their work-family balance and overall levels of well-being. Finally, researchers can aid these efforts by investigating boundary conditions and mediating processes that explain how work-life balance can be achieved and the mechanisms this would entail.

References

Ashforth BE, Kreiner GE, Fugate M (2000) All in a day's work: boundaries and micro role transitions. Acad Manage Rev 25:472–491

Baruch GK, Barnett R (1986) Role quality, multiple role involvement, and psychological well-being in midlife women. J Pers Soc Psychol 51:578–585

Berg P, Kalleberg A, Appelbaum E (2003) Balancing work and family: the role of high-commitment environments. Ind Relat 42:168–188

Blair-Loy. M (2001) Cultural constructions of family schemas: the case of women executives. Gender Soc 15:687–709

Blair-Loy M, Wharton A (2002) Employees' use of work-family policies and the workplace social context. Soc Forces 80:813–845

Burke RJ (2008) Preface. In Burke RJ, Cooper CL (eds) The long work hours culture: causes, consequences and choices (p. ix–xxvii). Emerald, Bingley

Clark SC (2000) Work/family border theory: a new theory of work/family balance. Hum Relat 53:747–770

Crosby F (1991) Juggling: the unexpected advantages of balancing career and home for women and their families. Free Press, New York, NY

Dalton DR, Mesch DJ (1990) The impact of flexible scheduling on employee attendance and turnover. Administr Sci Q 35:370–387

David J, Green P, Martin R, Suls J (1997) Differential roles of neuroticism, extraversion and event desirability on mood in daily life: an integrative model of top–down and bottom–up influences. J Pers Soc Psychol 73:149–159

Desrochers S, Hilton JM, Larwood L (2005) Preliminary validation of the Work-Family Integration-Blurring Scale. J Fam Issues 26:442–462

Diener E (1984) Subjective well-being. Psychol Bull 95:542–575

Diener E, Oishi S, Lucas RE (2003) Personality, culture, and subjective well-being. Ann Rev Psychol 54:403–425

Eby LT, Casper WJ, Lockwood A, Bordeaux C, Brinley A (2005) Work and family research in IO/OB: content analysis and review of the literature (1980–2002). J Vocat Behav 66:124–197

Edwards JR, Rothbard NP (2000) Mechanisms linking work and family: clarifying the relationship between work and family constructs. Acad Manage Rev 25:178–199

Friedman SD, Greenhaus JH (2000) Work and family—Allies or enemies? What happens when business professionals confront life choices. Oxford University Press, New York, NY

Friedman SD, Christensen P, DeGroot J (1998) Work and life: the end of the zero-sum game. Harv Bus Rev 76:119–129

Frone MR (2003) Work-family balance. American Psychological Association, Washington, DC

Glass J, Fujimoto T (1994) Housework, paid work, and depression among husbands and wives. J Health Soc Behav 35:179–191

Gove WR, Zeiss C (1987) Multiple roles and happiness. In Crosby F (ed) Spouse, parent, worker: on gender and multiple roles. Yale University Press, New Haven, CT, pp 125–137

Greenhaus JH, Beutell NJ (1985) Sources of conflict between work and family roles. Acad Manage Rev 10:76–88

Greenhaus JH, Collins KM, Shaw JD (2003) The relation between work-family balance and quality of life. J Vocat Behav 63:510–531

Grzywacz J, Marks N (2000) Reconceptualizing the work-family interface: an ecological perspective on the correlates of positive and negative spillover between work and family. J Occup Health Psychol 5:111–126

Hammer LB, Allen E, Grigsby TD (1997) Work-family conflict in dual-earner couples: within-individual and crossover effects of work and family. J Vocat Behav 50:185–203

Hammer LB, Bauer TN, Grandey AA (2003) Work-family conflict and work-related withdrawal behaviours. J Bus Psychol 17:419–436

Heller D, Watson D, Ilies R (2004) The role of person vs. situation in life satisfaction: a critical examination. Psychol Bull 130:574–600

Hill EJ, Yang C, Hawkins AJ, Ferris MA (2004) Cross-cultural test of the work-family interface in one corporation in 48 countries. J Marriage Fam 66:1300–1316

Hofstede G (2001) Culture's consequences: comparing values, behaviors, institutions, and organizations across nations. Sage, London

Ilies R, Schwind K, Wagner DT, Johnson M, DeRue DS, Ilgen DR (2007) When can employees have a family life? The effects of daily workload and affect on work-family conflict and social activities at home. J Appl Psychol 92:1368–1379

Ilies R, Schwind Wilson K, Wagner DT (2009) The spillover of daily job satisfaction onto employees' family lives: the facilitating role of work-family integration. Acad Manage J 52:87–102

Ilies R, Dimotakis N, De Pater IE (2010) Psychological and physiological reactions to high workloads: implications for well-being. Pers Psychol, 63, 407–436.

Jacobs J, Gerson K (2004) The time divide: work, family, and gender inequality. Harvard University Press, Cambridge

Janning M (2009) The efficacy of symbolic work-family integration for married professionals who share paid work– a descriptive study. J Hum Soc Sci 1:1–17

Jones F, Burke RJ, Westman M (2006) Work-life balance: key issues. In Jones F, Burke RJ, Westman M (eds) Work-life balance: a psychological perspective. Psychology Press, New York, NY, pp 1–9

Judge TA, Ilies R, Scott BA (2006) Work-family conflict, hostility, and guilt: effects on work and family outcomes. Pers Psychol 59:779–814

Kirchmeyer C (2000) Work-life initiatives: greed or benevolence regarding workers time. In Cooper CL, Rousseau DM (eds) Trends in organizational behavior. Wiley, West Sussex, pp 79–93

Kossek EE, Lautsch BA (2008) CEO of me: creating a life that works in the flexible job age. Wharton School Publishing, Philadelphia, PA

Kossek EE, Lee MD (2005) Making flexibility work: what managers have learned about implementing reduced-load work. Summary of Findings from Phase II of Alfred P. Sloan Grant # 2002-6-11. Michigan State University and McGill University

Kossek EE, Lautsch BA, Eaton SC (2006) Telecommuting, control, and boundary management: correlates of policy use and practice, job control, and work-family effectiveness. J Vocat Behav 68:347–467

Kreiner GE (2006) Consequences of work-home segmentation or integration: a person environment fit perspective. J Organ Behav 27:485–507

Lee MD, MacDermid SM, Williams ML, Buck ML, Leiba-O'Sullivan S (2002) Contextual factors in the success of reduced-load work arrangements among managers and professionals. Hum Resour Manage 41:209–223

Lobel SA, Googins BK, Bankert E (1999) The future of work and family: critical trends for policy, practice and research. Hum Resour Manage 38:243–254

Lykken DT, Tellegen A (1996) Happiness is a stochastic phenomenon. Psychol Sci 7:186–189

Merton RK (1957) Social theory and social structure. Free Press, New York, NY

Mesmer-Magnus JR, Viswesvaran C (2005) Convergence between measures of work-to-family and family-to-work conflict: a meta-analytic examination. J Vocat Behav 67:215–232
Michel JS, Hargis MB (2008) Linking mechanisms of work-family conflict and segmentation. J Vocat Behav 73:509–522
Milkie MA, Peltola P (1999) Playing all the roles: gender and the work–family balancing act. J Marriage Fam 61:476–490
Moen P, Sweet S (2003) Time clocks: couples' work hour strategies. In Moen P (ed) It's about time: career strains, strategies, and successes. Cornell University Press, Ithaca, NY, pp 17–34
Moen P, Yu Y (1998) Effective work/life strategies: working couples, work conditions, gender and life quality. Soc Probl 47:291–326
Nippert-Eng CE (1996) Home and work: negotiating boundaries through everyday work. University of Chicago Press, Chicago, IL
Nordenmark M (2002) Multiple social roles—a resource or a burden: is it possible for men and women to combine work with family life in a satisfactory way?. Gender Work Organ 9: 125–145
Olson-Buchanan JB, Boswell WR (2006) Blurring boundaries: correlates of integration and segmentation between work and non-work. J Vocat Behav 68:432–445
Osterman P (1995) Skill, training, and work organization in American establishments. Ind Relat 34:125–146
Pavot W, Diener E (1993) The affective and cognitive context of self-reported measures of subjective well-being. Soc Indic Res 28:1–20
Roehling PV, Roehling MV, Moen P (2001) The relationship between work-life policies and practices and employee loyalty: a life course perspective. J Fam Econ Issues 22:141–170
Rothbard NP (2001) Enriching or depleting? The dynamics of engagement in work and family roles. Adm Sci Q 46:655–668
Rothbard NP, Phillips KW, Dumas TL (2005) Managing multiple roles: work family policies and individuals' desires for segmentation. Organ Sci 16:243–258
Saltzstein AL, Ting Y, Saltzstein GH (2001) Work–family balance and job satisfaction: the impact of family-friendly policies on attitudes of federal government employees. Public Adm Rev 61:452–467
Strazdins L, Korda RJ, Lim LY, Broom DH, D'Souza RM (2004) Around-the-clock: parent work schedules and children's well-being in a 24-h economy. Soc Sci Med 59:1517–1527
Song Z, Foo MD, Uy M (2008) Mood spillover and crossover among dual-earner couples: a cell phone event sampling study. J Appl Psychol 94:1086–1094
Takeuchi R, Yun S, Teslu PT (2002) An examination of crossover and spillover effects of spouse and expatriate cross-cultural adjustment on expatriate outcomes. J Appl Psychol 85:655–666
Tellegen A, Lykken DT, Bouchard TJ Jr., Wilcox K, Segal N, Rich S (1988) Personality similarity in twins reared apart and together. J Pers Soc Psychol 54:1031–1039
Thoits P (1983) Multiple identities and psychological well-being: a reformulation and test of the social isolation hypothesis. Am Sociol Rev 48:174–187
Toossi M (2002) May). A century of change: the U.S. labor force, 1950–2050. Mon Labor Rev 25:5–28
Verbrugge LM (1986) Role burdens and physical health of women and men. Women Health 11: 47–78
Waldron I, Weiss CC, Hughes ME (1998) Interacting effects of multiple roles on women's health. J Health Soc Behav 39:216–236
Westman M (2001) Stress and strain crossover. Hum Relat 54:717–751
Westman M, Etzion D (2005) The crossover of work-family conflict from one spouse to the other. J Appl Soc Psychol 35:1936–1957
Westman M, Etzion D, Horovitz S (2004) The toll of unemployment does not stop with the unemployed. Hum Relat 57:823–844
Williams KJ, Alliger GM (1994) Role stressors, mood spillover, and perceptions of work-family conflict in employed parents. Acad Manage J 37:837–868
Zerubavel E (1991) The fine line: making distinctions in everyday life. Free Press, New York, NY

Part III
Mothers, Fathers, and Couples: Challenging Conventional Wisdom

Chapter 8
What to Expect When She's Expecting

Work-Family and Identity Integration Challenges and Opportunities of "Soon-to-Be" Working Professional Mothers

Jamie J. Ladge, Danna Greenberg, and Judith A. Clair

8.1 Introduction

There has been extensive writing both in the popular press and the academic literature on the unique work-life challenges of professional working mothers. Professional working mothers often must manage competing gender biases – being perceived as cold-hearted (Cuddy et al., 2004) or worse, bad mothers (Epstein et al., 1999) on one hand and less committed to their work roles (Correll et al., 2007) on the other. Despite recent figures that show women represent more than half the workforce and 40% of breadwinners or co-breadwinners in the United States, societal expectation that a mother's primary role is in the home remains strong (Boushey, 2009; Hayes, 1996; Johnston and Swanson, 2006; Williams, 2000). These societal expectations and competing gender biases create challenges for professional working mothers as they try to define who they are as professionals, as mothers, and as working mothers. How a woman integrates these identities has important implications for how she manages work-life demands.

The challenge of integrating these work and non-work identities begins when professional working women become pregnant. When a professional woman becomes pregnant she is likely to be well-established in her career.[1] As any individual gains more experience and competency in her career, psychological success stemming from work increases which leads to increased centrality of career identity for that individual (Harrington and Hall, 2007). As such, career identity is already highly central relative to these women's overall identity. Once pregnant, professional women must determine how their evolving pregnancy identity and maternal identity will integrate with their already established professional identity.

J.J. Ladge (✉)
Northeastern University, Boston, MA, USA
e-mail: j.ladge@neu.edu

[1] In general, women with advanced degrees in professional occupations are older first-time mothers (Martin, 2000; Amuedo-Dorantes and Kimmel, 2005).

S. Kaiser et al. (eds.), *Creating Balance?*, DOI 10.1007/978-3-642-16199-5_8,

The integration between these identities is a continual challenge and has implications for how women evaluate and reprioritize their roles and how they manage their work-family interface (Bailey, 1999; Greenberg et al., 2009; Ladge, 2008). In order to more fully understand how identity integration affects professional working mothers' work-family experiences, it is important that we begin to understand how the integration of career identity and maternal identity initially develops for these professional women when they become pregnant.

In this chapter, we focus on this issue of work-family and identity integration by examining how pregnant professional women experience changes in their professional self-conceptualization as they begin to integrate new maternal-centered identities with their established identities as working professionals. Using data from our research with working women in professional occupations that are pregnant for the first time, we highlight the questions that arise for women as they begin to appraise their professional identity in light of their transition into motherhood. We also look at how women reconcile their professional identity with their emerging maternal identity and explore the strategies women engage in as they re-craft their professional sense of selves in response to their changing non-work roles. In Sect. 8.4, we highlight the implications of this research for organizations and managers as they try to support pregnant professional women in the transition to becoming working mothers.

8.2 Overview of the Literature

While professions vary widely, there is a general code of conduct and series of traits expected of most professions in the United States. When professionals are trained in their respective trades, they are often socialized to adopt a code of conduct and set of characteristics that generally revolves around three primary goals: service quality, commitment and time. To fulfill these goals, professionals are expected to bring their whole selves to their work which often implies they will need to work long hours to prove their commitment and give the highest level of service to their organizations and customers. Because of these pressures, professionals often experience time or strain based role conflict when they have to balance work with competing out of work roles (Greenhaus and Beutell, 1985; Perlow, 1999).

While most American workers are working longer hours and are feeling greater stress with regards to work-life management (Schor, 1998), some scholars have suggested that professional workers may face even greater work-life stress (Milliken and Dunn-Jensen, 2005). Part of this burden arises from the extreme nature of professional work roles. In researching professionals at multi-national firms, Hewlett and Luce (2006) characterized the extreme nature of professional jobs as having an unpredictable work flow, a fast-paced environment, large amounts of travel, physical presence in the workplace for at least ten hours a day, and frequent attendance at work-related events outside of regular work hours. The extreme nature of professional jobs takes a toll on professionals' non-work lives and often has a negative effect on their relationships with their spouses, children and other family members.

In addition, the norms of ideal workers that pressure professionals to center their lives around work and to minimize home engagement is more typical in professional service type firms where the expectation of working long hours is imperative for meeting client demands and 24/7 operating hours (Williams, 2000). While fathers face these pressures and challenges, these issues are more pronounced for women who continue to take on the vast majority of childcare and household responsibilities (Bianchi, 2000; Mattingly and Bianchi, 2003).

8.2.1 Constructing Professional Identities and Motherhood

Despite the potentially harmful effects on nonwork roles, most professionals define themselves through their work. Professional identity is defined as a "self-definition as a member of a profession" (Chreim et al., 2007, p. 6) and is largely constructed as individuals are socialized into their profession during school when they learn their trade (e.g., law, business or medical school). These norms and expectations are further reinforced by peers, mentors and supervisors when professionals begin their careers. During this socialization process an individual begins to construct who they are and hope to be as a professional (Ibarra, 1999). Over time, individuals are able to establish their own professional identity.

Although one's sense of professional self typically stabilizes and strengthens with time, it can be altered as a result of changes in one's work environment or changes to one's personal life (Gibson, 2003; Pratt et al., 2006). Major life events such as the birth of a child are, more often than not, act as a driving force that prompts one to reconsider one's professional identity. When professional working women have children, they are traditionally more likely than men to experience significant changes to their professional sense of self as they must reconcile competing work and non-work demands and gendered expectations (Blair-Loy, 2003; Williams, 2000).

During pregnancy and/or after the birth of their first child, women often begin a process of recalibrating who they are as professional as well as questioning the value they place on their established identity vís a vís their new identity as a mother. For some women, their new identity as a mother even provides them with an opportunity to adjust an unsatisfactory professional working identity or reappraise a professional identity by altering career goals (Bailey, 1999).

While there have been some efforts by researchers to understand how professional identity is constructed (Gibson, 2003; Ibarra, 1999; Pratt et al., 2006), little is known about how women re-construct their professional identity in light of becoming a mother. In this study, we look at the issues that arise when women become pregnant for the first time and begin to question their professional identity. We focus on how women reconcile this self-questioning and begin to reconstruct their professional sense of selves in response to emerging identities as mothers. We conclude with implications for theory and practice. In the following section, we provide a brief description of the research that provides the basis for this study.

8.3 Overview of the Research Process

The data we present here was drawn from a qualitative study the authors conducted on first-time experiences being pregnant at work. Following Miles and Huberman (1994), we chose to use an inductive research approach, as we wanted to build an in-depth understanding of women's experiences of pregnancy from their own perspective. As such, our data gathering and data analysis were guided by an interest in identifying the commonalities and patterns that underlie pregnant women's experiences as they navigated their work experience.

We began our research with a pilot study in which we interviewed ten women about their experiences being pregnant at work. Based upon the pilot data and additional literature review, we designed a qualitative, inductive study in which we interviewed thirty women about their experiences being pregnant at work. Women were recruited for this study through a number of venues, including college alumni network groups, human resource professional networks, and local medical and health professionals who work with new mothers.

All research participants were pregnant with their first child at the time of their interview. We chose to study pregnant women because the issues women face surrounding the birth of their first child are more complex and ambiguous than those encountered by experienced mothers (Miller et al., 1996). All research participants were in professional work positions, had at least 3 years of full-time work experience, and intended to return to work after their maternity leave. Interviews lasted between 60 and 90 min and followed a semi-structured format. All interviews were transcribed, yielding approximately 1,000 pages of double-spaced text to analyze.

We began data analysis by reviewing the transcripts to identify the general themes that women were discussing regarding being pregnant at work. We then focused specifically on the data related to identity transition to identify the general themes that women discussed as they talked about their evolving sense of professional self and their emerging maternal identity (Glaser and Strauss, 1967). The authors met to discuss these themes and to review existing research that connected to the themes. The following discussion highlights the three themes that framed women's pregnancy experiences in an organizational context.

8.3.1 Becoming a Mother and the Reconstruction of Professional Identity

Our research shows there are three primary issues that arise as women begin to reconstruct their professional identity in light of becoming pregnant. The first concern involves a series of questions women ask themselves regarding who they are and hope to be as professionals. Second, we found that women begin to reconcile these questions about their professional identity with their future identity as working mothers. Finally, women begin a process of reconstructing their sense of professional self by making tentative commitments about who they hope to become as working mothers. We discuss each of these themes in more detail below.

8.3.2 Questioning Professional Identity

For most of the women we interviewed, their sense of self was focused primarily on being a professional. With an average age of 37, most of the women we interviewed had spent a significant portion of their lives building their careers and establishing their identities within their chosen professions. Becoming pregnant caused many of the women to begin to question what it meant to them to be a professional. Abby, a manager at a large nonprofit organization explained:

> I'm not a housewife at all, the weekend comes to an end and if I haven't been doing something special, I'm ready to get out of the house again. I don't want to look at the laundry, cleaning, I don't want to deal with it. Part of being home with the baby it's not stated but, it's going to be. . .so while you're home maybe you can take care of some of these things. I hate it. I just don't identify with it . . . If I'm home I really need to think about these things. I need to start being a mom and that's a huge change . . . I know I need to think about it but it just stresses me out. . . (004)

As the above quote illustrates, women were in a process of trying to discern who they were now that they were becoming mothers. A major aspect of this self-questioning involved determining what effects motherhood could have on their professional identity. As in any identity transition, there is a great deal of self-questioning that arises when one tries to reconcile an old identity with a new, conflicting identity (Corley and Gioia, 2004; Ibarra, 1999).

It was evident in our data that as women began to question what it meant to be a professional they began to contemplate changes that they would need to make in order to manage their work-family interface in the future. For example, Julie, a financial analyst, questioned how she was going to handle her professional responsibilities once her baby was born:

> So I think in some ways I'll probably still have to figure out how to cut down. There's no way I think I can handle sixty five to seventy five hours every single week. There's no way I can do that and then take care of a baby. (010)

Most of the women we interviewed understood that their professional lives would need to change with respect to their work hours and responsibilities. At the same time, women were unsure of what specifically had to change and how it would change.

8.3.3 How Do Women Reconcile Their New Identity as a Mother with Their Professional Identity?

In order to reconcile the questions pregnant women had about their changing professional identity, these women began a process of gathering information from a variety of sources. Often women looked to the advice and guidance from external sources to aid them in finding answers to their questions about their changing professional identity. These sources included role models and other mothers at their workplace, books and internet sites on working mother topics, and organizational policies and procedures.

One of the primary sources women engaged as they reconciled their professional identity with their maternal identity was to observe how other working mothers in their organizations balanced work and pregnancy and motherhood. Although most women highlighted the lack of working mother role models in their own organizations, these women could still identify one or two women through whom they learned about balancing work and motherhood. Cynthia, an associate in a management consulting firm explained,

> I have seen the partners do it, and I said to myself, you know, if they can do it in such a demanding profession, and – even, I have seen a couple of people that haven't worked on my team, but have worked within this organization, at levels below me, and they have been able to do it. It is definitely something that is achievable. So, yes, I would think it is more of an influential thing, of the people around me, kind of observing everyone else around me being able to do it, so – (. . .) (008)

Vicarious learning is one of the primary sources of self confidence (Bandura, 1986) as well as identity development (Gainor and Lent, 1998). By drawing on the experiences of others, pregnant women are able to begin to craft a vision for their own futures as working mothers. Moreover, other mothers become a primary source of information and play a significant role in shaping how women begin thinking about their identities as mothers (Deutsch et al., 1988) and as professionals. The women we interviewed sought out other working mothers as they established their plans for how they would reveal their pregnancy to their peers and supervisors, how they would negotiate maternity leave, and how they would balance work and motherhood after their babies were born. This is consistent with research on role transitions which states that individuals adopt a provisional self partly in response to the vicarious learning of others in similar roles (Ibarra, 1999).

Books and the internet provided another source of information that shaped how women reconciled their professional and maternal identities. Some of what women read focused on societal expectations and gender role ideologies on what it meant to be a good mother and good working mother. Many of the women we spoke with explained that they used these resources to think about their own situations and to resolve some of the questions that arose for them during their pregnancy and in their transition back to work.

Lastly, women looked to the work-family policies in their organizations as guidelines for how they might, or might not, adapt their professional selves. When women perceived their organizations to be family friendly, they had an easier time reconciling their identity as a mother with their professional identity. For Molly, a product manager at a large biotechnology firm explained:

> I think Biotech is very flexible with all of that. I know myself as a manager I've been very flexible. I know that the other managers in this building are very flexible, and Biotech as a whole is very flexible. I know a lot of the other sites where folks that don't have direct

> impact on day-to-day operations do work more of a flexible work schedule. They'll work four days, they'll work three days, two from home, one from home. And Biotech is very open to that. As a manager, we are given the option to telecommute from home. So that really lessens the need to be here 40-plus hours. And my boss is very open to that. If you want to work six hours a day and do the remainder at home, they're very open to that. So it's very flexible. And I don't think it's just me, I see it throughout the organization. (032)

Jenny, an academic administrator shared a similar reflection on her organization:

> I think that the job itself can be stressful. And, there are a lot of demands in the job. But, I do think that the organization is fairly flexible. There are several working mothers that do exactly what I do. (026)

Knowing that their organization was flexible and seeing others utilize that flexibility helped Molly and Jenny as well as others easily adapt to their changing professional and maternal identities. Conversely, we interviewed several women who did not feel their organizations were family friendly and were having a harder time envisioning how they were going to reconcile their professional and maternal identities. When professionals work in organizations which they perceive as not family friendly, they will be less likely to take advantage of family friendly programs for fear that they will be stigmatized (Allen, 2001; Thompson et al., 1999). Often this leads to increased feelings of flux, resentment to their organization, reduced commitment to their professional identity, and decreased overall job satisfaction. It will also likely lessen their commitment to their professional identity as they recognize they can't be mothers and successful professionals in these organizations.

In summary, organizations both in terms of the people within them and the policies they offer can have a strong influence on the degree to which a woman feels she is able to reconcile work and non-work identities (Ladge, 2008; Ladge et al., 2009). These influences can play a significant role in shaping how pregnant women begin to reconstruct themselves as professionals in light of motherhood.

In the last findings section, we will discuss the ways in which women begin to reconstruct their professional identity in response to these influences and their ability to resolve their professional and maternal identities.

8.3.4 Reconstructing Professional Identity

Once women begin to reconcile their changing identities, they then begin a process of reconstructing their professional sense of self. In other words, women re-craft the meaning and value they place on being a professional in light of their emerging maternal identity and make tentative commitments about who they hope to become as working mothers. This reconstruction happens as women consider their options for managing their professional identity and motherhood. The options women consider include leaving their jobs, redefining their current career track or their role, or simply trying to maintain the status quo with their professional identity. We discuss each of these options in more detail below.

8.3.4.1 Leaving

A very small number of women felt that it would be impossible to maintain their current professional role in light of their new responsibilities as a mother. This small group of women worked primarily in roles that could be categorized as extreme jobs (Hewlett and Luce, 2006). More often than not, the firms these women worked in did not offer a wide array of family friendly policies or declined proposals for flexible work arrangements. June's situation reflected this experience:

> And that ultimately did not get approved, which has kind of left a sour taste in my mouth because they've done it for other people, and now they're saying – oh, well, we're trying to set a precedent going forward. So, if you have any questions, you should come talk to us not to other people who have gone out on maternity leave and blah, blah, blah. So, that was a little upsetting. That's one of the reasons why I'm probably going to be looking for something closer to home once the baby is born because as I was saying with my 12 h day – that wouldn't allow me to do daycare drop-off or pickup, which number one is unfair to my husband – two, it's not fair to me, and I don't think it's fair to our baby either. The baby is not going to know any different the first year, so. (017)

8.3.4.2 Maintaining the Status Quo

Some women did not feel that having a baby was going to have much impact on their professional identity. Most women who had this view worked in firms or held positions which they perceived as family friendly. In these firms, women were more likely to report that they had a number of colleagues (and working mothers in particular) who utilized flexible work arrangements and had supervisors that supported them. In fact a small subset of the women we interviewed all worked in a progressive professional services firm which encouraged alternative career tracks to support working parents. These women had many role-models to follow in their own organizations and had the organizational resources to support their ability to easily maintain their professional identity in light of their transition to motherhood. For these women, becoming a mother had a negligible impact on their professional identity. For example, Tricia, a management consultant explained:

> I think that [I] have a kind of job where you can put in as many hours as you want to put into it. And I think the family – that's something that people want to put as many hours as you can put into it. So, I think it's finding that balance. (009)

It is important to note, we are not suggesting that these women were not struggling with their professional identity. They still were concerned about the compromises they were likely to make and they considered the impact their identity would have on their team members. In Tricia's case, she went on to explain, "On my own personal side, I think – I have a very supportive team that I work with, but it's a matter of you don't want to let anybody down – and once you take on the responsibility, it's not always easy to share it with somebody else." So regardless of having a supportive work environment, women like Tricia still struggled with fears about losing too much of their professional sense of self by taking advantage of reduced work hours and flexible work arrangements.

Other women explained that their career goals and aspirations had not changed much and they weren't approaching their professional identity any differently than before they were pregnant. For example, Beatrice explained that although she might have some apprehension over a new promotion, she was clear that her career goals had not changed:

> It hasn't affected my approach. I still do want to make that manager level. The only difference there is that if that position were to arise right now, it would be more difficult for me to accept it if it was offered to me. I would still accept it, but I would have some apprehension just because of the timing and the thought of what if I wanted to take three months' maternity leave, which is what I'm planning to do. But it really hasn't changed my perspective. I still want to get to that next manager level. (010)

8.3.4.3 Redefining Current Career Track

The aforementioned two scenarios do not represent the experience of how most professional women adjust their identity in response to motherhood. For most of the women in our study, becoming pregnant created the prospect of revising one's professional identity. Some women spoke about the desire and plan to ease up while others described pregnancy as the ultimate opportunity to reorient their lives in a way that provided them more balance. Monisa shared these sentiments:

> I think, you know, I think the most positive aspect is the opportunity it presents to me, to reorient my life. I am trying to look at that as a really positive thing. I worked for a long time. I, you know, on the one hand, do I want to, of course, go up the corporate ladder more? Yes, I do. On the other hand, it has been so hard for us to get pregnant and stay pregnant, I am really looking forward to being a mom, and working with my husband to try and figure out the best way for me to do that because I am not going to – it will be my most important priority. I want to make sure that I am set up in a way that I can be successful at that, first and foremost, and then figure out the work stuff secondarily. (012)

As women's attitudes about their careers change and as they think about how they will best meet the demands of their work and family roles, they will likely begin to reformulate the meaning and value they place on being a professional. Most women have the hopes of putting extensive psychological and physical time into their role as mothers which defies the expectations for professionals who need to be 100% committed to their jobs and to working long hours. This conflict has resulted in working mothers who have begun to re-craft their work roles in a way that enables them to have greater control over their personal lives. Inevitably, this means that changes will need to be made to the expectations of professionals and to the psychological contract these women have with their organizations (Millward, 2006).

The notion of letting go of even the smallest aspect of one's professional identity was challenging for most of the women we interviewed. Women like Julia, a management consultant, explained her internal struggles with feeling the pull away from one identity (professional) and push toward another (mother):

> I think that the hardest thing I'm finding now is my own internal struggle and then, now, being seen as not being able – I don't know whether it's not been as I won't be as dedicated

> and I don't know that it will be. I'll just be dedicated in a different way. But that's still a change for everybody, me and everybody else. So, I think that's the most difficult thing to be this far in your career and spend this long not having anything else pull you and then all of a sudden, to have that and people's natural impressions change but they're right to change. (007)

In summary, as a result of pregnancy and the transition to motherhood women begin a process of re-crafting their professional identity. This re-crafting process begins as women start to question their professional identity as a result of their pregnancy. Women reconcile these questions as they look to role models, written sources, and organizational practices, to help them begin to understand how they may need to alter their professional identity with their new role as a mother. Women eventually reconstruct their professional identity and make tentative commitments about easing up or leaving their jobs, by maintaining their established sense of professional identity, or by redefining their work and the importance they place on their professional identity.

8.4 Discussion and Conclusion

Most professional careers in the United States exist within work environments where demands for time and effort are high making it difficult for most professionals to put full effort into their roles outside of work. Individuals are socialized into such work environments and accept this as a normative state when they enter their given professional field. However, many professionals begin to question these norms particularly when significant external roles begin to compete with work roles. We have described in the chapter our examination of these dynamics in the context of professional women becoming pregnant for the first time. Becoming pregnant often prompts women to re-evaluate their identity as professionals. We don't view this as a negative. In fact, we view it as a unique opportunity for women to recalibrate their professional identity and redefine it in a way that is conducive to meeting non-work roles. Our analysis revealed that through this process of re-crafting their work roles, women are better positioned to gain greater control over their personal lives. In fact, this process of re-inventing one's professional identity is very much in line with what Hall defines as protean careers (Hall, 1976, 2001). The protean career is a process by which the person, not the organization, manages their own career. According to Hall, "the protean person's own personal career choices and search for self-fulfillment are the unifying or integrative elements in his or her life. The criterion of success is internal (psychological success), not external" (Hall, 1976, p. 201). The majority of the pregnant women we spoke with expressed a protean career attitude as they considered their future as working professionals. They expressed a strong desire to take control over their career and personal life, as opposed to having their organization do this for them.

We are not disillusioned by the fact that this re-evaluation will come at a cost for most women and is not an easy feat. As women set out to change the way they work and re-negotiate their psychological contracts with their organizations

(Millward, 2006) they are likely to have a hard battle fighting long established organizational work norms. While most modern professional organizations offer flexible work arrangements and other programs to facilitate better work-family harmony for their employees, their utilization rates are often low because most individuals think that taking advantage of them they will harm their professional image (Thompson et al., 1999). Additionally, managers often don't know how to evaluate employees when they don't follow a traditional career models (Bailyn, 1993). Finally, there can be negative career consequences that result from taking alternative career paths and from taking a family leave (Finkel et al., 1994; Judiesch and Lyness, 1999; Perlow, 1995).

While women aim to take greater control of their professional and overall career identity, organizations can and should play a role in facilitating this process. Organizations can offer programs that allow mothers (and fathers) to customize their careers in ways that are conducive to their unique life circumstances (Benko and Weisberg, 2007; Harrington and Ladge, 2009). These programs are necessary in all types of organizations and will facilitate greater retention of mothers (and fathers) in professional roles. Furthermore, organizations need to consider the inherent challenges that the professional career poses to a woman's ability to integrate professional and maternal identity. As our research shows, organizational norms and practices influence women's ability to re-craft their professional identity in light of their new maternal identity. Organizations need to consider the design and demands that they place on professionals and how these jobs may need to be re-crafted to accommodate working mothers and working fathers.

8.4.1 The Influence of National Culture on Women's Re-crafting of Professional Identity

In our analysis, we highlighted the influence that organizational context has on how pregnant women re-craft their professional identity. In organizational cultures that are more supportive of the work-family interface, women are likely to experience less torment as they adjust their professional identity. The reverse is likely to be true in organizations that are less supportive of employees' work-family demands.

Finally, we must note that our study, like most research on the work-life interface, is bound within the context of the United States (Poelmans et al., 2005). We believe that future research must think more broadly about our research findings and the influence that national culture may have on how women attempt to integrate their professional and maternal identity. Cultural norms towards working women are likely to influence how women construct their identity as mothers and as working mothers (Apter, 1994). National context will influence pregnant women's experiences re-crafting their professional identity as societal norms and values pertaining to work and family are likely to influence how women manage their work-life domains (Powell et al., 2009). Future research should consider how women in different national cultures re-craft their professional identity in light of becoming a mother. For example, future research may draw on comparisons of women's experiences across varying national cultures where acceptance of working mothers and cultural norms differ with respect to national assistance for new and working mothers. This line of research may encourage understanding of how national programs can influence corporate-sponsored programs and trigger shifts in gendered norms and expectations about pregnant working professionals and new mothers.

References

Amuedo-Dorantes C, Kimmel J (2005) The motherhood wage gap; the role of education and fertility. Rev Econ Household 3(1): 17–48; Guest Editor: Siv Gustafsson; earlier draft circulated as WMU Department of Economics WP # 03–07

Allen TD (2001) Family-supportive work environments: the role of organizational perceptions. J Vocat Behav 58:414–435

Apter T (1994) Working women don't have wives. St. Martin's Press, New York, NY

Bailey L (1999) Refracted selves? A study of changes in self-identity in the transition to motherhood. Sociology 33:335–352

Bailyn L (1993) Breaking the mold: women, men, and time in the new corporate world. Free Press, New York, NY

Bandura A (1986) Social foundations of thought and action: a social cognitive theory. Prentice-Hall, Englewood Cliffs, NJ

Benko C, Weisberg AC (2007) Mass career customization. Harvard Business School Press, Cambridge, MA

Bianchi SM (2000) Maternal employment and time with children: dramatic change or surprising continuity?. Demography 37:401–414

Blair-Loy M (2003) Competing devotions: career and family among women executives. Harvard University Press, Cambridge, MA

Boushey H (2009) Infographic: the importance of women breadwinners. A report by the Center for American Progress. http://www.americanprogress.org/issues/2009/04/women_breadwinners.html. Accessed 17 Feb 2010

Chreim S, Williams BE, Hinings B (2007) Inter-level influences on the reconstruction of professional role identity. Acad Manage J 50:1515–1539

Corley K, Gioia D (2004) Identity ambiguity and change in the wake of a corporate spin-off. Adm Sci Q 49:173–208

Correll SJ, Benard S, Paik I (2007) Getting a job: is there a motherhood penalty?. Am J Sociol 112:1297–1338

Cuddy AJ, Fiske ST, Glick P (2004) When professionals become mothers, warmth doesn't cut the ice. J Soc Issues 60:701–718

Deutsch FM, Ruble DN, Fleming A, Brooks-Gunn J, Stangor C (1988) Information-seeking and maternal self-definition during the transition to motherhood. J Pers Soc Psychol 55: 420–431

Epstein CF, Seron C, Oglensky B, Saute R (1999) The part-time paradox: time norms, professional life, family, and gender. Routledge, New York, NY

Finkel SK, Olswang S, She N (1994) Childbirth, tenure, and promotion for women faculty. Rev Higher Educ 17:259–270

Gainor K, Lent R (1998) Social cognitive expectations and racial identity attitudes in predicting the math choice intentions of black college students. J Couns Dev 45:403–413

Gibson DE (2003) Developing the professional self-concept: role model construals in early, middle and late career stages. Organ Sci 14:591–610

Glaser BG, Strauss AL (1967) The discovery of grounded theory: strategies for qualitative research. Aldine Publishing Company, Chicago, IL

Greenhaus JH, Beutell NJ (1985) Sources of conflict between work and family roles. Acad Manage Rev 10:76–88

Greenberg D, Ladge JJ, Clair JA (2009) Negotiating pregnancy at work: public and private conflicts. Negotiation Confl Manage Res 2:42–56

Hall DT (1976/2001) Careers in organizations. Goodyear, Pacific Palisades

Harrington B, Hall DT (2007) Career management and work-life integration: using self-assessment to navigate contemporary careers. Sage, Thousand Oaks, CA

Harrington B, Ladge JJ (2009) Work–life integration: present dynamics and future directions for organizations. Organ Dyn 38:148–157

Hayes S (1996) The cultural contradictions of mother- hood. Yale University Press, New Haven, CT

Hewlett SA, Luce CB (2006) Extreme jobs: the dangerous allure of the 70-h work-week. Harv Bus Rev 84:49–59

Ibarra H (1999) Provisional selves: experimenting with image and identity in professional adaptation. Adm Sci Q 44:764–791

Johnston DD, Swanson DH (2006) Constructing the "good mother": the experience of mothering ideologies by work status. Sex Roles 54:509–519

Judiesch MK, Lyness KS (1999) Left behind? The impact of leaves of absence on managers' career success. Acad Manage J 42:641–651

Ladge JJ (2008) Becoming a working mother: identity, efficacy and resocialization following re-entry. Unpublished doctoral dissertation, Boston College, Chestnut Hill, MA

Ladge JJ, Clair JA, Greenberg D (2009) Pregnancy in the workplace: identity and the transition to motherhood. Working Paper. Northeastern University, Boston, MA

Martin SP (2000) Diverging fertility among U.S. women who delay childbearing past age 30. Demography 37(4):523–33

Mattingly MJ, Bianchi SM (2003) Gender differences in the quantity and quality of free time: the U.S. experience. Soc Forces 81:999–1030

Miles MB, Huberman AM (1994) Qualitative data analysis. Sage, Thousand Oaks, CA

Miller VD, Jablin FM, Casey MK, Lamphear-Van Horn M, Ethington C (1996) The maternity leave as a role negotiation process. J Managerial Issues 8:286–309

Milliken FJ, Dunn-Jensen L (2005) The changing time demands of managerial and professional work: implications for managing the work-life boundary. In Kossek EE, Lambert SJ (eds) Work and life integration: organizational, cultural, and individual perspectives. Lawrence Erlbaum, Mahwah, NJ, pp 43–59

Millward LJ (2006) The transition to motherhood in an organizational context: an interpretative phenomenological analysis. J Occup Organ Psychol 79:315–333

Perlow LA (1995) Putting the work back into work/family. Group Organ Manage 20:227–239

Perlow LA (1999) The time famine: towards a sociology of work time. Adm Sci Q 44:57–81

Poelmans S, O'Driscoll M, Beham B (2005) An overview of international research on the work-family interface. In Poelmans S (ed) Work and family: an international research perspective. Lawrence Erlbaum, Mahwah, NJ, pp 3–46

Powell GN, Francesco AM, Ling Y (2009) Toward culture-sensitive theories of the work-family interface. J Organ Behav 30:597–616

Pratt MG, Rockmann K, Kaufmann J (2006) Constructing professional identity: the role of work and identity learning cycles in the customization of identity among medical residents. Acad Manage J 49:235–262

Schor J (1998) The overspent American: upscaling, downshifting and the new consumer. Basic Books, New York, NY

Thompson CA, Beauvais LL, Lyness KS (1999) When work–family benefits are not enough: the influence of work–family culture on benefit utilization, organizational attachment, and work–family conflict. J Vocat Behav 54:392–415

Williams JC (2000) Unbending gender: why family and work conflict and what to do about it. Oxford University Press, Oxford

Chapter 9
Fatherhood and Flexible Working: A Contradiction in Terms?

Simon Burnett, Caroline Gatrell, Cary Cooper, and Paul Sparrow

[A] constellation of ideas around what it means to speak of 'good fatherhood' [emerged] after the election of the New Labour government in 1997 . . . focus[ing] on. . . recent developments around the idea of 'work-life' balance.
(Collier, 2001, p. 520)

[W]hile there may have been a shift in discourses of fatherhood, spatially separated employment and family lives make it difficult for fathers to engage in the practices that this implies.
(Halford, 2006, p. 387)

9.1 Introduction

This chapter derives from a desire to investigate the issues of modern fatherhood in the context of flexible working practices and gender; offering a broad-scale analysis as to the extent to which predominantly Western, employed, white-collar, middle-class fathers are embracing flexibility and under what familial and workplace circumstances. In so doing, we highlight the tensions and correlations between traditional notions of fathers as economic providers who go "out" to work and the emergence of late modern ideas concerning the dynamism between fatherhood and more flexible modes of working. Our focus on the work-life situation of fathers in senior roles is both timely and relevant to the growing body of research on work-life balance and male working patterns, which identifies white-collar fathers as under-researched and insufficiently catered for within policy. The topic of fatherhood and flexibility is also important because existing research identifies professional fathers as likely to be highly conflicted due to employers' expectations that they should work increasingly long hours, such pressures being at odds with contemporary paternal desires to invest more time in children's upbringing. Such conflict is especially likely to affect fathers of infant children, who often work longer hours when their offspring are very young than at any other point in their career trajectory (Gatrell and Cooper, 2008).

S. Burnett (✉)
Lancaster University Management School, Lancaster, UK
e-mail: s.burnett1@lancaster.ac.uk

S. Kaiser et al. (eds.), *Creating Balance?*, DOI 10.1007/978-3-642-16199-5_9,

We begin our analysis of fatherhood and flexibility with the observation that flexible working practices have been identified by contemporary governments as key elements within family and employment policy, and demonstrate that they are commonly interpreted and enacted by politicians, organizations and employees in relation to desires to facilitate improved work-life balance. We then assess the history, impact and shifts of culturally accepted discourses regarding paternity – described here as "idealised images" – in order to discuss the presently emerging rhetorical and practical interactions between fatherhood and flexibility; paying particular regard to the conflicts between long-hours work cultures based on "presenteeism" and fatherly desires to be free to engage with their children. We ultimately offer a summation as to the relationships between flexible working policies, employment practices and fatherhood problematizing their traditional, gendered nature, and suggest "flexible" fatherhood may, contradictorily, still ultimately require mothers to clean up the mess.

9.2 The Age of Flexibility

> Hierarchy and three-piece suits give way to first names, shirtsleeves, hoopla, and project-based flexibility. (Peters and Waterman, 1995, p. xxv)

The concept of "work-life balance" became a priority in trans-government policy initiatives throughout the 1990s, in response to the work-family pressures experienced by dual earner couples where both parents were employed (Lewis and Cooper, 1999; Swan and Cooper, 2005). This was partly because research on health and employment had begun to establish links between poor work-life balance and stress (Cartwright and Cooper, 1997) wherein long hours and work intensification for working parents were shown to be impacting negatively on relationships with their partners and children. These pressures were increasingly demonstrated to be linked to stress-related illness and stress-inducing behaviours (Worrall and Cooper, 1999).

While interpretations differ across cultures, concerns about parenting, employment and work-life balance continue to be prominent within international political, consultancy and research agendas, as the demands of life in the neo-liberal economy exerted upon working parents increasingly heighten (cf. Gatrell and Cooper, 2008; Lewis and Cooper, 2005; Lewis et al., 2007). Policies to address this issue are regularly formulated on the precept of offering staff the chance to be employed on a "flexible" basis in order to manage the competing pressures of paid work and families. In theory, the notion of "flexibility" includes a plethora of working practices such as job shares, "flexi-space" and "flexi-time" [the re-organization of individuals' spatial and temporal needs] (see Hill et al., 2001), and home-working. In practice however, a number of organizations tend to offer flexibility in predominantly one form only: the double-edged opportunity of reducing employees' contractual hours to part-time working (Lewis and Cooper, 2005; Swan and Cooper, 2005). The range of options on offer to employees is hence, ironically, often rather inflexible.

Many such initiatives are also gendered. This is because the concept of flexible working is hard to disentangle from socially embedded assumptions regarding the embodied roles which fathers and mothers are oft-expected to play (Lewis et al., 2007). Although organizational and governmental policy statements about work-life balance regularly refer to "parents" and are presented as if gender neutral, the majority of the ideas about flexible working were developed with women in mind and are associated primarily with mother- rather than fatherhood (ibid). Equal opportunities agendas were also originally formulated on the basis of gendered expectations that mothers would usually be second income earners and they, rather than men, would likely utilise offers of flexibility and part-time working in order to facilitate meeting their family commitments (Gatrell and Cooper, 2008; Lewis and Cooper, 2005). Moreover, fathers [especially those in professional and managerial roles] were expected to demonstrate high work orientation through working long hours and maintaining embodied presence in the office (presenteeism) well beyond the official "end" of the working day (Gatrell and Cooper, 2008).

As a result, opportunities to work less than whole-time have been available for women, across a range of occupations, for over two decades. This solution has though been far from unproblematic, as this form of [gendered] part-time or flexible working has been symptomatically linked to discrimination, low pay and reduced opportunities for promotion (see Blair-Loy, 2003; Gatrell, 2005, 2007a; Gatrell and Cooper, 2008; Williams, 2000). However, while part-time working arrangements may be potentially disadvantageous to women in career terms, such flexible opportunities, despite their burgeoning political popularity and assumed affinity, are often either not offered to or taken up by fathers (Gatrell, 2005; Hochschild, 1997). Homeworking, for example, is reported as being only offered to a "privileged few": "over 60% are male, located predominantly towards the top of the occupational hierarchy in managerial and professional occupations, and 58% have dependent children" (Halford, 2006, p. 387). The reasons for men's more limited access to flexible working vary: some fathers, especially those in senior positions who might wish to work from home or part-time, can face barriers from organizations reluctant to offer such practices (cf. Fleetwood, 2007) or alternatively, and as focused upon in this chapter, many may be discouraged due to a personal sense that high presenteeism in full-time paid work is a driving source of their masculinity and, equally, a necessity in securing continued employment (Hochschild, 1997). As a result, opportunities to work flexibly are often more readily accessible to mothers in practice, while many professional fathers, either conducive with or regardless of their own desires, often sustain a customary and highly visible presence at work due to traditional cultural expectations (Collinson and Collinson, 2004).

9.3 The Idealised Imageries of Fatherhood

> Views about what it means to be a father and the roles of fatherhood are constructed over many years . . . boys become fathers to boys who will become fathers in the future. (Cabrera et al., 2000, p. 131)

Despite the maternally-oriented origins of flexibility and the indicated difficulties experienced by many contemporary fathers in practically engaging with its methods the association of fatherhood and flexible working practices is regardless presently becoming an increasingly politically celebrated and culturally substantiated ideal (cf. Brandth and Kvande, 2001; Cabrera et al., 2000; Halford, 2006). However, whilst for many whether within "nuclear" of more variegated family settings, the biological and physical preconditions of fatherhood have effectively endured unchanged throughout history. Its trajectory as a social, discursive concept – the manners in which fatherhood is widely conceived of, thought, written and acted upon – towards its current "flexible" incarnation, in the West at least, have undergone a variety of marked, developmental shifts. In order to understand the present then it can be of great benefit to address that which has past, and in particular consider the ideas and events characteristic of previous eras that arose from and reflect some of their contemporaneously commonly shared views, anxieties and images (cf. Burnett, 2008; Skirbekk and Gilje, 2001). For the practicalities of constructing and articulating a reasonable argument, any theoretical and socio-historical account must though necessarily be culturally and temporally selective. To such an end, those selections employed concerning the changing values attached to fatherhood in the following discussion, and displayed in Fig. 9.1 below, are made in an attempt to demonstrate in brief certain periodic transitions in common

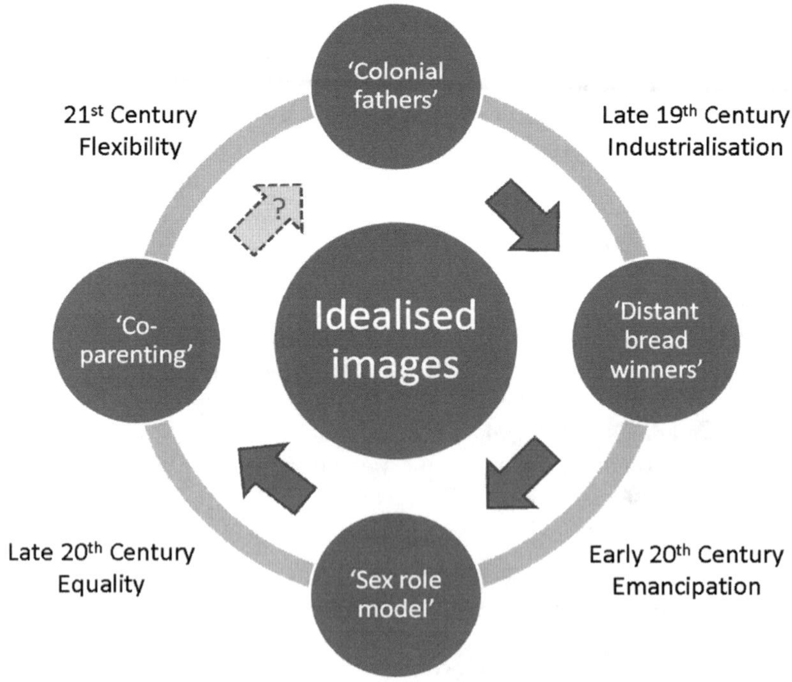

Fig. 9.1 The cultural transitions of fatherhood

[predominantly Western] perceptions of what fatherhood is alongside the associated historic and cultural definitions of what it should be. This is not intended as an exhaustive account spanning all geo- and epochal boundaries or individuals within any given timeframe; but more so a narrative to convey how the shifting social enterprise of fatherhood has been, and thereby presently is, widely understood.

As depicted below, since the circa 1800s, fatherhood has been beset by a "deep-seated ambivalence" of images regarding what the role "should be" (Pleck, 1997, p. 351). This is indicated by the dynamic and modifying significance, theories and even labels attached to it. Amongst others, Cabrera et al. suggest that since this period generic social perceptions of fatherhood can be traced as having evolved through stages of the "colonial father", to "distant breadwinner", to the "sex role model" and to "co-parent" (Cabrera et al., 2000, p. 127; cf. also Pleck, 1997). This analysis will now address each of these ideals, concluding that they have indeed progressed so and have engineered the environmental conditions for a fifth, the "flexible father", which is infused with their legacies and argued herein to have emerged in late modernity.

9.3.1 The "Colonial Father"

For the late eighteenth and early nineteenth century there is little disagreement that mothers, if not employed female carers or "nannies", as with their current counterparts, tended to provide a majority of the hands-on care essential for infants and young children. However, fathers were typically viewed, similarly on both sides of the Atlantic in "colonial" America and Victorian England, as resolutely embodying the over-arching position of "moral overseer" and holding the ultimate responsibility for, and influence on, their offspring (Pleck, 1997, p. 351). Patriarchy – in its "precise meaning" of "father-rule" (Tosh, 2007, p. 3) – was thereby hegemonic in the constitution of family life. These societies were reportedly "merciless in pillorying men" who did not reflect this image, ostracising them socially as well as financially: "It is not surprising that political thinkers held that the authority relations of the household were a microcosm of the state" (ibid).

According to Pleck (1997) a father's duty under such social conditions was, if literate himself, to educate his children in the skills of reading and writing, advise them as to what God and the world required of them, and guide them into an appropriate "occupational calling". He would even play a meaningful role in the arranging and ratifying of the courtship and marriage of both sons and daughters (ibid, p. 352). An epitomising exemplar from the period of familial relations under the "colonial father" appears in GW Cottell's "The Token of Friendship; or Home, The Centre of Affections" (1844):

> The father gives his kind command,
> The mother joins, approves;
> The children all attentive stand,
> Then each obedient moves.
> (cited in Pleck, 1997, p. 352)

Such a construction of the paternal position was hence embedded in the contemporaneous politico-cultural conception of the difference between and projected roles of the sexes. Marriage and the homestead were central to masculinity in the period, with a man's lineage and "place in posterity" being secured through the reproduction and instruction of [preferably] sons who would carry on the family name (Tosh, 2007, p. 2, 4). It is argued that in Britain the Victorians, especially, placed a higher value on family life than any generation before or since (ibid, xi) and that the church, a paramount social institution of the time, very much supported and reinforced these beliefs. It is then perhaps in more modern times difficult to appreciate "the power and depth" (Pleck, 1997, p. 358) of this so widespread mentality in which fathers were reified to such a foremost and hegemonic social position.

9.3.2 The "Distant Breadwinner"

In the second half of the nineteenth century however the church had somewhat waned in terms of its social prowess and influence as mass Industrialisation swept through Western society. These impacts affected virtually all facets of social life, as, in the US particularly, many fathers "left their small farms and businesses" to seek gainful employment in the newly expanding cities and industrial centres. Such a shift resulted in a new model of fatherhood as the "distant breadwinner", earning a crust for the family at home, and socially sanctioned the relocation of the authority for making decisions regarding the rearing of children to the mother (Cabrera et al., 2000, p. 127; Pleck, 1997, p. 358). The imagery of the constant and central presence of mothers as children's primary caregivers, and fathers' predominant absence for vast periods of the day, contributed to the germination of the increasingly popular assumption by many developmental theorists of the time that father-child relationships had little to no impact on children's advancement (Cabrera et al., 2000, p. 127); significantly countermanding the "colonial" rhetorical dominance of before. Tosh (1996) reports how this social context customary within professional, middle-class, family life was very similar in North America and Britain; most notably definable by the growing separation of home and work and with it the relocated derivation of masculinity. The expected station for [good] fathers was now in the workplace (cf. Halford, 2006).

A progressive shift towards a greater perceived role for the mother, in terms of both embracing responsibility and failure for how a child was deemed to have been brought up, and a more indirect and primarily financial role for the father is unmistakable in this industrial period of the late eighteen hundreds (Pleck, 1997). This predominant construction of fatherhood was accompanied by a parallel shift in gender ideology overall, with the conceived "purity of the female sphere" [i.e. the home] elevating woman above man in terms of their suitability for rearing the young (ibid). This was reflected in the wider, socio-structural establishment of both the presumption of maternal custody in the event of divorce and the introduction of female teachers into [American] elementary schools (ibid, p. 353, 4).

9.3.3 The "Sex Role Model"

By the early twentieth century, the context and idealised imagery of fatherhood had somewhat shifted again. The effects of 1940s war-time production necessity and emancipation movements saw many more women joining the ranks of the employed (cf. Cabrera et al., 2000) and the toll of the war on populations, with hundreds upon thousands of fathers going to fight and a great number not returning forced many mothers to learn either temporary or permanent independence. In part resulting from the increase in paternal absence and maternal self-sufficiency, the notion of "excessive mothering" consequently became an "issue of concern" by the post-war years, with a number of prominent military psychiatrists through the 1940s and 1950s even blaming soldiers' battle breakdowns on over-caring [or over-bearing] mothering. This contributed greatly to the [re-]ushering in of the impression of a father's direct importance to raising a child as a "sex role model" (Pleck, 1997, p. 356). Whilst this conception did not become all-pervasive, the impression of the distant father-breadwinner still remained, it is nonetheless "historically important" as it represented the first positive image to have exerted significant cultural gravitas toward advocating more involved fatherhood since that of the Victorian or colonial "moral overseer" (ibid). Significantly though, this "new" view encouraged paternal participation within limited parameters: fathers should teach sons how to be men by way of example. The post-World War Two father hence came to be seen as a "towering figure" of importance in the life of his children, the newly returned, iconical "modern involved dad" (Cabrera et al., 2000, p. 127), but principally through the potential for detrimental effects levied by their absence, rather than much that was directly positive from their presence (Pleck, 1997).

9.3.4 The "Co-parent"

The aloof-albeit-present "sex role model" imagery of the father soon came to be softened in the latter half of the twentieth century however, in conjunction with a greater societal emphasis on gender equality and as the noted increase in maternal employment crept ever higher. At the midway point just after the Second World War, 12% [of North American] women with preschool age children were in the workforce. By 1997 this had risen in real terms to two-thirds (Cabrera et al., 2000, p. 127). Simultaneously, it is reported that the proportion of two-parent families in which the father acted as the sole "breadwinner" had noticeably declined (ibid, p. 128). It was not though solely economic responsibilities that had come to be more widely shared. Statistics indicate that in the 1960s and 1970s, fathers tended to spend "about 30% to 45% as much time with their children as mothers did"; whereas [in the 1990s] this figure had increased to 67% on weekdays and up to 87% on weekends (ibid). Barnett and Baruch concur, stating from their research that: "fathers with employed wives and... who hold less traditional attitudes ... did proportionately more feminine home chores" (Barnett and Baruch, 1987, p. 35). This suggests, as discussed in more depth anon, that higher [although still not necessarily entirely equal] paternal participation, the idealised image of "co-parenting", arose, once again, in conjunction with a politically and culturally conducive environment.

9.3.5 The "Flexible Father"

With such general societal changes in the division of familial workforce participation indicated, it is unsurprising that, as discussed in the introduction to this chapter, recent related shifts in the patterns and organization of work in the neoliberal, globalised economy have also become more common place. Reflecting on such conditions, Toffler states:

> Watching masses of peasants scything a field three hundred years ago, only a madman would have dreamed that the time would have come when the fields would be depopulated, when people would crowd into urban factories to earn their daily bread... Today it takes an act of courage to suggest that our biggest factories and office towers may, within our lifetimes, stand half empty, reduced to use as ghostly warehouses or converted into living space. (1980, p. 210, cited in Hill et al., 1996, p. 293)

This is though seemingly an end towards which the theory of contemporary flexible modes of work and production are engineered: "a return to cottage industry on a new, higher, electronic basis with its new emphasis on the home as the centre of society" (ibid). There is then leeway to argue, as posited above, that the emerging predominant idealised conceptualisation of being a father – or more specifically, being a good father – within the realm of contemporary employed, white-collar, Western, middle class society, is a discursive notion progressive from "breadwinner" and "co-parent" iconoclasm, predicated upon utilising modern employment practices: dad should be flexible enough to both earn a wage and be able to help fix dinner and read a bedside story (cf. Cartwright and Cooper, 1997; Gatrell and Cooper, 2008; Hill et al., 2001). There is an ingrained, unitaristic assumption amongst its advocates that a workable consensus can be established between all parties with a vested interest in such work-life patterns: fathers, mothers, employers, and politicians; and that this itself is an a priori social good (Collier, 2001, p. 541). Significantly, this is also somewhat indicative of a return to colonial ideals of fatherhood, albeit without the overtones of patriarchy and hence the broken arrow in Fig. 9.1; as Tosh reports how Victorian middle-class men were [also] beset by conflicting expectations, between their responsibilities towards "domestic concerns and regular fireside evenings" and generating capital (Tosh, 1996, p. 49).

As with those which preceded it, this ideal and neologism of the "flexible father" is not presented as a society-wide representation of the activities and idiosyncrasies of all modern family life; but rather as an historically emergent, culturally supported and produced rhetorical archetype, upon which a number of employment and political policies and family and organisational practices are, since the mid-nineties, arguably becoming based. In support of this assertion, a review into "Mental Capital and Wellbeing in the twentyfirst century" commissioned as part of the UK Government's Foresight Project, states: "Findings show that [flexible work arrangements] enable individuals to integrate work and family responsibilities within their personal time and space and are instrumental in achieving a healthy work-family balance" (Westman, 2008, p. 2). Evidence can [and below will] be provided to demonstrate though that flexible working practices have not seamlessly drip-fed from theory into reality, acting as a "magic-bullet" to provide families, and particularly working fathers, with the additional temporal and spatial pliancy envisioned. Furthermore, we show how domestic and child rearing duties have not subsequently become equally divided but often remain the responsibility of mothers – even when they are also in full-time employment – as an axiomatic consequence of the ascendance of the image of the "flexible father".

9.4 Modern Men and Flexible Working

> [G]lobalisation and the growing intensification of work will combine to increase workers' levels of stress and anxiety, and affect their health and efficiency. Changes in the nature of work will also interact with changes at home, such as growing numbers of two-earner households . . . thereby creating pressures on families. Maintaining and improving well-being in the face of these trends will be a major challenge. (Foresight Mental Capital and Wellbeing Project, Executive Summary, 2008, p. 28)

It is widely reported in both academic and news media that British employees, in particular, frequently work more hours, take shorter lunch breaks and fewer holidays compared to the vast majority of their European counterparts (cf. Giga et al., 2008, p. 4; International Labour Organization, 2007; Trade Union Congress, 2007). According to 2003 statistics, approximately a quarter of the UK's entire working population exceed the limit imposed by the Working Time Regulations of 48 h per week, and 9% report regularly work over 60 h per week (Giga et al., 2008, p. 4). Two thirds of those who breached the 48 h threshold had deliberately "opted out" of the directive's protection, and, more ominously, one in four indicated they had been given little to no choice to do so. Furthermore, a study published by the British government's Department of Trade and Industry found that male employees with children tended to work the longest hours of all (Chartered Institute of Personnel and Development, 2009); whereby, on average, "UK fathers with children under 14 years of age work for 46 h a week and working evenings, nights and/or weekends has become commonplace for many, with as many as 39% of fathers working more than 60 h a week" (Halford, 2006, p. 387).

Within this context, Eggebeen and Knoester provide supportive evidence that, due to their familial responsibilities, contemporary white-collar, middle class fathers tend to exhibit greater attachment and commitment to the labour force and their careers than men with no children (Eggebeen and Knoester, 2001, p. 384; cf. Halford, 2006). Ironically though, despite the reputed heightened focus on work, their senses of happiness, subjective well-being and life satisfaction are deemed to draw from their family rather than work roles (Eggebeen and Knoester, 2001, p. 382). Such home- and workplace circumstances would seem conducive to encouraging fathers to embrace and request flexible working practices where available: a legal right now extended to many working parents in Britain. Hill et al. indeed suggest that such actions can reduce the time and stress of commuting, afford increased freedom as to where to live, enable synchronicity with children's school schedules, and relinquish more quality time to be spent at both home and work (Hill et al., 2001). It is even proposed that the implementation of flexibility programmes sufficient to improve employee perception of their flexibility by one point on a 5-point scale is: "statistically equivalent to reducing workload by 11 h per week per employee" (ibid, p. 56). It would seem logical then to assume that an array of policies intended to elicit such freeing effects would likely receive a notable degree of popularity. However, as previously indicated, despite the impressive array available, their politically-driven promotion and innate theoretical appeal, flexible working practices have not to date received a ubiquitous uptake amongst Western, white-collar, middle-class, employed fathers (Giga et al., 2008). The archetype of the "flexible father" is currently seemingly somewhat more of a rhetorical than actual reality.

As testimony to this, Brandth and Kvande provide an exemplar piece of work on modern fatherhood and flexibility in Norway, describing the "Scandinavian Welfare State" as being "at the 'cutting edge' in... creating a more egalitarian society" (Brandth and Kvande, 2001, p. 251). Centering upon paternal leave allocations awarded at the birth of a child, Norwegian political and social welfare efforts have been made to ensure that workplace benefits are shared between couples in order to facilitate equality and "strengthen the father-child relationship" (ibid, p. 252). Fathers are eligible to choose between two systems: an inflexible, non-transferable "paternity quota" of a single block of 4 weeks sanctioned leave, provided the mother works, so the father can take part in early childcare; or to utilise a more flexible "time account", whereby negotiations between parents, their employers and the state are entered into in order to share the parental leave through reducing work hours for each over a longer period as is deemed most appropriate. It was widely expected that the "time account" system would be incredibly popular as fathers would not need to be entirely absent from work for a protracted length of time. Counter-intuitively though, it was found that the quota of a 4 week block was more appealing to the vast majority, for whilst: "fathers' use of leave depended on sharing... with the mother it worked poorly" (ibid, p. 261). This was based upon many fathers reasoning that their jobs were too "demanding" to work reduced hours across a prolonged period and the unappealing, associated impacts of reduced income, presenteeism and status (ibid, p. 262):

> Mothers returning to work before maximum leave time is over are met with some suspicion... And fathers risk being defined as part-time workers. (ibid, p. 263)

Whilst clearly popular as a political and theoretical device, intended to improve fatherly engagement with children and simultaneously maintain their ability to work in an uninterrupted and continuous fashion; in reality the initiatives were, arguably, introduced in a cultural climate still steeped in traditional and gendered "breadwinner"-type perceptions of family roles and the importance and validity of long hours and strong workplace presenteeism. As shown by the relatively successful practical emergence and embracement of all the preceding idealised images of fatherhood identified in Fig. 9.1 ["colonial", "breadwinner", "sex role model" and "co-parent"], successful shifts in ideologies must occur in unison with a pronounced degree of cultural receptivity to become accepted comportments of actual behaviour. This is not to say that Toffler's comments, shown above and taken from Hill et al. (1996, p. 293), regarding how our biggest factories may, within our lifetimes, stand half empty will not necessarily occur, but rather that such a conceptual shift has not yet been widely accommodated and reified into practice.

In addition to this theorising upon why middle class, white-collar fathers may resist embracing the emergent and politicised "flexible" imagery, and in accordance with the implication that wider cultural support is still lacking, there is reason to suggest that a number of organizations are similarly hesitant (cf. Hill et al., 2001, p. 51). This is because the erosion of work and home-life boundaries by practices intended to engineer employees' flexibility can actually result in a perceived "threat" to companies' productivity and control (Halford, 2006). Whilst the primary focus of this chapter is on the embracement of flexible working practices from the perspective of fathers, organizations clearly and unavoidably contribute to the influential environmental conditions in which modern fathers find themselves. For many employers, the "ideal professional [male] worker" remains reminiscent of the "distant breadwinner", who fulfils his family obligations through a strong commitment to paid employment, leaving responsibility for childcare to the maternal realm. Fleetwood concurs, stating that in neo-liberal market economies such as the UK and the USA, any policy which is developed to improve the situation for workers "is likely to achieve only limited success" (Fleetwood, 2007, cited in Gatrell and Cooper, 2008, p. 73). This is because in a market-led economy: "employers will usually seek to find ways of re-shaping "worker friendly" policies" to their own advantage and, quite possibly, to the detriment and expense of those whom they were originally intended to benefit (ibid). This argument thus enables the supposition that flexible working practices that directly benefit organizations [such as, for example, flexi-space initiatives that expect staff to always be available through email and mobile phone access, or flexi-time policies that demand working atypical or "annualized' hours] can be decidedly "employee and family unfriendly", and often act as hindrances to the establishment of an effective work-life balance (Fleetwood, 2007).

Halford articulates that whilst individual fathers are of course able to actively participate in negotiating the agency of their approach to fatherhood within such organizational and cultural constraints, there is an overbearing weight of contemporary empirical evidence that demonstrates: "even amongst those fathers who endorse new [flexible] representations of fatherhood, the embodied practices of fathering children take place outside the time-space of the working day" (Halford, 2006, p. 387). Even amongst "modern" fathers then it seems, flexibility is often either shunned, or forced upon them in a form reconstituted for the benefit of the employer, not the employee.

9.5 Conclusion: Are Flexible Dads Founded Upon Mums?

> [I]t is clear that [flexible] working is not producing equality in parents' child-rearing responsibilities and roles ... new practices of fatherhood linked to home-working still depended on a traditional gender division of labour to police the boundaries of public and private... This confirms ... that fathering is mediated by wives. (Halford, 2006, pp. 398–399)

As articulated at length above, the contemporary relationship within the realm of white-collar, middle-class employment between flexible working initiatives, their utilisation in workplace policies and the besieged imagery and practices of fatherhood betwixt them, is far from having reached an "ideal" mutually beneficial and societal-embraced consensus. Whilst notable technological, political and social developments in recent years have indeed seen a significant increase in the amount of people who are "organizational employees and also work from home" (Halford, 2006, p. 387); the archetypal "flexible father" is arguably very much mired in a gendered and culturally unreceptive environment that is stalling his further deep-seeded appropriation. The presently popular representations of men as workers and fathers are thus ironically resulting in a potentially family-damaging paradox: "While marriage and fatherhood may be linked to career development, career success means more demanding work and longer hours; that is, more time spent away from the family" (ibid, p. 386). Social pressures and personal desires for fathers to engage more with their children and to function in a particularly troublesome economy are hence rendering the dualistic or Janus-image of the flexible employed-yet-engaged father somewhat difficult to sustain. In response to such tensions, across political, organisational and personal spectrums numerous policy debates and cultural artefacts, such as books, films, newspaper and magazine articles, and television programmes, are frequently lamenting how it is now: "an increasingly hard time to be a father" (Collier, 2001, p. 531).

However, in addition to the culturally and organisationally-generated conditioning influences exerted upon fathers' interaction [of lack their of] with flexible working practices, a range of empirical research studies highlight that domestic and paternal engagement with their children is also commonly and profoundly mediated by the children's mothers (ibid). As already indicated, since the mid-1990s in conjunction with "third way" neo-liberal economic governmental policies concerning employer and employee flexibility and attempts to achieve sustainable fiscal growth, has been the deliberate heightening of values and assumptions regarding social cohesion, emancipation and the equal sharing of familial responsibilities. Such societal notions have progressed developmentally from the preceding cultural climates that begot the ideologies of "colonial", "breadwinner", "sex role model" and "co-parent" fathering, to the present age in which we ideally see families: "marked by the qualities of emotional and sexual equality" (ibid, p. 527).

Recalling the progression of the previously discussed ideals of fatherhood from Fig. 9.1, some mothers welcome "involved" paternal/child relationships believing such contact with children to be "healthy", whilst other women find this more difficult to deal with. Consequently, the reluctance on the part of some mothers to share parent-child involvement equally with fathers relates to maternal desire to maintain the more traditional, post-"colonial" role of lead parent. It is recognised though that even in [particularly heterosexual] families that have utilised flexibility to enhance joint-parental engagement, tensions often arise regarding the amount of time spent by fathers undertaking childcare (Gatrell, 2007b). This appears to be the case both within intact relationships and post separation/divorce (Gatrell, 2007b; Smart et al., 2001). Gatrell argues that these tensions emerge because mothers and fathers can interpret the notion of parental care and involvement differently (ibid). Fathers tend to define "involved fatherhood" as spending quality time with children in addition to investing in paid work, but, most notably, as excluding domestic labour. Mothers, on the other hand, are more likely to interpret paternal involvement with children in terms of fathers sharing the more unappealing domestic chores [such as household laundry and food preparation] which do not comprise direct contact with children.

At present, while there is then evidence that some fathers are striving to work flexibly and seeking a better work-life balance, and are hence more likely to be involved with children's upbringing on a day-to-day basis than in the past, the same still does not apply to domestic labour: working mothers in heterosexual relationships still undertake substantially more domestic chores than their male partners. This applies not only to women who are in part-time paid work (Dryden, 1999), as Scott (1999, p. 73) notes that even where men and women in dual-earner households devote the same amount of time to their employment, women undertake around 9 h per week more housework than men. It is further compounded by Gatrell (2005) that while men may want equality in terms of parenting, they are unlikely to desire equal responsibility for housework and domestic labour, preferring to cast themselves in the role of "helper". As Delphy and Leonard observe:

> Employed women therefore generally end up working very long days, at the least favoured aspects of household work . . . even when they have "good" husbands. (Delphy and Leonard, 1992, p. 240)

Sullivan likewise concludes that the free time available for employed women with infant children is less than that available to their husbands/partners and notes: "women's time is . . . more pressured in terms of intensity" (Sullivan, 1997, p. 235). As a result, some mothers feel resentful that paternal engagement with children is not beset by housework and may be regarded as "quality time", while maternal time with children is compromised by household labour (Gatrell, 2007b).

Mothers may find it distressing then if more "involved" flexible fatherhood' still allows men to partake in unencumbered contact with children, whilst they are obliged to "mop up" the additional domestic chores. So, whilst it may indeed be a particularly hard time to be a father, it seems the pressures associated with the budding desires to flexibly navigate the cultural expectations of being a good employee and a good dad are very much predicated upon the expectation of, unsung, maternal support.

Acknowledgements The authors would like to acknowledge their research partners Working Families (Chief Executive Officer Sarah Jackson (OBE); Policy and Research Officer Jonathan Swan), and to express their gratitude to the Lottery Research Fund for financing this project.

References

Barnett RC, Baruch GK (1987) Determinants of fathers: participation in family work. J Marriage Fam 49:29–40

Blair-Loy M (2003) Competing devotions: career and family among women executives. Harvard University Press, Cambridge

Brandth B, Kvande E (2001) Flexible work and flexible fathers. Work Empl Soc 15:251–267

Burnett SB (2008) The happiness agenda: the sources, sites and protagonists of a modern obsession. PhD Thesis, University of Lancaster

Cabrera NJ, Tamis-LeMonda CS, Bradley RH, Hofferth S, Lamb ME (2000) Fatherhood in the twenty-first century. Child Dev 71:127–136

Cartwright S, Cooper C (1997) Managing work-place stress. Sage, Thousand Oaks, CA

Chartered Institute of Personnel and Development (2009) Does the UK have a long hours culture? (Part of 'Working Hours in the UK'). http://www.cipd.co.uk/subjects/hrpract/hoursandholidays/ukworkhrs. Accessed on 6 Oct 2009

Collier R (2001) A hard time to be a father?: reassessing the relationship between law, policy, and family (practices. J Law Soc 28:520–545

Collinson D, Collinson M (2004) The power of time: leadership, management and gender. In Epstein CF, Kalleberg AL (eds) Fighting for time: shifting the boundaries of work and social life. Russell Sage Foundation, New York, NY, pp 219–246

Delphy C, Leonard D (1992) Familiar exploitation: a new analysis of marriage in contemporary Western societies. Polity Press in association with Blackwell, Oxford

Dryden C (1999) Being married, doing gender. Routledge, London

Eggebeen DJ, Knoester C (2001) Does fatherhood matter for men?. J Marriage Fam 63: 381–393

Fleetwood S (2007) Why work-life balance now?. Int J Hum Resour Manage 18:351–360

Foresight Mental Capital and Wellbeing Project (2008) Final project report – executive summary. The Government Office for Science, London. http://www.foresight.gov.uk/Mental%20Capital/Mental_capital_&_wellbeing_Exec_Sum.pdf Accessed 17 Feb 2010

Gatrell C (2005) Hard labour: the sociology of parenthood. Open University Press, Maidenhead

Gatrell C (2007a) A fractional commitment? Part-time employment and the maternal body. Int J Hum Resour Manage 18:462–474

Gatrell C (2007b) Whose child is it anyway? The negotiation of paternal entitlements within marriage. Sociol Rev 55:353–373

Gatrell C, Cooper CL (2008) Work-life balance: working for whom?. Eur J Int Manage 2:71–86

Giga SI, Jain AK, Cooper CL (2008) State-of-science review: SR-C7 – Working longer: hours of work and health. The Government Office for Science, London. http://www.foresight.gov.uk/Mental%20Capital/SR-C7_MCW.pdf. Accessed on 17 Feb 2010

Halford S (2006) Collapsing the boundaries? Fatherhood, organization and home-working. Gender Work Organ 13:383–402

Hill EJ, Hawkins AJ, Miller BC (1996) Work and family in the virtual office: perceived influences of mobile telework. Fam Relat 45:293–301

Hill EJ, Hawkins AJ, Ferris M, Weitzman M (2001) Finding an extra day a week: the positive influence of perceived job flexibility on work and family life balance. Fam Relat 50:49–58

Hochschild A (1997) The time bind: when work becomes home and home becomes work. Henry Holt, New York

International Labour Organization (2007) Press release: working time around the world (7 June 2007). http://www.ilo.org/global/About_the_ILO/Media_and_public_information/Press_releases/lang–en/WCMS_082827. Accessed 17 Feb 2010

Lewis S, Cooper CL (1999) The work-family agenda in changing contexts. J Occup Health Psychol 4:382–393

Lewis S, Cooper CL (2005) Work-life integration: case studies of organizational change. Wiley, Chichester

Lewis S, Gambles R, Rapoport R (2007) The constraints of a work-life balance approach: an international perspective. Int J Hum Resour Manage 18:360–374

Pleck JH (1997) American fathering in historical perspective. In Hansen KV, Garey AI (eds) Families in the US: kinship and domestic politics. Temple University Press, Philadelphia, PA, pp 351–362

Peters T, Waterman RH (1995) In search of excellence: lessons from America's best-run companies. HarperCollins, London

Skirbekk G, Gilje N (2001) A history of Western thought: from ancient Greece to the twentieth century. Routledge, London and New York

Smart C, Neale B, Wade A (2001) The changing experience of childhood: families and divorce. Polity Press, Cambridge

Sullivan, O (1997) Time waits for no (wo)man: an investigation of the gendered experience of domestic time, Sociology, 31: 221–239

Swan J, Cooper CL (2005) Time, health and the family : what working families want. Working Families, London

Tosh J (1996) Authority and nurture in middle-class fatherhood: the case of early and mid-Victorian England. Gender Hist 8:48–64

Tosh J (2007) A man's place: masculinity and middle class home in Victorian England. St Edmundsbury Press Ltd, Bury St. Edmunds

Trade Union Congress (2007) Press release: real, but oh so slow, progress on long hours in London (20 February 2007). http://www.tuc.org.uk/work_life/tuc-12966-f0.cfm. Accessed on 17 Feb 2010

Westman M (2008) State-of-Science Review: SR-C4 – Flexible working time arrangements and their impact on work-family interface and mental wellbeing at work. The Government Office for Science, London. http://www.foresight.gov.uk/Mental%20Capital/SR-C4_MCW.pdf. Accessed on 17 Feb 2010

Williams JC (2000) Unbending gender: why family and work conflict and what to do about it. Oxford University Press, Oxford

Worrall L, Cooper CL (1999) Working patterns and working hours: their impact on UK managers. Leadersh Organ Dev J 20:6–10

Chapter 10
Dual-Career Couples: Specific Challenges for Work-Life Integration

Andrea E. Abele and Judith Volmer

10.1 Introduction

10.1.1 Overview: The Development of Dual-Career Couples

Over the past 40 years, industrialized nations have experienced major changes in their labor market characteristics. Of special importance is the steady increase of women's participation in the paid workforce. In the United States (US), 72.3% of all women aged between 25 and 54 (US Bureau of Labor Statistics, 2009) were in employment, and 59.1% of women aged between 15 and 64 in the European Union (EU) (EUROSTAT, 2009). Due to women nowadays having educational attainments equal to men's (representing, for example, 59% of university graduates in the EU (Rusconi and Solga, 2008)) and due to changed values on women's and men's roles in society, traditional family models have been steadily replaced by more egalitarian partnership constellations. Women's roles are no longer confined to family and home-keeper roles, and men are no longer the sole breadwinners in the family.

A relatively new partnership constellation is one where both partners are highly educated, have a high upward career orientation, and work full time in a demanding job. This particular partnership arrangement has been termed dual-career couple (DCC)[1] and it has existed for about 40 years in noteworthy numbers. The traditional career was a "male" one in which the husband was a successful professional and the wife supported her husband, either without employment or with employment that was clearly subordinated to the husband's career. The central concern of DCCs, in contrast, is to arrange both partners' careers and upward career aspirations with a happy family life.

A.E. Abele (✉)
University of Erlangen-Nuremberg, Erlangen, Germany
e-mail: abele@phil.uni-erlangen.de

[1] Whereas dual-earner couples have been observed for a longer time, the novelty of DCCs pertains to the fact that both partners not only work full time but additionally they have highly demanding professional positions and both partners have high career aspirations.

S. Kaiser et al. (eds.), *Creating Balance?*, DOI 10.1007/978-3-642-16199-5_10,

Since the 1960s, research has paid increasing attention to DCCs. In their seminal work, Rapoport and Rapoport (1969) defined DCCs as families where both partners pursue a professional career. Other definitions emphasize a high career orientation together with a high partnership orientation for DCCs. Boehnke (2007) defines DCCs as couples with or without children where both partners are highly educated (university degree or comparable), work full-time in a challenging position, and live together in one household for at least 5 years. Behnke and Meuser (2003a, b) describe DCCs as partnerships wherein both pursue a professional career, without necessarily working full time. Although each of these definitions includes slightly different criteria, all concur that DCCs are well educated couples who are highly committed to their jobs and show respect for and interest in their partner's career. DCCs are considered to be "avant-garde" (Walther and Lukoschat, 2008) as they influence societal images of a modern family and strive towards gender and value equality (Apostal and Helland, 1993).

10.1.2 Aim and Structure of the Chapter

The aim of this chapter is to give a comprehensive overview of the topic of DCCs. We will first provide information on the frequency of DCCs in Europe and the US. We will then discuss the potential advantages and disadvantages of this partnership arrangement; and we will review findings on DCCs in the area of stress and well-being; career and mobility issues; partnership and parenthood issues; as well as gender related questions. We will then turn to strategies for facilitating the work-life integration of DCCs on the individual level, on the level of couples, and on the level of organizations.

10.2 Dual-Career Couple Arrangements in Professionals: Frequency

Because scholars use different definitions of DCC, it is difficult to estimates the frequency of such arrangements. Rusconi and Solga (2008) estimate that about 30% of couples live in dual-career arrangements and that this number is similar in different countries (e.g., Germany, Italy, Switzerland, and the US). For example, they report a study showing that 28% of North American middle-class couples live in dual-career constellations (Williams and Han, 2003). For Germany, estimates for DCC households range from 1 to 8% for all education levels (Boehnke, 2007) to approximately 30% for academically educated couples (Rusconi and Solga, 2008) and up to 80% for scientists (Abele, 2009; Rusconi, 2003). DCC might be stable or a temporary partnership constellation, being altered into a more traditional partnership constellation like a one-and-a-half-earner constellation after the birth of a child (e.g., Gerhard et al., 2003).

10.3 Theoretical Models

The research basically draws on three theoretical models when explaining why couples choose to live in DCCs. First, the economic theory of the family (e.g., Becker, 1981; Blossfeld and Drobnic, 2001) argues that partners bring together their individual resources with the aim of maximizing their joint economic utility (Bernadi, 1999). According to this view, households maximize their shared unitary utility when both partners engage in and specialize in activities in which they are highly qualified. Partners can engage in paid work and/or housework. When both partners are highly educated, the highest joint economic utility will be achieved when both partners work full time in career-oriented jobs. From a society's point of view, it is also wise to transform the money invested in both partners' education into full-time paid work. Choosing the dual-career path also increases flexibility in times of economic insecurity and can be regarded as a highly adaptive strategy (Oppenheimer, 1997). If one of the partners (temporarily) loses his/her job, there is still some family income.

Second, social-exchange theory (Emerson, 1976) argues that partners not only rely on economic exchange but also on social exchanges. Partners invest social behavior into the relationship. They reciprocate this investment which ensures the long-term nature of relationships. According to social exchange theory, partners who make more investments also have more power in the relationship. When both partners are highly educated, similar investments can be expected and, hence, similar power will result.

Third, gender approaches argue that gender equality is highly valued – especially by women – in modern Western societies and that in DCC arrangements there is a good chance that the roles of men and women in such partnerships are characterized by gender equality (Abele, 2009; Apostal and Helland, 1993; Behnke and Meuser, 2003a, b; Boehnke, 2007; Walther and Lukoschat, 2008). Gender approaches build on value shifts that accompanied societal changes from the pre-industrial to the post-industrial era. In the industrial era, the work and family domains were strictly separated with a sharp role differentiation between men and women. In the post-industrial era, partnership, equality between partners, and shared responsibilities have become more important. Since then, the number of working women increased steadily together with the number of couples in which both partners pursued a career. Increasing equality in partnership was followed by changing family structures with a concentration on the nuclear family, role negotiations in partnerships, and more egalitarian relationships. The distinction between "sex" and "gender" is crucial in respective theorizing. Whereas "sex" is related to mainly biological and socio-demographic aspects, gender is related to both psychological aspects (gender identity, gender self-concept) and to social aspects, notably the social construction of gender (such as gender roles, gender roles expectations; e.g., Abele, 2000; Deaux and LaFrance, 1998; Eagly, 1987). Egalitarian conceptualizations of gender roles emphasize that men and women should be free to choose the roles they want to perform and that there should be no power gap between men's and women's roles. Working full-time increases both partners' autonomy, offers opportunities for recognition and contribution to society, and shows acknowledgment and appreciation of each partner's profession.

To summarize, the DCC is a relatively new form of partnership arrangement that has emerged as a consequence of women's increasingly better education and higher workforce participation in the last few decades. It reflects considerations of joint economic utility but it particularly reflects value shifts towards more equality in gender roles.

10.4 Work-Life Integration in DCCs: Specific Challenges

10.4.1 The Pioneer Research of Rapoport and Rapoport

Pioneer research on work-life integration of DCCs was conducted by Rapoport and Rapoport (1969). They conducted an interview study with 16 dual-career families to investigate whether these families experience specific challenges and dilemmas from their lifestyle and how they cope with them. The authors found five major forms of dilemma: overload dilemma, personal norm dilemma, dilemma of identity, social network dilemma, and role cycling dilemma. The couples interviewed reported that recognition of joint interests and pursuing the goal of optimizing career versus family decisions was most important to deal with these dilemmas.

- *Overload dilemmas* refer to the issue that full-time working couples have less time and fewer resources for housework. Study participants reported that they delegate some responsibilities to other people in order to deal with overload dilemmas.
- *Normative dilemmas* arose from discrepancies between personal and social norms. Working mothers were particularly confronted with the social norm expecting mothers to stay at home and take care of their children and considering working-full time as being egoistic and at the expense of children's well-being. To resolve normative dilemmas, interviewees reported that they try to separate themselves from such social norms whenever possible.
- *Identity dilemmas* concerned the switch between different role requirements. For example, women reported that they would switch roles by showing aggressive behavior when necessary at work and caring behavior when required at home.
- *Social network dilemmas* were also reported. Time constraints kept DCCs from frequent interactions with relatives and friends. Moreover, friendships with couples with different role arrangements became difficult because they triggered the normative dilemma. Participants reported choosing friendships very carefully and with couples who shared their values and life style and engaging in couple-based friendships.
- Finally, DCCs were confronted with two types of *role-cycling conflicts*, specifically the career-family dilemmas and the dual-career cycling dilemmas. The first

type of role-cycling conflict refers to a decision whether to curtail career engagement in favor of family commitment. The latter refers to decisions regarding job offers and the difficulty of accommodating the demands of two careers.

Rapoport and Rapoport's (1969) study on dual-career families was a guideline for further research. All the dilemmas reported by these authors can still be observed in DCC arrangements. Moreover, the results from their interview study show that whether couples are successful and feel comfortable with this partnership constellation or not largely depends on their joint interests and on their optimism and skills in coping with the challenges arising from the DCC arrangement.

10.4.2 Later Research

Researchers have pointed out that making the choice to live in a dual-career arrangement offers numerous advantages but at the same time creates several sources of conflict and stress (e.g., Green and Zenisek, 1983). The advantages of DCCs include, for example, the optimization of family income (see above, joint utility), similar power and influence between partners (see above, social exchange), gender equality, personal growth, high autonomy, and mutual respect (Abele, 2009; Behnke and Meuser, 2003a, b; Boehnke, 2007). The potential disadvantages concern first of all the above mentioned "overload" dilemma (Rapoport and Rapoport, 1969), e.g., high stress levels. They further concern difficulties in arranging two careers, and possibly reduced career progress for both partners. Finally, the partnership itself could suffer under the "overload dilemma" and/or DCCs could refrain from becoming parents. All these aspects can be discussed with respect to possibly different effects on men and women. In the following, we will review empirical findings on these topics. Although a detailed consideration of country-specific employment and family structures is beyond the scope of this chapter, it is important to mention that cultural differences can also shape women's and men's experiences in dual-career constellations (Esping-Andersen, 1999).

10.5 Empirical Findings on the Specific Challenges of DCCs

10.5.1 Stress and Well-Being

Dual-career couples have to fulfill multiple roles and are confronted with demands and stressors from the work *and* private life domains (Domsch and Ladwig, 2007). They are highly committed to their job, respect their partner's career, and strive towards a satisfactory partnership and private life (family, leisure, friends, etc.). Yet, the weekly time budget is limited and much of this time budget is already dedicated to the work domain, leaving only a limited amount of remaining time. Researchers have considered role strain and role conflict, work-family conflict, and mediating variables (e.g., social support) and its effects on DCCs' stress versus well-being.

Many studies have shown that dual careers frequently provide couples with several benefits and compensations but that quantitative and qualitative overload may cause stress. Elloy and Smith (2003), for instance, found that DCCs experienced more stress, work-family conflict, family conflict, role ambiguity, role conflict and overload than single-career couples. Parasuraman et al. (1992) differentiated between work and family role stressors and job and family satisfaction, and found that the effects of the within-domain stressors on well-being were stronger than between-domain relationships. Job stressors had a negative impact only on job satisfaction and family stressors decreased only family satisfaction. However, stressors from different life domains added to overall life stress. Parasuraman et al. (1992) further found that work support was associated with increased job satisfaction, while spouse support was associated with more family satisfaction. A study on DCCs by Aryee and Luk (1996) showed that women experienced higher levels of work-family conflict than men. Crossfield et al. (2005) found a strong positive association between women's work stressors and the anxiety and depression reported by their male partners, but they found only modest crossover from men's work stressors to women. To summarize, most of the studies have supported the view of DCCs being stressed and overloaded. Study findings show that the effects of within-domain stressors are stronger than between-domain stressors.

A different line of research, however, emphasizes the positive effects the engagement in two domains can have and suggests that experiences from both domains can be mutually rewarding (Greenhaus and Powell, 2006). For example, individuals with high job autonomy and a strong network have been found to be more satisfied with childcare and have healthier children. Vice versa, positive family experiences and partner support were shown to be positively related to work success (Friedman and Greenhaus, 2000). Future research is warranted to determine under which conditions positive or negative effects of living in a DCC arrangement on stress and well-being occur.

10.5.2 Career and Mobility Issues

Geographical mobility is an important precondition for career development, especially for high potentials (Ackers, 2004; Challiol and Mignonac, 2005; Robert and Bukodi, 2002). Early studies on mobility issues showed that people relocate for mainly two reasons; for career enhancement or to stay employed (Bielby and Bielby, 1992). Some research revealed that women were less willing to accept job transfers than men (Landau et al., 1992). Men's careers were often given priority and therefore women decided not to relocate (Abele, 1996; Behnke and Meuser, 2003a, b). Relocation decisions in DCCs are especially complicated as compatibility issues have to be considered (e.g., Challiol and Mignonac, 2005). Even DCCs often give priority to the man's career, suggesting that traditional gender roles still exist (Valcour and Tolbert, 2003). Ackers (2004) investigated the mobility of EU DCCs working in science. They found that not only parenthood but even partnering had a strong impact on career decision making. Specifically, the results of that study show that living in a dual scientist partner constellation had a detrimental effect on a female's career success. Women more often than men made compromises in favor of their partner's career. Robert and Bukodi (2002) investigated the effects of spouse's resources on career success in the former communist Hungary. Their findings demonstrated that the spouse's occupational and informational resources exerted a positive impact on the other partner's upward career moves, supporting advantage accumulation (i.e., education and profession) and status similarity frameworks. Working in the same discipline (Burkett et al., 1999) or with the same employer (Moen, 2002) can also be advantageous for career progression.

10.5.3 Marital Satisfaction and Partnership Stability

Green and Zenisek (1983) reviewed DCC literature with respect to marital satisfaction. They found negative effects in older studies (e.g., Axelson, 1963). Partners were less satisfied with their dual-career partnership in the early days of this form of relationship. This may be due to the above reported "normative dilemma", perhaps also to the "identity dilemma" and "social network dilemma". Later studies revealed no effects or even positive effects of DCC on marital satisfaction (e.g., Yogev, 1981). Green and Zenisek (1983) term this trend "an evolution of adjustment" (p. 174). We guess that the normative dilemma became smaller, and the identity dilemma also became less pronounced when DCC was not "exotic" any more but became "normal" in the sense of more frequent. Parasuraman et al. (1992) found that spouse support (i.e., informational and emotional support) predicts family satisfaction of DCCs, and that this effect is more pronounced for women than for men. Burley (1995) found a negative effect of work-family conflict on marital adjustment, whereas respondents had higher marital adjustment levels when they experienced career social support from their partners and felt that household duties were distributed equally. In our own research we found that DCCs were less satisfied with how much leisure time they could spend together with their partner than other couples. However, DCCs were more satisfied with their partnership than other couples (Weerda, 2003).

To summarize, we assume that partnership satisfaction of DCCs is not per se higher or lower than partnership satisfaction in non-dual-career couples, but that there are rather chances and risks in both forms of partnership as we have outlined above. DCCS may suffer from time constraints in several respects (partner, family, friends, leisure, etc.) but they also gain from more money, shared values and interests, autonomy, and gender and power equality.

10.5.4 Parenthood

Rounding out the above "dilemma" list provided by Rapoport and Rapoport (1969) one could add the important "parenthood dilemma" for DCC couples. The high time constraints under which DCCs live, the difficulties in arranging two careers, and the detrimental effects discontinuity has on a person's career success, make it extremely difficult to combine DCC with parenthood. DCCs practice different strategies for dealing with this "parenthood dilemma". First, one member of the DCC, usually the woman, interrupts her career or at least reduces her work-load when a child was born. Research on career patterns revealed that women more often than men interrupt their careers (Abele and Spurk, in press; Gattiker and Larwood, 1990; Melamed, 1995, 1996; Tharenou et al., 1994; Williams and Han, 2003). This arrangement was described as a return to more traditional gender roles (Abele, 2009; Künzler, 1994). Lower career success of the partner staying home is the consequence (Abele and Spurk, in press). Second, the decision to become parents is postponed until a later age (Swiss and Walker, 1993). Partners assume that their careers will be more settled than at career entry and having a child will be less "detrimental" for the mother's (or the father's) career. The effect is that professionals are becoming older and older when they have their first child (Abele, 2009). In some cases postponing parenthood also results in involuntary childlessness due to medical reasons. Third, the dual-career partners often decide to stay childless. Several findings suggest that the general decrease in birth-rates in Western countries is especially pronounced for highly educated women living in DCC arrangements (Blossfeld and Drobnic, 2001; Blossfeld and Timm, 2003). When childlessness is due to pragmatic considerations of compatibility of careers and family, people experience these decisions as painful (Solga and Wimbauer, 2005).

10.5.5 Gender and Dual-Career Couples

Women more often live in DCC arrangements than men. This is due to gender differences in partner selection. Highly educated women almost always live with partners who are as educated as they are and who almost always work full time, which is called assortative mating. Highly educated men also often practice assortative mating, but they also live with partners who are less highly educated than they are and who do not work full-time (e.g., Blossfeld and Timm, 2003). Gender roles have also changed more with respect to women than with respect to men (Abele, 2000, 2003; Diekman and Eagly, 2000). Finally, value shifts regarding gender equality primarily concern women, as women's status had to be made equal to men's status.

As has been outlined above, DCC arrangements often reflect – among others – such a gender-equality concern of both partners. However, even in DCC arrangements, gender differences in, for instance, the distribution of labor in the household as to the time spent for child care or elderly care are such that women engage in these activities more than men (e.g., Burley, 1995). Also so-called "compatibility management" is performed more by women than by men (Ackers, 2004; Behnke and Meuser, 2003a). Women are more concerned than men with coordinating their careers with those of their partners and more often than men (temporarily) reduce their career commitment to place priority on the partner's career. For example, women represent 74% of "second hires" (the second partner hired in a couple recruitment) at universities (Schiebinger et al., 2008). Thus, although dual-career partnership arrangements are highly gender-equalized women are still the ones who perform more family duties, show more social investment in compatibility management, and are more willing to prioritize the partner's career over their own (see also Rusconi and Solga, 2008; Solga and Wimbauer, 2005). The dilemma created by norms expecting women to take care of the family and the housework could still be more salient for women than for men and these gender role expectations might lead women to more engagement in compatibility management, household activities, and the decision to place priority on their partner's career. Recent studies on gender stereotypes have shown that societal changes have led to changes in gender role concepts (Spence and Buckner, 2000; Twenge, 2001), suggesting that partnerships could become more gender-equalized in the future.

10.6 How to Facilitate the Work-Life Integration of DCC

Discussing DCC arrangements, Moen (2003) talks of "linked lives" which means that dual-career partners' lives are embedded with and influenced by each other and that the achievement or failure of dual career arrangements is dependent on a social-relational process in which the interdependence of both partners' lives has to be coordinated and negotiated. Processes that lead to better or worse work-life integration of DCCs and – more generally – to success or failure of this partnership arrangement have to be analyzed on three levels; the individual level, the couple level, and the external level (Rusconi and Solga, 2008). We discuss issues on the individual level (time and stress management, goals etc.), the couple level (negotiation, communication, etc.) and the external level which we conceptualize as organizational human resource management regarding DCCs.

10.6.1 Individual Level

As the partnership arrangement of a DCC is often associated with overload dilemmas, individuals should strive towards effective time management. Time management strategies include planning one's short-term and long-term activities, setting goals, assessing the time used for specific tasks, making to-do lists, and monitoring one's behavior. Time management strategies increase the awareness of time use, show limits of time capabilities, give feedback loops, and guide the investment of effort. Meta-analytical research has demonstrated that time management behaviors relate positively to perceived control of time, job satisfaction, and health, and negatively to stress (Claessens et al., 2007). Training is available for people in order to improve their time management skills (Claessens et al., 2007; Green and Skinner, 2005).

Of course, not all overload dilemmas can be solved by means of time management. However, when individuals are nonetheless stressed and under time pressure, stress management training is recommended. Stress management training can be classified as primary (e.g., job redesign, autonomy change), secondary (e.g., progressive muscle relaxation, cognitive reframing), or tertiary (e.g., recovery interventions, burn-out assistance programs). Especially cognitive-behavioral programs that belong to secondary interventions have been shown to be effective in reducing stress (Richardson and Rothstein, 2008). Cognitive-behavioral programs teach people to change their cognitions about challenging situations from pessimistic and distorted to more optimistic and realistic thoughts. Reframing techniques are effective for both the work and family domain.

We further propose that goals are crucial in facilitating the work-life integration of DCCs. Goal setting theory (Locke and Latham, 1990) suggests that people should set specific and difficult goals for themselves. Thus, especially for DCCs with the demand of balancing two careers, it is important to reflect at an early stage upon one's goals and priorities in work and private life. Selecting goals and implementing them will lead to optimal outcomes. Self-management in the form of self-set goals and goal-pursuing behavior has been shown to be positively associated with career and life satisfaction (Abele and Wiese, 2008; Riediger et al., 2005). Sometimes, however, when one cannot implement self-set goals, adaptation of goals will be the best strategy. Adapting one's goals – for example, by means of goal disengagement – was shown to have positive effects on well-being and a variety of other outcomes (Brandstätter and Renner, 1990; Heckhausen et al., 2001; Wrosch et al., 2007).

10.6.2 Couple Level

Often the partnership arrangement of a DCC has not been planned out ahead, but has rather developed. Often, the future partners come to know each other while they are at university and quite often they have even graduated in the same field (Blossfeld and Timm, 2003). Such a partnership arrangement requires time management and a distribution of duties which is perceived as fair by both partners. The "dilemmas" (Rapoport and Rapoport, 1969) arising from such an arrangement are still manageable. If the partners appreciate each other and their DCC arrangement, when they have enough joint interests and are generally optimistic, then they will be able to negotiate the reciprocal duties and especially to talk about the mutual expectations.

Things become more complicated when relocation decisions have to be made and/or when the DCC wants to have a child. Rapoport and Rapoport (1969) termed these the dual-career cycling dilemma and the career-family dilemma. Again it is of utmost importance that the couple communicates intensely about each other's goals and wishes and that the process of "compatibility management" takes place such that each partner experiences it as adequate and fair. It may be helpful to frame this compatibility management in terms of a negotiation. Social psychological research has provided a number of tools for successful negotiations (e.g., Fisher et al., 1981; Thompson, 2005) like, for instance, a "win-win" orientation, optimal preparation, overcoming a fixed-pie bias, expanding the pie, etc. The aim of such a negotiation is to create as many options as possible and then to come to the best solution for both partners. This best solution is specific to the respective couple and there are no predetermined best outcomes.

Looking at the other dilemmas outlined above, some of them seem more easily manageable than others. The overload dilemma, for instance, can be reduced by "outsourcing" quite a few housekeeping tasks. The economic situation of DCCs is often such that the couple can afford this help and they should not hesitate to do so. Practical skills in time management are also helpful (see above). The overload dilemma could also be reduced when partners experience each other as supporting, not only on the practical level of concrete help, but also on the level of emotional support (see above, Friedman and Greenhaus, 2000; Parasuraman et al., 1992).

Normative dilemmas and identity dilemmas have become less during the last 40 years, but they have not disappeared. Seeking social support from the partner and from similar others, together with clearly knowing one's own goals and values, is helpful for coping with these dilemmas.

Finally, social network dilemmas are still as significant as they were 40 years ago; perhaps these dilemmas have even become larger. Time constraints keeping DCCs from frequent interactions with relatives and friends still persist. Due to high mobility requirements and frequent relocation decisions, DCCs not only have too little time to cultivate their friendships but also they are often far apart from their friends and relatives and thus there are also geographical restrictions on the cultivation of social relationships. We regard these social network dilemmas as highly important; more research should be devoted to this issue.

10.6.3 Organizational Level

Organizations have to strive for their goals in a competitive market. In order to be successful organizations have – among other factors – to manage their human resources which means recruiting candidates who are best suited for the position in question as well as developing people working already in the organization. Organizations are not per se interested in managing their human resources such that they will provide tools to help DCCs to arrange their "linked lives" successfully. Rather, organizations will react to the specific needs of DCCs when this group provides a competitive advantage (Kölbl, 2008). There are several advantages conceivable for organizations to invest in human resource management of DCCs. People living in DCC arrangements are highly motivated and want to perform especially well. Furthermore, those people are highly committed to the organization, as one of the most important obstacles to a successful DCC arrangement is finding appointments at the same place or at least nearby.

On the recruiting level, dual-career policies begin with the realization that in order to attract the best candidates employers cannot regard candidates as single people without social ties, but that it is important to consider the candidate's social ties very early in the recruitment process because otherwise the candidate might not accept the job offer. In Germany, for instance, several universities have installed specific dual-career services, because they have acknowledged that "recruiting the best" often means "recruiting the best couple" (e.g., Corpina, 1996; Domsch and Krüger-Basener, 2003; Harvey and Buckley, 1998; Pierce and Delahaye, 1996;

Wolf-Wendel et al., 2003). Specific legal prescriptions have been developed in recent years for increasing the flexibility in hiring couples. One of the most important strategies is twin assignment or dual hiring. Given that women's career decisions are more strongly influenced by partner employment decisions than men's career decisions, twin assignment and dual hires can be considered as enabling a more egalitarian and diverse workforce (Schiebinger et al., 2008). Further strategies are, for instance, support of the partner's job search, flexible work arrangements (e.g., virtual collaboration, flexible working hours, sabbaticals, year wise working hours accounts, etc.), support for child care (e.g., company-run day care for children of different age levels) and support for domestic duties (Domsch and Krüger-Basener, 1999). Universities are becoming more and more aware of the emerging number of academics living in DCCs and the necessity to support them. Having a competitive dual-career program that meets applicants' needs and expectations influences whether qualified candidates can be attracted, since career decisions are strongly influenced by employment opportunities for partners. In the US, dual hires at university faculties have significantly increased from 3% in the 1970s to 13% in the 2000s (Schiebinger et al., 2008). Providing effective dual-career programs can strengthen international competitiveness and prevent "brain-drain" (Backhaus et al., 2002).

Networking is central for dual-career support and includes networks within and between universities and to external institutions (Funk and Gramespacher, 2008). In Germany, for example, some dual-career services have built regional networks of several universities wherein each university has a dual-career service and offers consulting and career advice for scientists (Funk and Gramespacher, 2008).

On the level of retaining and developing staff with high potential, dual-career policies aim at helping the couples both to improve their professional skills further and to arrange their work and private lives. Strategies are again flexible work arrangements (e.g., virtual collaboration, e-learning, flexible working hours and family-friendly meeting hours, shared appointments), and support for childcare and domestic duties (Domsch and Krüger-Basener, 1999). Some best practice examples illustrate how organizations facilitate the arrangement of work and private life. For example, a well-known European furniture company offers part-time and job-sharing models for executives, gives diversity seminars for executives with an emphasis on the acknowledgement of employees' different needs and requirements, holds yearly feedback sessions on work and private situation, offers re-employment advice after parental leave, etc. Some other companies also provide best practice examples like flexible working hours, job-splitting, leadership in part-time employment, web-access, work-life balance seminars, parent network meetings, intranet information on respective issues, etc. Most of these initiatives address women, and organizations place emphasis on topics such as part-time employment, re-employment after parental leave, and childcare.

No doubt, these respective initiatives are helpful for balancing work and family life. However, human resource management with respect to DCCs must be broader, because DCCs' necessities are not confined to work-life balance issues. The partnership arrangement of DCCs can be regarded as a new and alternative career model that challenges the traditional thinking and acting patterns of decision makers. Consequently, organizations have to develop new management concepts that allow employees to combine their high career orientation with a high partnership orientation. Flexible work arrangements have to be changed from mere employment concepts to career-oriented concepts and traditional career concepts have to be adapted such that they fit to modern partnership constellations.

Generally, the management of DCCs on the level of organizations requires flexibility on the part of human resource managers as well as on the part of the couples. Dual career management as a specific component of human resource management will only be successful if two preconditions are met. First, the chief executive officers have to be convinced that dual career management really helps to attain the organizational goals. Second, the organizational culture has to be such that gender equality in career and family is highly valued (Kölbl, 2008). It is important to mention that dual-career programs facilitate dual hires at the same organization or institution (or at least nearby) but that the understanding of these programs is that the final decision of hiring the applicant's partner has to depend on his or her expertise (Schiebinger et al., 2008). Excellence is the criterion for deciding upon hiring the partner, and dual-career programs should be a "win-win" situation for both the employer and the employee.

10.7 Conclusion

Throughout this chapter we have been concerned with dual career couples as if these were a homogeneous group. This is, of course, not the case. Even if we deal with highly educated professionals, DCCs differ with respect to their field of employment and in some of these fields it may be easier to arrange both partners' careers than in others. The upward career orientation may also differ between DCCs. Some may strive to the "top" of career development achievable in their respective profession; some may be career-oriented up to a certain limit. Some couples may perceive their "linked lives" as a unit; other couples may perceive each other as autonomous individuals. One may even think of DCCs as a continuum with – always high educated – dual-earner couples with a clear priority of one partner's career at one end of the continuum and highly upward oriented dual career couples with equal priority of both careers at the other end of the continuum. Compatibility management strategies on the level of the couple are always necessary. Societal support (child care, parental leave, DCC-friendly tax systems, etc.) and societal appreciation of respective ways of living are also always necessary. Organizational strategies of dual career management help a lot and are the more necessary the more the respective DCC belongs to the highly upward oriented couples with equal priority of both careers.

References

Abele AE (1996) Berufliche Laufbahnentwicklung von Akademikerinnen und Akademikern der Universität Erlangen-Nürnberg (BELA-E). Erste Erhebung der prospektiven Längsschnittstudie. University of Erlangen-Nuremberg, Nuremberg, Internal Paper

Abele AE (2000) The dual impact model of gender on career related processes – and the reciprocal impact of career on gender. In Eckes T, Trautner H-M (eds) The developmental social psychology of gender. Lawrence Erlbaum, Mahwah, NJ, pp 361–388

Abele AE (2003) The dynamics of masculine-agentic and feminine-communal traits. Findings from a prospective study. J Pers Soc Psychol 85:768–776

Abele AE (2009) Doppelkarrierepaare – Entstehung und Relevanz der Thematik. In Funk J, Gramespacher E, Rothäusler I (eds) Dual Career Couples in Theorie und Praxis. Verlag B. Budrich, Opladen

Abele, AE, Spurk D (in press) The Dual Impact of Gender on Objective Career Success: Longitudinal Effects of the Gender-Self-Concept and the Timing of Parenthood. International Journal of Behavioral Development.

Abele AE, Wiese B (2008) The nomological network of self-management strategies and career success. J Organ Occup Psychol 81:733–749

Ackers L (2004) Managing relationships in peripatetic careers: scientific mobility in the European union. Women's Stud Int Forum 27:189–201

Apostal RA, Helland C (1993) Commitment to and role changes in dual career families. J Career Dev 20:121–129

Aryee S, Luk V (1996) Balancing two major parts of adult life experience: work and family identity among dual-earner couples. Hum Relat 49:465–487

Axelson LJ (1963) The marital adjustment and marital role definitions of husbands of working and non-working wives. Marriage Fam Living 25:189–195

Backhaus B, Ninke L, Over A (2002) Brain drain – brain gain. Eine Untersuchung über internationale Berufskarrieren. Stifterverband für die Deutsche Wirtschaft. http://www.stifterverband.org/publikationen_und_podcasts/positionen_dokumentationen/braindrain_brain gain_2002.pdf. Accessed on 30 Sept 2009

Becker GS (1981) A treatise on the family. Harvard University Press, Cambridge, MA

Behnke C, Meuser M (2003a) Vereinbarkeitsmanagement. Die Herstellung von Gemeinschaft bei Doppelkarrierepaaren. Soziale Welt 54:163–174

Behnke C, Meuser M (2003b) Doppelkarrieren in Wirtschaft und Wissenschaft. Zeitschrift für Frauenforschung und Geschlechterstudien 21:62–74

Bernardi F (1999) Does the husband matter?. Eur Soc Rev 15:285–300

Bielby WT, Bielby DD (1992) I will follow him: family ties, gender-role beliefs, and reluctance to relocate for a better job. Am J Sociol 97:1241–1267

Blossfeld H-P, Drobnic S (2001) Careers of couples in contemporary societies. Oxford University Press, Oxford

Blossfeld H-P, Timm A (2003) Who marries whom? Educational systems as marriage markets in modern societies. Kluwer, Dordrecht

Boehnke M (2007) Männerwelten – Frauenwelten. Dual Career Couples im deutsch-deutschen Vergleich. VDM, Saarbrücken

Brandtstädter J, Renner G (1990) Tenacious goal pursuit and flexible goal adjustment: explication and age-related analysis of assimilative and accommodative strategies of coping. Psychol Aging 5:58–67

Burkett SL, Lusth JC, Ruhala LA, Ruhala RJ, Vrbsky S, Hyslop B (1999) Solving the dual-career dilemma: Three case-studies. Paper presented at the 1999 ASEE Annual Conference and Exposition: Engineering Education to Serve the World; Charlotte, NC (June 1999)

Burley KA (1995) Family variables as mediators of the relationship between work-and-family conflict and marital adjustment among dual-career men and women. J Soc Psychol 135: 483–497

Challiol H, Mignonac K (2005) Relocation decision-making and couple relationships: a quantitative and qualitative study of dual earner couples. J Organ Behav 26:247–274

Claessens BJC, van Eerde W, Rutte CG, Roe RA (2007) A review of time management literature. Pers Rev 36:255–276

Corpina P (1996) Laufbahnentwicklung von Dual-Career Couples. Gestaltung partnerschaftsorientierter Laufbahnen. Difo-Druck, Bamberg

Crossfield S, Kinman G, Jones F (2005) Crossover of occupational stress in dual-career couples. Commun Work Fam 8:211–232

Deaux K, LaFrance M (1998) Gender. In Gilbert DT, Fiske ST, Lindzey G (eds) The handbook of social psychology, vol 1. McGraw-Hill, New York, NY, pp 788–827

Diekman AB, Eagly AH (2000) Stereotypes as dynamic constructs: women and men of the past, the present and the future. Pers Soc Psychol Bull 26:1171–1188

Domsch ME, Krüger-Basener M (1999) Personalplanung und -entwicklung für Dual Career Couples (DCCs) In: Rosenstiel L; Regnet E, Domsch ME (eds) Führung von Mitarbeitern. Handbuch für erfolgreiches Personalmanagement. Schäffer-Poeschel, Stuttgart, pp 547–558

Domsch ME, Krüger-Basener M (2003) Personalplanung und -entwicklung für Dual Career Couples (DDCs). In von Rosenstiel L, Regnet E, Domsch ME (eds) Führung von Mitarbeitern. Handbuch für erfolgreiches Personalmanagement. Schäffer-Poeschel, Stuttgart, pp 475–488

Domsch ME, Ladwig DH (2007) Doppelkarrierepaare – Eine Herausforderung für die betriebliche Familienpolitik. In Dilger A, Gerlach I, Schneider H (eds) Betriebliche Familienpolitik. Potenziale und Instrumente aus multidisziplinärer Sicht. Verlag für Sozialwissenschaften, Wiesbaden, pp 75–85

Eagly A (1987) Sex differences in social behaviour. A social role interpretation. Lawrence Erlbaum, Hillsdale, NJ

Elloy D, Smith C (2003) Patterns of stress, work-family conflict, role conflict, role ambiguity and overload among dual-career and single-career couples: an Australian study. Cross Cult Manage 10:55–66

Emerson R (1976) Social exchange theory. Ann Rev Sociol 2:335–362

Esping-Andersen G (1999) Social foundations of postindustrial economies. Oxford University Press, Oxford

Eurostat (2009) European statistical data support. http://epp.eurostat.ec.europa.eu/tgm/table.do?tabl=table&init=1&plugin=1&language=en&pcode=tsiem010. Accessed on 27 Aug 2009

Fisher R, Patton B, Ury W (1981) Getting to yes. Negotiating agreement without giving in. Houghton Mifflin & Co, Boston, MA

Friedman SD, Greenhaus JH (2000) Work and family – allies or enemies? What happens when business professionals confront life choices. Oxford University Press, New York, NY

Funk J, Gramespacher E (2008) Netzwerke bilden. Wie Hochschulen Dual Career Couples fördern können. Forschung und Lehre 10:691

Gattiker UE, Larwood L (1990) Predictors of career achievement in corporate hierarchy. Hum Relat 43:703–726

Gerhard U, Knijn T, Weckwert A (2003) Erwerbstätige Mütter. Ein europäischer Vergleich. Beck, Munich

Green D, Zenisek T (1983) Dual career couples: individual and organizational implications. J Bus Ethics 2:171–184

Green P, Skinner D (2005) Does time management training work? An evaluation. Int J Train Dev 9:124–139

Greenhaus JH, Powell GN (2006) When work and family are allies: a theory of work-family enrichment. Acad Manage Rev 31:72–92

Harvey MG, Buckley MR (1998) The process for developing an international program for dual-career couples. Hum Resour Manage Rev 8:99–123

Heckhausen J, Wrosch C, Fleeson W (2001) Developmental regulation before and after a developmental deadline: the sample case of "biological clock" for childbearing. Psychol Aging 16:400–413

Kölbl S (2008) Entwicklung eines Konzepts für das Management hoch qualifizierter Humanressourcen. University Press, Kassel

Künzler J (1994) Familiale Arbeitsteilung. Die Beteiligung von Männern an der Hausarbeit. Kleine, Bielefeld

Landau JC, Shamir B, Arthur MB (1992) Predictors of willingness to relocate for managerial and professional employees. J Organ Behav 13:667–680

Locke EA, Latham GC (1990) A theory of goal setting and task performance. Prentice-Hall, Englewood Cliffs, NJ

Melamed T (1995) Career success: the moderating effect of gender. J Vocat Behav 47:35–60

Melamed T (1996) Career success: an assessment of a gender-specific model. J Occup Organ Psychol 69:217–242

Moen P (2002) Two careers, one employer: couples working for the same corporation. J Vocat Behav 61:466–483

Moen P (2003) It's about time. Couples and careers. ILR Press, Ithaca, NY

Oppenheimer VK (1997) Women's employment and the gain to marriage: the specialization and trading model. Ann Rev Sociol 23:431–453

Parasuramanan S, Greenhaus JH, Granrose CS (1992) Role-stressors, social support, and well-being among two-career couples. J Organ Behav 13:339–356

Pierce J, Delahaye BL (1996) Human resource management implications of dual-career couples. Int J Hum Resour Manage 7:905–923

Rapoport R, Rapoport RN (1969) The dual-career family: a variant pattern and social change. Hum Relat 22:3–30

Richardson KM, Rothstein HR (2008) Effects of occupational stress management intervention programs: a meta-analysis. J Occup Health Psychol 3:69–93

Riediger M, Freund AM, Baltes PB (2005) Managing life through personal goals: intergoal facilitation and intensity of goal pursuit in younger and older adulthood. J Gerontol Psychol Sci 60:84–91

Robert P, Bukodi E (2002) Dual career pathways. The occupational attainment of married couples in Hungary. Eur Soc Rev 18:217–232

Rusconi A (2003) Alters- und Erwerbskonstellationen in Akademikerpartnerschaften – Empirische Analysen mit dem Mikrozensus 1997. http://www.diejungeakademie.de/pdf/Rusconi_Mikrozensus.pdf. Accessed on 24 Aug 2009

Rusconi A, Solga H (2008) A systematic reflection upon dual career couples. Discussion Paper, Social Science Research Center Berlin (WZB), Berlin

Schiebinger L, Henderson AD, Gilmartin S (2008) Dual-career academic couples: what universities need to know. Stanford University, Palo Alto, CA

Solga H, Wimbauer C (2005) "Wenn zwei das gleiche tun". Ideal und Realität sozialer (Un-)Gleichheit in Dual Career Couples. Verlag B. Budrich, Opladen

Spence JT, Buckner CE (2000) Instrumental and expressive traits, trait stereotypes, and sexist attitudes. Psychol Women Q 24:44–62

Swiss D, Walker J (1993) Women and the work/family dilemma. How today's professional women are finding solutions. Wiley, New York, NY

Tharenou P, Latimer S, Conroy D (1994) How do you make it to the top? An examination of influences on women's and men's managerial advancement. Acad Manage J 37:899–931

Thompson LL (2005) The mind and heart of the negotiator. Prentice Hall, New Jersey

Twenge JM (2001) Changes in women's assertiveness in response to status and roles: a cross-temporal meta-analysis, 1931–1993. J Pers Soc Psychol 81:133–145

US Bureau of Labor Statistics (2009) Household data. Employment status of the civilian, noninstitutional population by age, sex, and race. http://www.bls.gov/web/cpseea13.pdf. Accessed on 27 Aug 2009

Valcour M, Tolbert P (2003) Gender, family and career in the era of boundarylessness: determinants and effects of intra- and inter-organizational mobility. Int J Hum Resour Manage 14:768–787

Walther K, Lukoschat H (2008) Kinder und Karrieren: Die neuen Paare. Bertelsmann, Gütersloh

Weerda K (2003) Persönliche Ziele, Berufserfolg und subjektives Wohlbefinden. Untersuchung an Unternehmensberatern. Dissertation,University of Erlangen-Nuremberg

Williams S, Han SK (2003) Career clocks: forked roads. In Moen P (ed) It's about time. Couples and careers. ILR Press, Ithaca, NY, pp 80–97

Wolf-Wendel L, Twombley SB, Rice S (2003) The two-body problem: dual-career-couple hiring policies in higher education. John Hopkins University Press, Baltimore, MD
Wrosch C, Miller GE, Scheier MF, de Pontet SB (2007) Giving up on unattainable goals: benefits for health?. Pers Soc Psychol Bull 33:251–265
Yogev S (1981) Do professional women have egalitarian marital relationships?. J Marriage Fam 43:865–871

Part IV
Corporate Strategies Facilitating Work-Life Integration

Chapter 11
Corporate Work-Life Balance Initiatives: Use and Effectiveness

T. Alexandra Beauregard

Managers and professionals are uniquely placed to experience difficulties combining work responsibilities with non-work commitments, due to long hours of work and extensive use of communications technologies enabling work tasks to be performed at any time, anywhere. In response to a workforce increasingly concerned with maintaining a work-life balance, organisations now offer a range of initiatives designed to facilitate the integration of work and non-work domains. These initiatives usually take the form of flexible work options, family or personal leave, and organizational assistance with childcare or eldercare (Estes and Michael, 2005). Common initiatives include flexible work hours, telework (working from home or a satellite location), job sharing (sharing a full-time job between two employees), voluntary reduced work hours, parental leave, and financial and/or informational assistance with childcare and eldercare services.

This chapter will examine the antecedents to and outcomes of corporate work-life balance initiative usage among professionals, as well as equity concerns and "backlash" among non-users of such initiatives. Interconnecting influences such as gender, diversity, values, and dispositional characteristics will be explored, and their role in the use and effectiveness of corporate work-life balance initiatives analysed. The chapter will seek to highlight some of the implications of the research for policy implementation, with an aim to improving the likelihood of work-life balance initiatives achieving their intended purpose among managers and professionals: enhancing work-life balance and strengthening performance on the job.

11.1 Who Uses Work-Life Balance Initiatives?

First of all, it is worth noting that work-life balance initiatives frequently experience low levels of take-up among staff eligible to use them (Pocock, 2005). There are a number of reasons why this is the case. Some employees have no perceived need or desire to use the initiatives. Some employees are concerned about the potential repercussions of using the initiatives, as will be discussed later in this chapter. Some

T.A. Beauregard (✉)
London School of Economics, London, UK
e-mail: a.beauregard@lse.ac.uk

S. Kaiser et al. (eds.), *Creating Balance?*, DOI 10.1007/978-3-642-16199-5_11,

employees, however, are simply unaware that the initiatives exist, or that they are entitled to use them (Kodz et al., 2002; Lewis et al., 2000). It is not always clear to staff what initiatives are available, or exactly what these initiatives entail (Duxbury et al., 2003).

Women, however, are more likely both to have knowledge of the work-life initiatives offered by their employer (Blair-Loy and Wharton, 2002; Prottas et al., 2007) and to take up these initiatives (Houston, 2005), as are those with longer tenure in the organization (Lambert et al., 2008). Managers and professionals tend to have greater access to work-life balance initiatives than non-supervisory white collar employees and blue collar workers (Swanburg et al., 2005), and managers have also been found to use these initiatives to a greater extent than employees with no supervisory responsibilities (Lambert et al., 2008). Two important predictors of usage appear to be personal commitments outside of work, and the climate for initiative use within the organization. Individuals with young children, those who provide care to elderly, ill, or disabled relatives, and those who have non-family related commitments outside of work are more likely to use flexible work arrangements, leave, and referral services (Blair-Loy and Wharton, 2002; Lambert et al., 2008). In their study of academics, Shockley and Allen (in press) found a positive relationship between need for occupational achievement and use of flexible work hours when family responsibility was high; similarly, for academics with high family responsibility, their need to segment work from other life roles played a less important role in predicting their use of telework. As for the importance of the social context, managers whose colleagues use flexible work arrangements are more likely to take up these practices themselves (Kossek et al., 1999; Lambert et al., 2008). Research by Allen (2001) found that workers who perceived their organizations to be more supportive of family concerns were more likely to use the work-life balance initiatives available to them.

Although work-life balance initiatives are often targeted at highly skilled workers in order to improve recruitment and retention (Gray and Tudball, 2003), managers and professionals don't always feel capable of using these initiatives due to organisational pressures for long work hours and the domination of workplace values negating work-life balance, such as availability and presenteeism (Blair-Loy, 2001; Fried, 1998; Perlow, 1997). Blair-Loy and Wharton (2002) note that organizations offering work-life balance initiatives enjoy enhanced legitimacy and a reputation as desirable employers, but that these organizations may also dissuade staff from using the initiatives; their study of professionals and managers at an international bank renowned for being "family-friendly" revealed that staff received ambiguous and contradictory messages about using the wide assortment of initiatives on offer. Long work hours and high levels of dedication were demanded of managers and professionals, and as a result, nearly two-thirds of the study sample reported that taking an extended parental leave or setting limits on the hours they spent at work would hurt their career advancement. Here again we see evidence of how significant the social context of the work group can be. Blair-Loy and Wharton (2002) found that managers and professionals are more likely to use work-life balance initiatives when they are protected from perceived negative career consequences by the buffering effect of working with powerful supervisors and colleagues – i.e., work groups with longer tenure, composed primarily of men, with fewer family responsibilities.

Concerns regarding career penalties feature heavily in much of the research on work-life balance initiative usage. Shockley and Allen's (in press) study of academics found that a workplace emphasis on "face time" predicted low use of flexible work hours, and for professionals with few family responsibilities, the need for occupational achievement was also negatively linked to flexible work hour use. Butler et al. (2004) found that more positive work outcome expectancies, i.e., no harm to status or career progression, were related to higher rates of work-life balance initiative use. The latter authors also found a significant effect of worker disposition: employees with higher work-family self-efficacy – a stronger perceived ability to handle the conflicting demands of work and family – reported greater intentions to use the initiatives on offer.

Another recurring issue in this area is gender, and this issue is connected to employee concerns regarding career penalties. Compared to women, few men take up the work-life balance initiatives offered by their employing organizations, instead using vacation or sick leave when family commitments necessitate time away from work (Berry and Rao, 1997; Bygren and Duvander, 2006; Pleck, 1993). In Bygren and Duvander's (2006) study of American couples, the women in the sample were the ones who made the most changes to their work schedules to accommodate family responsibilities, in large part because the men had fewer work-life balance initiatives available to them and because their wages and benefits were higher and thus unpaid leave was not an appealing option. The authors' research found that men employed in the private sector, at small workplaces, and in workplaces dominated by men were less likely to use parental leave, as were those employed by organizations where other men had not previously taken up their parental leave entitlement to a large extent. Fried's (1998) study revealed that men were informally discouraged from taking family-related leave, as were women working in male-dominated positions.

One of the main explanations for why men do not use a larger share of the leave available to them is that they find it difficult to be absent from their workplaces (Brandth and Kvande, 2001, 2002; Haas et al., 2002). Liff and Cameron (1997) suggest that both men and women with career aspirations will be reluctant to use work-life balance initiatives due to fear of being perceived as less committed to their organization. This proposition has received support in a number of studies. Brandth and Kvande's (2002) study of Norwegian working fathers found that as men progress up the managerial career ladder, they exhibit a reduced tendency to use the paternity leave to which they are entitled. Bittman et al. (2004) case studies of two large Australian organizations showed that employees at all levels of the organizations thought that taking leave or reducing work hours could irreparably damage men's careers. Nord et al.'s (2002) case studies of two American firms revealed employee impressions that use of the companies' work-life balance initiatives would lead to being perceived by peers as "slackers", or as poor team players, and that men, more than women, felt that using the initiatives would have a negative impact on their career progression.

Given these findings, it is unsurprising that work-life balance initiatives are often under-utilized by men, as well as by single employees and career-oriented women (Bailyn et al., 1997; Whitehouse and Zetlin, 1999). As Fried (1998) points out, organizational culture often produces different behaviour in men and women even when work-life balance initiatives are ostensibly gender-neutral. The perception often exists in organizations that work-life balance initiatives are intended primarily for the benefit of working mothers. As a result, when men take up these initiatives, their use can be seen as unusual and subject to question (Daly et al., 2008). In Bittman et al. (2004) Australian case studies, among the workplace barriers discouraging men's take-up of work-life balance initiatives were the novelty of men's use of such initiatives, and doubts about the legitimacy of men's claims to family responsibilities. Men are rarely regarded as primary caregivers for children or the elderly; consequently, their desire to take time away from work for family reasons can be seen as an indulgence (Daly et al., 2008).

Clearly, there are a number of factors predicting managers' and professionals' use of work-life balance initiatives. Chief among these are commitments outside of work, usually in the form of caregiving responsibilities, and organizational climate for initiative use, which manifests itself most frequently in employee concerns regarding the effects of use on career progression. In an ironic twist, anxiety regarding negative career repercussions of initiative use has been linked to higher levels of work-life conflict (Anderson et al., 2002; Thompson et al., 1999), the very phenomenon work-life balance initiatives were designed to eradicate. In the next section, research on the outcomes of initiative use will be reviewed: Does using these initiatives enhance work-life balance for managers and professionals, and what are the consequences for their performance on the job?

11.2 What Are the Outcomes for Managers and Professionals Using Work-Life Balance Initiatives?

According to Fleetwood (2007) corporate work-life balance initiatives are simply employer-friendly work practices in disguise, and do little to improve work-life balance for users. While examining the effects of these initiatives on organizational performance is beyond the scope of this chapter (see Beauregard and Henry, 2009 for a review of how firms benefit from offering work-life balance practices), we will now examine evidence for the assertion that initiative usage is of little help to managers and professionals struggling to combine work responsibilities with non-work commitments. Much of the research investigating the effects of corporate work-life balance initiatives examines availability rather than use, rendering the results more helpful to organizations pondering the advantages of offering initiatives than to individual managers and professionals pondering the advantages of using them. There have, however, been a few studies focused specifically on initiative use.

11.2.1 Effects of Initiative Use on Work-Life Balance

Using flexible work hours has been linked to lower levels of work-to-life conflict (Anderson et al., 2002; Hill et al., 2001) and non-directional work-life conflict (Lee and Duxbury, 1998; Saltzstein et al., 2001). In a study of IBM teleworkers, Hill et al. (2003) found that working from home was a significant predictor of work-life balance and perceived success in one's personal and/or family life. In Kossek and Nichol's (1992) study of a corporate onsite childcare centre, users of this service reported a greater ability to balance multiple roles than that reported by non-users.

Aside from these main effects of initiative use on work-life outcomes, there is also evidence of indirect effects. Research by O'Driscoll et al. (2003) showed that the relationship between employee use of work-life balance initiatives and work-to-life conflict was mediated by perceptions of the organization as being supportive of family concerns. In Shockley and Allen's (2007) study of female managers and professionals, use of flexible work hours and ability to work from home was associated with reduced work-to-life conflict and life-to-work conflict for those with greater family responsibilities. For those with few family responsibilities, however, use of these work-life balance initiatives was linked to higher levels of work-life conflict.

It would seem, then, that using work-life balance initiatives can help employees balance their work and non-work demands. But what is the impact of initiative use on managers' and professionals' ability to meet work demands? In the next section, we will inspect the links between initiative use and job performance, focusing on telework, flexible work hours, childcare provision, and voluntary reduced hours as the initiatives most commonly investigated in the research literature.

11.2.2 Effects of Initiative Use on Job Performance

Telework, in which employees work from locations other than the office (usually at home), has a mixed reputation when it comes to employee performance. On the one hand, there are a number of research reviews showing positive links between use of telework and productivity (Kurland and Bailey, 1999; Pitt-Catsouphes and Marchetta, 1991). For instance, studies employing self-report measures of productivity frequently show positive effects of telework use upon performance (Callentine, 1995; Hill et al., 1998), and both quantitative and qualitative research has found that participation in telework programs was associated with higher performance ratings from supervisors (Frolick et al., 1993; Kossek et al., 2006). These positive effects may be dependent upon who was responsible for initiating the telework arrangements. Hartman et al. (1991) found that when supervisors imposed telework arrangements upon employees, more time spent teleworking was associated with lower productivity. Higher productivity levels were reported by those participating in employee-initiated or mutually-initiated telework programs.

On the other hand, there is evidence to suggest that telework can present a number of drawbacks to users of this work-life balance initiative. In Nord et al. (2002) qualitative study of management consultants, those participating in telework practices reported experiencing isolation, reduced social interaction with their colleagues, and difficulties in achieving team synergy due to their reliance on remote modes of communication such as e-mail, instant messaging, and the telephone. Professional isolation has been identified elsewhere in the research as a pitfall of telework, as have reduced informal learning, less participation in mentoring relationships, decreased teamwork, and negative effects on professional development activities like interpersonal networking (Cooper and Kurland, 2002; Kurland and Bailey, 1999).

Flexible work hours have also been associated with greater self-reported effort and productivity (Chow and Keng-Howe, 2006; Nord et al., 2002), although in their review of studies carried out by Dunham et al. (1987) and Pierce and Newstrom (1982, 1983), Kossek and Ozeki (1999) suggested that more limited amounts of flexibility, wherein workers specified in advance which hours they would work, may produce higher levels of performance than varying schedules on an ad-hoc basis. In addition to increased productivity, using flexible work hours has also been linked to reduced absenteeism (Dalton and Mesch, 1990; Halpern, 2005; Krauz and Freibach, 1983). This relationship may be strongest when employees are not required to obtain approval from their managers for the flexibility required (Pierce and Newstrom, 1983).

Childcare provision, like telework, has received mixed reviews when it comes to employee performance. A study of on-site childcare by Kossek and Nichol (1992) found no effects of use on employees' performance as rated by supervisors, although users of the childcare program rated the quality of their performance more highly than non-users did of their own performance. While some studies have found that using on-site childcare centres reduces levels of absenteeism (Auerbach, 1990; Milkovich and Gomez, 1976), others have found no relationship between childcare centre use and absence at work (Goff et al., 1990; Kossek and Nichol, 1992; Thomas and Ganster, 1995). This discrepancy in findings might be explained by the role of work-life conflict as a mediator in the relationship between childcare use and absenteeism. Users of on-site childcare centres will only experience lower levels of work-life conflict if the childcare provided is the most satisfactory of all childcare options available to the employee, and only then will absence at work be reduced (Goff et al., 1990).

Voluntary reduced hours of work have not often been studied in isolation (i.e., apart from other work-life balance initiatives), so it is difficult to determine their effect on managers' and professionals' performance at work. Research by Lewis (1997) found that chartered accountants working reduced hours on a voluntary basis reported greater productivity and efficiency on the job. Nord et al. (2002), however, found that reduced hours of work, accompanied by reduced salaries, did not necessarily result in reduced work or reduced goals. Research participants identified a number of instances in which their firms' human resources policies did not make sufficient adjustments to evaluation or compensation of work in the context of fewer contracted hours.

In addition to studies considering these four work-life balance initiatives separately, there exists research that examines bundles or groupings of practices and their joint effects on employee performance. Some of this research has linked use of work-life balance initiatives to improved self-reported performance, in the form of greater focus, concentration, and motivation (Raabe, 1996; Williams et al., 2000). The perceived usefulness of work-life balance initiatives has been associated with greater participation in organizational citizenship behaviours (Lambert, 2000), and the use of a range of initiatives (including healthcare, fitness, and education benefits) has been related to both increased task performance and contextual performance (as represented by interpersonal facilitation and job dedication), mediated by affective commitment to the employing organization (Muse et al., 2008).

11.2.3 Effects of Initiative Use on Organizational Commitment

The increased commitment of initiative users to the initiative providers is a recurring theme in the work-life balance literature. In general, employees who use onsite childcare centres, referral services, and other family-supportive initiatives report higher levels of commitment to their employing organization (Goldberg et al., 1989; Grover and Crooker, 1995; Orthner and Pittman, 1986; Youngblood and Chambers-Cook, 1984). Using flexible work hours has also been linked to greater organizational commitment, as well as decreased intentions to leave the organization (Aryee et al., 1998; Halpern, 2005; Houston and Waumsley, 2003; Nord et al., 2002).

Some indirect relationships between work-life balance initiatives and organizational commitment have also been found. In a study of knowledge workers in Scotland, flexible work hours were positively related to organizational commitment and extrinsic job satisfaction, mediated by trust in the organization (Scholarios and Marks, 2004). Casper and Harris (2008), meanwhile, found that for the men in their study, the availability of flexible work hours was positively related to organizational commitment only when use of flexible work hours was high. When use was low, the availability of this initiative was negatively associated with organizational commitment, suggesting that increased loyalty to the employer may only be generated among users and future users of initiatives, rather than all employees.

Some caveats exist to the link between initiative use and organizational commitment. Eaton's (2003) study of professional and technical employees in biopharmaceutical companies found that the availability of work-life balance initiatives led to enhanced organizational commitment, but only when employees felt that using the initiatives would not result in negative work outcomes, such as career penalties or deterioration in relationships with co-workers. Nord et al. (2002) observed that even where initiative use was supported by the organization, difficulties could arise for employees when the initiatives were not compatible with the organization's external environment, particularly in terms of client expectations for uninterrupted access to the management consultants in their study. A number of Nord et al.'s study participants also expressed concern about the effects of work-life balance initiative use on their prospects for promotion and career advancement. Are career concerns such as these justified? The next section will examine the evidence for negative consequences of initiative use on career progression.

11.2.4 Effects of Initiative Use on Career Advancement

Due to the time and workload pressures inherent in many managerial and professional jobs, decreased-workload options such as voluntary reduced hours are not often available to employees in the upper echelons of their occupation. Researchers have argued that when these types of work-life balance initiatives are made available to managers and professionals, usage is frequently associated with negative career outcomes (Raabe, 1996). Work-life balance initiatives such as telework, flexible work hours, and family leave contribute to reduced visibility of the employees using them. This can present problems when time spent in the workplace is used as a key indicator of employee commitment and contributions to the organization. As a result, the use of work-life balance initiatives that remove managers from their regular workplace has been linked to fewer promotions, poorer performance evaluations, and reduced salary increases (Bailyn, 1997; Judiesch and Lyness, 1999; Perlow, 1995). Results such as these have been found in both experimental and observational research. For example, experiments have shown employees using work-life balance initiatives are perceived by co-workers as being less committed to the organization, less ambitious, and less suitable for promotion, despite no differences being perceived in their capability to do the job effectively (Allen and Russell, 1999; Rogier and Padgett, 2004). In the field, research has demonstrated that accountants using flexible work hours are perceived as less likely to be promoted and more likely to leave the organization, while female engineers using flexible work hours may be excluded from decision-making roles or from career advancement opportunities (Cohen and Single, 2001; Watts, 2009).

The impact of work-life balance initiative use on career prospects can also be seen indirectly. Wayne and Cordeiro (2003) investigated the effects of taking family leave on co-workers' perceptions of employees' organizational citizenship behaviours. They found that men who took family leave were assessed as being less likely to work overtime, less likely to be punctual, and less likely to help their colleagues, despite identical performance ratings to men having taken no family leave. As individuals' citizenship behaviours have been shown to influence managerial decisions regarding their performance appraisals and promotability (see Organ et al., 2006, for a review), results such as these reflect another way in which work-life balance initiative use has the potential for negative repercussions. In Nord et al. (2002) study of management consultants, some participants cited instances of colleagues stigmatizing those using flexible work hours and telework initiatives with comments such as "Nice of you to show up today" and "What is he doing all day?" (p. 229). Remarks such as these reveal an underlying resentment toward those using work-life balance initiatives, characterized in the popular press as "backlash". An account of the effects of initiative use on the outcomes of managers and professionals would not be complete without an examination of this phenomenon, presented in the next section.

11.3 Equity Concerns and "Backlash" Among Non-users

Despite recommendations for approaching work-life balance from a perspective of inclusion, many organizations and individual employees continue to see work-life balance initiatives as intended for parents of young children, and particularly for mothers (Visser and Williams, 2006). In some cases, this may be an accurate perception; practices such as on-site childcare centres and financial assistance with dependent care are necessarily targeted at the subset of the employee population with caregiving responsibilities, usually for children. In other cases, even "caring-neutral" practices such as flexible work hours and telework may be marketed to employees as "family-friendly" programs designed primarily for those with family responsibilities, and consequently may be used predominantly by working parents.

According to Adams's (1965) Equity Theory, individuals compare their ratio of outcomes to inputs to that of others in a similar situation, or to a given standard. In other words, they assess the positive or negative returns they receive from a relationship relative to the contributions or investments they make in that relationship. If their inputs exceed those of another individual with the same outcomes, or if their outcomes are less than those of another individual with the same inputs, they will perceive that they are under-rewarded and that inequity exists (Adams, 1965). Responses to inequity include distress, anger, and behavioural reactions such as reduction of inputs (Hegtvedt et al., 2002). Perceptions of injustice in organizations have been found to predict deviant workplace behaviour among under-rewarded employees, as well as emotional exhaustion, depression, and anxiety (Aquino et al., 1999; Tepper, 2001). Some argue that work-life balance initiatives create inequities in the workplace, as only a subset of employees – usually parents of young children – are provided with these extra options. For instance, Burkett (2000) argues that on-site childcare centres are financed by all members of the organization, but used primarily by white, middle-class parents. Those who do not (or cannot) use work-life balance initiatives may bear an increased workload (Young, 1999), such as covering for co-workers who take time off to care for sick children, or serving as representatives at early-morning or late-afternoon meetings eschewed by those using flexible work hours. Employees who are unable to make use of the extra benefits provided by work-life balance initiatives may thus perceive themselves as being under-rewarded, and so might those who perform extra duties so that their co-workers may use work-life balance initiatives to attend to family concerns (Galinsky et al., 1997; Hegtvedt et al., 2002).

Research demonstrates that organizations offering work-life balance initiatives can indeed incur perceptions of injustice among single employees and/or those without children (Kirby and Krone, 2002; Parker and Allen, 2001). In Nord et al. (2002) study, management consultants who were not using any work-life balance initiatives reported that their own work-life balance needs were neglected, and those without children remarked that their organizations' work-life balance initiatives led to unjust burdens on employees with fewer family responsibilities (for instance, requiring them to travel for business more often than consultants with children). Similarly, participants in Kirby and Krone's (2002) study reported that single workers without children were required to travel more often on business than their colleagues with families. Perceptions of unfairness can sometimes be attributed to self-serving bias. Researchers often find that employees who use or who would in future use work-family benefits tend to perceive them as more fair, as do employees with spouses and/or children (Casper et al., 2007; Parker and Allen, 2001). In contrast, forty percent of respondents in the 1997 National Study of the Changing Workforce agreed that they would resent their employers' provision of work-life balance practices that did not help them personally (Galinsky et al., 1997). However, as the population of the Western world ages and the trend toward and importance of eldercare increases, the continuing emphasis of work-life balance initiatives on childcare issues may prove especially frustrating for workers with caregiving responsibilities for elderly relations, contributing to greater inequities and resentment on their part.

The way in which organizations frame their provision of work-life balance initiatives can also influence employees' perceptions of how equitable those practices are. According to Lewis et al. (2000), work-life balance initiatives can be construed by organizations as favours rather than as entitlements. As a result, these initiatives are often viewed by both employers and employees as a cost to the organization that cannot be afforded in times of economic difficulties (Lewis, 1997). This too can result in feelings of inequity among those who do not have family responsibilities and/or those who are not using the initiatives on offer.

Aside from equity issues surrounding employees' familial commitments and the improved access to work-life balance initiatives that these commitments might entail, eligibility for use of work-life balance initiatives can also be dependent upon employees' position within the organization and their immediate supervisors. Due to the relatively high levels of autonomy inherent in their work, managers and professionals tend to enjoy greater access to telework and flexible work hours than support staff, clerical employees, and manufacturing workers (Brewer, 2000; Casper et al., 2007; Golden, 2001; McDonald et al., 2005). Despite these advantages, managers and professionals are still subject to problems associated with the implementation of work-life balance initiatives. In a large-scale study of over 30,000 Canadian workers, Duxbury et al. (2003) found evidence of inconsistent application of initiatives across departments within the same organization. For example, some departments provided access to voluntary reduced hours and paid time off for children's medical appointments, while others did not. The authors also discovered that the initiatives were subject to a large amount of interpretation by individual supervisors. Supervisors often act as gatekeepers to an organization's work-life balance initiatives, adding to the potential for inequitable distribution of practices among employees. Drawing on evidence from case studies of four companies in the Scottish financial sector, Bond and Wise (2003) reported that despite managerial discretion being built into a number of work-life balance initiatives and codified in staff handbooks, awareness of statutory family leave provisions was variable and often quite poor among managers, who frequently had limited training in work-life balance-related human resource policies. Research has demonstrated that factors completely unrelated to employees' requests to use work-life balance initiatives can have a profound influence on the likelihood of those requests being granted. For example, female managers are more likely than male managers to grant requests for alternative work arrangements (Powell and Mainiero, 1999). Supervisors with greater parental responsibility have been found to exhibit more flexibility in helping employees balance their work and home commitments, while supervisors with a greater need for control have been found to display less flexibility in this regard (Parker and Allen, 2002).

11.4 Conclusions

From a review of the literature, it seems clear that work-life balance initiatives have the potential to ease the pressures experienced by busy managers and professionals trying to combine challenging, time-consuming jobs with family responsibilities and personal commitments outside of work. Users of initiatives appear to enjoy improved work-life balance and greater productivity on the job. According to the research, however, these gains can too easily be offset by organizational climates that fail to fully support the initiatives on offer. Managers and professionals using work-life balance initiatives may encounter resentment from colleagues who feel they are shouldering an unfair workload, professional isolation and reduced networking opportunities, a reputation among peers and superiors for being less committed to the firm, and ultimately, damaged prospects for career advancement.

How can these issues be addressed? Organizations may need to pay more attention to aligning workload distribution and compensation with the different schedules and hours put in by individual workers. For instance, single employees who are assigned the lion's share of business travel could be compensated with time off in lieu or the equivalent of overtime pay, thus reducing perceptions of inequitable treatment. Those working voluntary reduced hours could receive task assignments and goals commensurate with the hours they are paid to work, rather than the hours worked by their colleagues. A concerted effort could be made to include individuals using telework in more activities involving social interaction, such as weekly face-to-face team meetings or participation in organization-level committees or conferences. Improved organizational communication regarding the aim of work-life balance initiatives – to help individuals balance their work and non-work responsibilities in order to facilitate optimal performance in each domain – might help to reverse perceptions that users of these initiatives lack commitment to their careers and to their employers. This, in turn, may help create a more work-life balance friendly organizational climate, and turn the tide against the negative career consequences suffered by some users of work-life balance initiatives.

The research reviewed in this chapter indicates that the beneficial outcomes of work-life balance initiatives for managers and professionals may not be realized if the initiatives are implemented in an organizational climate unsupportive of work-life balance issues. As employees with greater access to work-life balance initiatives and greater levels of power relative to those at lower levels of the organizational hierarchy, managers and professionals have a responsibility to themselves and to their subordinates to take an active role in changing perceptions of work-life balance initiatives and their users. Unapologetic use of initiatives combined with sincere efforts to allay colleagues' concerns regarding workload distribution may go some way toward normalizing initiative use, especially for men. Managing client and colleague expectations can help assuage others' unease regarding the availability of the work-life balance initiative user; according to management consultants interviewed by Nord et al., 2002, p. 232), clients are "fine if you just tell them" which days the consultants will be working, and from what location. Role modelling effective management of employees using work-life balance initiatives is vital to changing perceptions among peers that initiative users are "a pain in the neck to have. . .working for you" (Nord et al., 2002, p. 234). By ensuring these employees are equitably treated with regard to task distribution, compensation, appraisal, and promotion opportunities, and by placing an emphasis on efficiency and results rather than face time and hours worked, managers and professionals can dispel apprehension among subordinates that work-life balance initiative use spells career derailment. Finally, providing emotional and instrumental support for subordinates who seek to use work-life balance initiatives is key to ameliorating the outcomes of initiative use for employees at all levels of the organization.

Taking on this responsibility for change is obviously challenging for managers and professionals operating in an environment where the use of work-life balance initiatives is associated with a reputation for low commitment to the organization and reduced prospects for career advancement – not least because of the implications for their own reputation and career prospects. Waiting patiently for organizations to change from the very top down is not, however, a viable option. Managers and professionals must play a role in effecting this transformation themselves if work-life balance initiatives are ever to achieve their full potential.

References

Adams JS (1965) Inequity in social exchange. In Berkowitz L (ed) Advances in experimental and social psychology. Academic, San Diego, CA, pp 267–299

Allen TD (2001) Family-supportive work environments: the role of organizational perceptions. J Vocat Behav 58:414–435

Allen TD, Russell JE (1999) Parental leave of absence: some not so family friendly implications. J Appl Soc Psychol 29:166–191

Anderson SE, Coffey BS, Byerly RT (2002) Formal organizational initiatives and informal workplace practices: links to work-life conflict and job-related outcomes. J Manage 28:787–810

Aquino K, Lewis MU, Bradfield M (1999) Justice constructs, negative affectivity, and employee deviance: a proposed model and empirical test. J Organ Behav 20:1073–1091

Aryee S, Luk V, Stone R (1998) Family-responsive variables and retention-relevant outcomes among employed parents. Hum Relat 51:73–87

Auerbach J (1990) Employer-supported child care as a women-responsive policy. J Fam Issues 11:384–400

Bailyn L (1997) The impact of corporate culture on work-family integration. In Parasuraman S, Greenhaus JH (eds) Integrating work and family: challenges and choices for a changing world. Quorum, Westport, CT, pp 209–219

Bailyn L, Fletcher JK, Kolb D (1997) Unexpected connections: considering employees' personal lives can revitalise your business. Sloan Manage Rev 38:11–19

Beauregard TA, Henry LC (2009) Making the link between work-life balance practices and organizational performance. Hum Resour Manage Rev 19:9–22

Berry JO, Rao JM (1997) Balancing employment and fatherhood: a systems perspective. J Fam Issues 18:386–402

Bittman M, Hoffmann S, Thompson D (2004) Men's uptake of family-friendly employment provisions. Policy Research Paper No. 22. Australian Government Department of Family and Community Services, Canberra, ACT

Blair-Loy M (2001) Cultural constructions of family schemas: the case of female finance executives. Gender Soc 15:687–709

Blair-Loy M, Wharton AS (2002) Employees' use of work-family policies and the workplace social context. Soc Forces 80:813–845

Bond S, Wise S (2003) Family leave policies and devolution to the line. Pers Rev 32:58–72

Brandth B, Kvande E (2001) Flexible work and flexible fathers. Work Empl Soc 15:251–267

Brandth B, Kvande E (2002) Reflexive fathers: negotiating parental leave and working life. Gender Work Organ 9:186–203

Brewer AM (2000) Work design for flexible work scheduling: barriers and gender implications. Gender Work Organ 7:33–44

Burkett E (2000) The baby boom: how family-friendly America cheats the childless. Free Press, New York, NY

Butler A, Gasser M, Smart L (2004) A social-cognitive perspective on using family-friendly benefits. J Vocat Behav 65:57–70

Bygren M, Duvander AZ (2006) Parents' workplace situation and fathers' parental leave use. J Marriage Fam 68:363–372

Callentine LU (1995) The ecology of the mobile workplace: influence of household composition and home workspace on satisfaction, stress, and effectiveness. Unpublished master's thesis, Cornell University, Ithaca, NY

Casper WJ, Harris CM (2008) Work-life benefits and organizational attachment: self-interest utility and signaling theory models. J Vocat Behav 72:95–109

Casper WJ, Weltman D, Kwesiga E (2007) Beyond family-friendly: the construct and measurement of singles-friendly work culture. J Vocat Behav 70:478–501

Chow IH, Keng-Howe IC (2006) The effect of alternative work schedules on employee performance. Int J Empl Stud 14:105–130

Cohen JR, Single LE (2001) An examination of the perceived impact of flexible work arrangements on professional opportunities in public accounting. J Bus Ethics 32:317–328

Cooper CD, Kurland NB (2002) Telecommuting, professional isolation, and employee development in public and private organizations. J Organ Behav 23:511–532

Dalton DR, Mesch DJ (1990) The impact of flexible scheduling on employee attendance and turnover. Adm Sci Q 35:370–387

Daly K, Ashbourne L, Hawkins L (2008) Work-life issues for fathers. In Korabik K, Lero DS, Whitehead DL (eds) Handbook of work-family integration: research, theory, and best practices. Academic, London, pp 249–264

Dunham RB, Pierce JL, Castaneda MB (1987) Alternative work schedules: two field quasi-experiments. Pers Psychol 40:215–243

Duxbury L, Higgins C, Coghill D (2003) Voices of Canadians: seeking work-life balance. Human Resources Development Canada, Hull

Eaton SC (2003) If you can use them: flexibility policies, organizational commitment, and perceived performance. Ind Relat 42:145–167

Estes SB, Michael J (2005) Work-family policies and gender inequality at work. In Kossek E, Pitt-Catsouphes M (eds) Work and family encyclopedia. Sloan Work and Family Research Network, Chestnut Hill, GA. http://wfnetwork.bc.edu/encyclopedia_entry.php?id=1230&area=All. Accessed on 22 Oct 2009

Fleetwood S (2007) Why work-life balance now?. Int J Hum ResourManage 18: 387–400

Fried M (1998) Taking time: parental leave policy and corporate culture. Temple University Press, Philadelphia, PA

Frolick MN, Wilkes RB, Urwiler R (1993) Telecommuting as a workplace alternative: an identification of significant factors at home in American firms' determination of work-at-home policies. J Strateg Inf Syst 2:206–222

Galinsky JT, Bond E, Swanberg JE (1997) Synthesis of findings: the 1997 national study of the changing workforce. Families and Work Institute, New York, NY

Goff SJ, Mount MK, Jamison RL (1990) Employer supported child care, work/family conflict, and absenteeism: a field study. Pers Psychol 43:793–809

Goldberg WA, Greenberger E, Koch-Jones J, O'Neil R, Hamill S (1989) Attractiveness of child care and related employer-supported benefits and policies to married and single parents. Child Youth Care Q 18:23–37

Golden L (2001) Flexible work schedules: what are we trading off to get them?. Mon Labor Rev 124:50–67

Gray M, Tudball J (2003) Family-friendly work practices: differences within and between workplaces. J Ind Relat 45:269–291

Grover SL, Crooker KJ (1995) Who appreciates family-responsive human resource policies: the impact of family-friendly policies on the organizational attachment of parents and non-parents. Pers Psychol 48:271–288

Haas L, Allard K, Hwang P (2002) The impact of organizational culture on men's use of parental leave in Sweden. Commun Work Fam 5:319–342

Halpern DF (2005) How time-flexible work policies can reduce stress, improve health, and save money. Stress Health 21:157–168

Hartman RI, Stoner CR, Arora R (1991) An investigation of selected variables affecting telecommuting productivity and satisfaction. J Bus Psychol 6:207–225

Hegtvedt K, Clay-Warner J, Ferrigno E (2002) Reactions to injustice: factors affecting workers' resentment toward family-friendly policies. Soc Psychol Q 65:386–400

Hill EJ, Ferris M, Martinson V (2003) Does it matter where you work? A comparison of how three work venues (traditional office, virtual office, and home office) influence aspects of work and personal/family life. J Vocat Behav 63:220–241

Hill EJ, Hawkins AJ, Ferris M, Weitzman M (2001) Finding an extra day a week: the positive influence of perceived job flexibility on work and family life balance. Fam Relat 50: 49–58

Hill EJ, Miller BC, Weiner SP, Colihan J (1998) Influences of the virtual office on aspects of work and work/life balance. Pers Psychol 51:667–683

Houston D (Ed) (2005) Work-life balance in the 21st century. Palgrave Macmillan, Basingstoke/New York

Houston DM, Waumsley JA (2003) Attitudes to flexible working and family life. Policy Press, York

Judiesch MK, Lyness KS (1999) Left behind? The impact of leaves of absence on managers' career success. Acad Manage J 42:641–651

Kirby EL, Krone KJ (2002) "The policy exists but you can't really use it": communication and the structuration of work-family policies. J Appl Commun Res 30:50–77

Kodz J, Harper H, Dench S (2002) Work-life balance: beyond the rhetoric. Institute for Employment Studies Report 384. IES, London

Kossek EE, Barber AE, Winters D (1999) Using flexible schedules in the management world: the power of peers. Hum Resour Manage 38:33–49

Kossek EE, Lautsch BA, Eaton SC (2006) Telecommuting, control, and boundary management: correlates of policy use and practice, job control, and work–family effectiveness. J Vocat Behav 68:347–367

Kossek EE, Nichol V (1992) The effects of on-site child care on employee attitudes and performance. Pers Psychol 45:485–509

Kossek EE, Ozeki C (1999) Bridging the work-family policy and productivity gap: a literature review. Commun Work Fam 2:7–32

Krauz M, Freibach N (1983) Effects of flexible working time for employed women upon satisfaction, strains and absenteeism. J Occup Psychol 2:155–159

Kurland NB, Bailey DE (1999) When workers are here, there, and everywhere: a discussion of the advantages and challenges of telework. Organ Dyn 28:53–68

Lambert SJ (2000) Added benefits: the link between work-life benefits and organizational citizenship behavior. Acad Manage J 43:801–815

Lambert AD, Marler JH, Gueutal HG (2008) Individual differences: factors affecting employee utilization of flexible work arrangements. J Vocat Behav 73:107–117

Lee CM, Duxbury L (1998) Employed parents' support from partners, employers, and friends. J Soc Psychol 138:303–322

Lewis S (1997) 'Family friendly' employment policies: a route to changing organizational culture or playing about at the margins?. Gender Work Organ 4:13–24

Lewis S, Kagan C, Heaton P (2000) Managing work-family diversity for parents of disabled children: beyond policy to practice and partnership. Pers Rev 29:417–430

Liff S, Cameron I (1997) Changing equality cultures to move beyond "women's problems". Gender Work Organ 4:35–46

McDonald P, Guthrie D, Bradley L, Shakespeare-Finch J (2005) Investigating work-family policy aims and employee experiences. Empl Relat 27:478–494

Milkovich GT, Gomez LR (1976) Day care and selected employee work behaviors. Acad Manage J 19:111–115

Muse L, Harris SG, Giles WF, Field HS (2008) Work-life benefits and positive organizational behavior: is there a connection?. J Organ Behav 29:171–192

Nord WR, Fox S, Phoenix A, Viano K (2002) Real-world reactions to work-life balance programs: lessons for effective implementation. Organ Dyn 30:223–238

Organ DW, Podsakoff PM, MacKenzie SB (2006) Organizational citizenship behavior: its nature, antecedents, and consequences. Sage, London

Orthner DK, Pittman JF (1986) Family contributions to work commitment. J Marriage Fam 48:573–581

O'Driscoll M, Poelmans S, Spector P, Kalliath T, Allen T, Cooper C, Sanchez J (2003) Family-responsive interventions, perceived organizational and supervisor support, work-family conflict and psychological strain. Int J Stress Manage 10:326–344

Parker LB, Allen TD (2001) Work/family benefits: variables related to employees' fairness perceptions. J Vocat Behav 58:453–468

Parker LB, Allen TD (2002) Factors related to supervisor work/family sensitivity and flexibility. In: Society for Industrial and Organizational Psychology, 17th Annual SIOP Conference, Toronto, ON, 12–14 April 2002

Perlow LA (1995) Putting the work back into work/family. Group Organ Manage 20: 227–239

Perlow LA (1997) Finding time: how corporations, individuals and families can benefit from new work practices. ILR Press, Ithaca, NY

Pierce JL, Newstrom JW (1982) Employee responses to flexible work schedules: an inter-organization, inter-system comparison. J Manage 8:9–25

Pierce JL, Newstrom JW (1983) The design of flexible work schedules and employee responses: relationships and process. J Occup Behav 4:247–262

Pitt-Catsouphes M, Marchetta A (1991) A coming of age: telework. Boston University, Center on Work and Family, Boston, MA

Pleck JH (1993) Are "family-supportive" employer policies relevant to men?. In Hood JC (ed) Men, work, and family. Sage, Newbury Park, CA, pp 217–237
Pocock B (2005) Work-life 'balance' in Australia: limited progress, dim prospects. Asia Pac J Hum Resour 43:198–209
Powell GN, Mainiero LA (1999) Managerial decision making regarding alternative work arrangements. J Occup Organ Psychol 72:41–56
Prottas DJ, Thompson CA, Kopelman RE, Jahn EW (2007) Work-family programs: factors affecting employee knowledge and accuracy. Pers Rev 36:163–189
Raabe PH (1996) Constructing pluralistic work and career arrangements. In Lewis S, Lewis J (eds) The work-life challenge: rethinking employment. Sage, London, pp 128–141
Rogier SA, Padgett MY (2004) The impact of utilizing a flexible work schedule on the perceived career advancement potential of women. Hum Resour Dev Q 15:89–106
Saltzstein AL, Ting Y, Saltzstein GH (2001) Work-family balance and job satisfaction: the impact of family-friendly policies on attitudes of federal government employees. Public Adm Rev 61:452–467
Scholarios D, Marks A (2004) Work-life balance and the software worker. Hum Resour Manage J 14:54–74
Shockley KM, Allen TD (2007) When flexibility helps: another look at the availability of flexible work arrangements and work-family conflict. J Vocat Behav 71:479–493
Swanburg JE, Pitt-Catsouphes M, Drescher-Burke K (2005) A question of justice: disparities in employees' access to flexible schedule arrangements. J Fam Issues 26:866–895
Tepper BJ (2001) Health consequences of organizational injustice: tests of main and interactive effects. Organ Behav Hum Decis Process 86:197–215
Thomas LT, Ganster DC (1995) Impact of family-supportive work variables on work-family conflict and strain: a control perspective. J Appl Psychol 80:6–15
Thompson CA, Beauvais LL, Lyness KS (1999) When work-life benefits are not enough: the influence of work-life culture on benefit utilization, organizational attachment, and work-life conflict. J Vocat Behav 54:392–415
Visser F, Williams L (2006) Work-life balance: rhetoric versus reality?. The Work Foundation, London
Watts JH (2009) Leaders of men: women 'managing' in construction. Work Empl Soc 23:512–530
Wayne JH, Cordeiro BL (2003) Who is a good organizational citizen? Social perception of male and female employees who use family leave. Sex Roles 49:233–246
Whitehouse G, Zetlin D (1999) Family friendly policies: distribution and implementation in Australian workplaces. Econ Lab Relat Rev 10:221–239
Williams ML, Ford LR, Dohring PL, Lee MD, MacDermid SM (2000, August) Outcomes of reduced load work arrangements at managerial and professional levels: perspectives from multiple stakeholders. Academy of Management, Annual Meeting, Toronto, ON, 7–9 August 2000
Young MB (1999) Work-family backlash: begging the question, what's fair?. Ann Am Acad Pol Soc Sci 562:32–46
Youngblood S, Chambers-Cook K (1984) Child care assistance can improve employee attitudes and behavior. Pers Adm 29:45–47
Shockley KM, Allen TD (2010) Investigating the missing link in flexible work arrangement utilization: an individual difference perspective. J Vocat Behav 76:131–142

Chapter 12
Temporal Workplace Flexibility and Associated Work-Life Outcomes for Professionals

E. Jeffrey Hill, Sarah June Carroll, Blake L. Jones, Lydia A. Buswell, Tamara A. Fackrell, and Adam M. Galovan

12.1 Introduction

An increasing proportion of workers are dual-earners in professional occupations, actively seeking rewarding, full-time careers while at the same time engaging in satisfying home and personal life. However, the demands of today's global economic environment often make work-life integration problematic (Hill et al., 2006). One adaptive strategy to deal with work-life conflict is to choose to be flexible about when one works and opting to reduce work hours, especially when demands in the home are great (Allen and Shockley, 2009; Grzywacz, 2009; Voydanoff, 2007). This adaptive strategy is called temporal workplace flexibility.

The purpose of this chapter is to review scholarly literature that examines the relationship between temporal workplace flexibility (e.g., schedule flexibility, compressed work-weeks, regular part-time employment, job sharing, transition from leave, and phased retirement) to work-life integration and associated outcomes (e.g., work-life conflict, work-life fit, work-life harmony, job satisfaction, marital satisfaction, and mental health). We give primary emphasis to those studies comprised of professional employees. These employees typically have higher educational levels, more job responsibility, higher income, and longer work hours than non-professionals. Though most scholarly literature in this domain has its origin in the United States, we have sought to include research from other countries as well.

12.1.1 Definition of Temporal Workplace Flexibility

Workplace flexibility is defined as "the ability of workers to make choices influencing when, where, and for how long they engage in work-related tasks" (Hill et al., 2008a, p. 152). Temporal workplace flexibility focuses on the when and for how long components of this definition. Several important concepts are embedded here. First, the ability of workers to choose when and for how long they work is central. If the organization required employees to work part-time or to vary their

---

E.J. Hill (✉)
School of Family Life, Brigham Young University, Provo, UT, USA
e-mail: jeff_hill@byu.edu

S. Kaiser et al. (eds.), *Creating Balance?*, DOI 10.1007/978-3-642-16199-5_12,

work schedules according to business needs, this would not be categorized as temporal workplace flexibility. Only if the flexible options were chosen by the employee would they qualify as temporal workplace flexibility. Substantial research demonstrates the value of workers' control or agency in relation to workplace flexibility in promoting optimal outcomes for the individual, family, and workplace (Voyandoff, 2007). Next, the when element of the definition refers to choosing to arrange work schedules within certain constraints according to personal and business needs. This type of flexibility includes schedule flexibility and compressed work weeks (where an employee works full-time hours in less than 5 days), and to some degree job sharing, regular part-time employment, transition from leave, and phased retirement. Finally, the for-how-long element of the definition refers to the choice of reducing the work hours below a standard full-time work week and work year. Some examples of this type of flexibility are job sharing, regular part-time employment, transition from leave, and phased retirement. As can be seen, some types of temporal workplace flexibility combine both the when and the for-how-long elements of the definition.

12.1.2 Theoretical Perspective

A useful theoretical perspective in studying temporal workplace flexibility is person-environment (P-E) fit theory. This theory emphasizes the demands individuals deal with vis-à-vis the resources available to meet those demands. Thus, it provides a useful perspective for linking temporal workplace flexibility to work-life integration and associated outcomes (see Edwards, 1996; Grzywacz et al., 2008; Voydanoff, 2007). Temporal workplace flexibility provides numerous options for employees to choose when and for how long they work. These choices may be seen as resources to better enable employees to deal with the numerous demands in both the work and family domains. By better managing the work and family demands, temporal workplace flexibility is hypothesized to be associated with less work-to-family and family-to-work conflict, which in turn enhances work-life integration (or work-life fit and work-life harmony) and associated outcomes (such as job satisfaction, marital satisfaction, and mental health).

12.2 Literature Review

The remainder of the chapter will review six major forms of temporal workplace flexibility: schedule flexibility, compressed work-week schedules, regular part-time employment, job sharing, transition from leave, and phased retirement. For each form of temporal workplace flexibility we will offer a definition, explain its usage, and then summarize research linking it to work-life integration and associated outcomes with studies that, as much as possible, utilize professional employees.

12.2.1 Schedule Flexibility

Definition. Schedule flexibility is defined as a type of flexible work arrangement that allows employees to vary their work schedules, within certain ranges and dimensions established by the employer, according to their differing needs (Ronen, 1981). Employees choose individual beginning and ending work times, lunch hours, and breaks, so long as the required number of hours is met within a certain time frame (usually 40 h in 1 week). Often employers establish "core hours" when the employees are required to be present (10 am to 3 pm for example), or require each day to total 8 h (Baltes et al., 1999).

Usage. Schedule flexibility is increasingly more common. Using a nationally representative sample of 1,100 employers in the United States, Galinsky et al. (2008) found that 79% of employers allowed employees to periodically change starting and quitting times within some range of hours, which is an increase from only 68% in 1998. Using a sample representing 312,459 employees in twenty-five US organizations, most of whom were professionals, Hill et al. (2008b) found differences by gender and life stage in the usage of flextime. For those without young children, the proportion using flextime did not meaningfully differ by gender or life stage. However, when a young child was present in the home (age 5 or less) women were almost twice as likely as men to use schedule flexibility.

Findings. Most research reveals that professional employees benefit from schedule flexibility. This flexibility is generally shown to be negatively associated with work-family conflict and positively associated with work-life integration and associated outcomes. Using a large multi-industry sample of US workers comprised mostly of professionals ($N = 19{,}704$), Grzywacz et al. (2008) found that using flextime was associated with less individual stress and burnout. Using a US sample of professionals, Hill et al. (2001) found that schedule flexibility was positively related to work-family balance. Employing a creative breakpoint analysis, they found that professional employees with high schedule flexibility were able to work 59 h per week before reporting work-life difficulty compared to 51 h per week for those with low schedule flexibility. The benefit was most pronounced for women professionals with pre-school children. Their breakpoint was 41 h per week with schedule flexibility, but only 31 h per week without schedule flexibility, which is a benefit of 10 h per week. In this sample, only 30% with high schedule flexibility reported work-life difficulty, compared to 49% with low schedule flexibility (Hill et al., 2001).

Most research related to schedule flexibility has been undertaken in the United States. However, truly multi-national studies are emerging. One such study (Hill et al., 2009) gathered data from professional employees in 75 countries ($N = 24{,}436$) and examined the relationship of schedule flexibility and work-at-home to work-life conflict. They divided their sample into four groups of countries based on cultural (individualistic vs. collectivist), economic (average country GDP), and governmental policy factors (degree of work-life legislation). Schedule flexibility was universally associated with much lower work-life conflict in every area of the world. Using break-point analysis, they found that those with schedule flexibility were able to work an average of 54 h per week before reporting work-life difficulty, compared to 37 h per week for those without schedule flexibility, an astounding benefit of 17 h per week. They also found that only 13% of the respondents with high schedule flexibility reported work-life conflict, compared to 35% of the respondents with low schedule flexibility. Finally, their study found that schedule flexibility was more strongly related to lower work-life conflict than was work-at-home.

Other research reveals that offering schedule flexibility may relate to positive outcomes for everyone, not just for those who use the flexibility (Hayman, 2009). Jones et al. (2008), using a nationally representative sample from Singapore ($N = 1601$), found that perceived flexibility predicted work-family fit even better than used flexibility did. The implication is that actually using temporal workplace flexibility is not a prerequisite for enhancing work-life integration and associated outcomes. "Indeed, just the perception that the flexibility would be available when needed appears to be sufficient" (p. 781). Similarly, Grzywacz et al. (2008) found that 30–50% of the differences in flextime could be accounted for by perceived flexibility. They summarized, "... participation in formal arrangements that involve flextime promotes a sense among workers that they have the discretion to fit job-related responsibilities into their broader lives, and this discretion contributes to less stress and burnout ..." (p. 210). A study in the U. K. found that perceived accessibility to family-friendly workplace options seemed to be more important than actual workplace availability in helping employees feel that their workplace was supportive of family life (Budd and Mumford, 2006). Also, in a study of professional women in Hong Kong, Lo (2003) found that, next to salary, schedule flexibility was the second leading descriptor of the ideal job. In addition, married professional women in Singapore indicated that it was the number of hours worked per week, rather than schedule inflexibility, which positively predicted work-family conflict (Aryee, 1992). Finally, results from their landmark meta-analysis (Baltes et al., 1999) revealed that schedule flexibility was most strongly related to lower employee absenteeism. It was also positively related to employee productivity, job satisfaction, and satisfaction with work schedule. However, they also found that these relationships were weakened or disappeared for managers and professionals.

12.2.2 Compressed Work-week Schedule

Definition. A compressed work-week schedule is a form of temporal workplace flexibility where employees work longer hours for fewer days of the week rather than the traditional 8-h workday 5 days a week. Most organizations with compressed schedules have implemented what are known as 4/40 (four 10-h days per week), 3/36 (three 12-h days per week), or 9/80 (80 h of work in 9 days over 2 weeks) schedules (Wadsworth, 2009). The major benefit of compressed work-week schedules is having one or more additional days at home per week, while continuing to work full-time. This may be particularly beneficial to professional employees who are typically in dual-earner households, have long commutes, or struggle to have the time at home to attend to household responsibilities.

Usage. The proportion of companies offering compressed work-week schedules appears to be increasing in the United States. According to the National Study of Employers (Galinsky et al., 2008), 41% of employers allow employees to compress their work-week by working longer hours on fewer days for at least part of the year, compared to 37% in 1998. While the focus of this chapter is on the employee's choice of temporal flexibility usage, it should be noted that some employers and agencies have recently increased their implementation or allowance of compressed work-week schedules as a means of saving energy and other costs to the company.

Findings. Results of studies relating compressed work-week schedules to outcomes associated with work-life integration are generally positive. Using a large multi-industry sample of US workers, comprised mostly of professionals ($N =$ 19,704), Grzywacz et al. (2008) found that using a compressed work-week was associated with less individual stress and burnout. Likewise, using a sample of professionals working with information systems personnel, working a 3 day per 38 h (3/38) week, Latack and Foster (1985) found that the compressed work-week schedule was most favorably received by those employees who had participated in the decision to implement this schedule. The compressed work-week schedule was associated with reductions in sick time, overtime, and personal leave time. More recently, positive family outcomes were associated with a compressed work-week schedule. In a sample of city government employees, some of which were professionals, Facer and Wadsworth (2008) compared those working 4/40 schedules to those working a different hourly schedule. Employees working a 4/40 schedule had lower levels of work-family conflict and greater productivity compared to those who do not work 4/40 schedules; however, job satisfaction did not differ according to work schedule. In a study of professional and technical workers with flexible work schedules, Eaton (2003) noted that the perception of flexibility is what really makes the difference. Control over time, flexibility, and pace of work were positively related to job commitment and job productivity. In their seminal meta-analysis, Baltes et al. (1999) found that compressed work-week schedules were positively related to job satisfaction and satisfaction with work schedule, but were unrelated to absenteeism and productivity. It should be noted that this study was not limited to professional employees.

While much of the research relating to compressed work-weeks has focused on positive outcomes, some researchers have reported negative outcomes and drawbacks of compressed work-week schedules. In a Belgian sample not limited to professional employees, those working compressed work week schedules reported up to 40% more health complaints, lower psychological performance, and lower quality of sleep than employees with non-flexible work schedules (Martens et al., 1999). Some potential drawbacks involve the greater physical and mental exertion that is required for working consecutive 10–12 h shifts, which can drain stamina and focus. This can then lead to health and safety risks in the workplace, particularly for jobs which require high levels of precision, sustained attention, or close proximity to hazardous materials or machinery (Human Resources and Skills Development

Canada, n.d.). It should be noted that some of these potential risks are not inherently associated with many types of professional jobs, but would more likely be problematic for industrial and manufacturing positions or manual labor. Additional studies have shown that compressed work week schedules were related to other negative outcomes as well. For example, researchers found specific problems with job performance due to decreased reaction times, alertness, or cognitive functioning, brought on with increases in fatigue (Armstong-Stassen, 1998; Rosa, 1991). In another study, women who used compressed work-week schedules reported higher levels of time stress than women with traditional work schedules (Fast and Frederick, 1996).

12.2.3 Regular Part-Time Employment

Definition. Regular part-time employment is a flexible work arrangement involving reduced hours and reduced workload compared to full-time employment. A common cut-off point is 30 h per week at one's main job (OECD.Stat Extracts, 2009). Usually pay and benefits for part-time employees are pro-rated according to the employee's time commitment. Recently, many companies have offered professional employees reduced-hour, high-status, career-oriented options that contrast with traditionally lower status part-time positions and maintain pro-rated professional salaries and benefits (Barnett, 1998; Barnett and Gareis, 2000). Hill et al. (2004a) define this as "new-concept" part-time employment. Offering work schedules that fit professional employees' personal/family needs is necessary to "winning the war for talent" by retaining "professional employees and managers with key skills" (Barnett and Hall, 2001, p. 192). Since many professionals are "time poor" rather than "money poor" regular part-time work arrangements may be a good fit.

Usage. The percentage of all employees who work 30 h per week or less at their main job varies greatly throughout the world (OECD.Stat Extracts, 2009). The largest proportion is 36% in the Netherlands and the smallest is 3% in Slovakia. Other rates include Korea (9%), United States (12%), European Union (18%), Japan (20%), Germany (22%), United Kingdom (23%), and Australia (24%). Universally, women are more likely than men to work part-time. In the Netherlands, 60% of employed women work part-time compared to 16% of employed men. In the European Union the figures are 29% for women and 7% for men. In the United States, 17% of employed women work part-time and 8% of employed men.

Using their large multi-organization sample, comprised mostly of professionals in the United States, Hill et al. (2008b) found differences by gender and life stage in the usage of part-time employment. Very few men (1%) and women (1%) worked part-time in Life Stage 1 (age less than 35 and no children). However, in Life Stage 2 (oldest child less than age 6) some women (16%), but very few men (1%), worked part-time. In Life Stage 3 (at least one child under age 6 and at least one child age 6 or older) 16% of the women, but none (0%) of the men, worked part-time.

Findings. Generally, the review of literature supports regular part-time employment as having a beneficial influence on outcomes associated with work-life integration. In a study of female professionals with young children, Hill et al.

(2004a) compared those working full-time with those working regular part-time schedules. The regular part-time group reported significantly lower work-to-family interference, work-to-family strain, job travel, and unnecessary work, and greater work-family fit, job flexibility, and family success. Raabe (1998) found that 86% of part-time managers reported an increase in the quantity of time available for family/personal life and hence less work-family conflict. A common perceived drawback is that working reduced-hours leads to reduced career opportunities – a so-called "mommy track" (Hill et al., 2004b). However, in a study of professionals and managers working in part-time positions, Lee and MacDermid (1998) found that 35% of the participants were actually promoted during the time of the study.

Browning et al. (2000) suggested a model which could apply to dual-professional families in their quest for work-life integration. In this model both parents engage in paid employment, but the combined work hours of the mother and the father do not exceed 60 h per week. These hours could be divided 30/30, 40/20, or in any combination the couple deems best. Browning et al. suggest that "the happiest families are those in which both husband and wife have some paid employment, share household chores and child care, and work less than full-time positions" (p. 317). Hill et al. (2006) tested this hypothesis by comparing three groups of married couples with children: full-time/full-time (FT/FT); at least one partner part-time (60-h); and full-time/not employed (FT/NE). In this sample of IBM professional employees in the United States, 60-h couples reported significantly greater job flexibility, improved work-family fit, enhanced family satisfaction, and less work-to-family conflict than either the FT/FT or the FT/NE groups. In a separate study, using a different group of IBM professional employees in the United States, working part-time was found to be strongly associated with work-family balance (Hill et al., 2004b).

Not all research related to regular part-time employment for professionals is positive. In a study of part-time physicians, Barnett and Gareis (2002) found that "physicians working longer hours reported higher marital-role quality than those working fewer hours to the extent that they performed fewer low-schedule-control household tasks than did their reduced-hours counterparts" (p. 364). In addition, in dual-earner households with children, where the mother works part-time, there tends to be a more gendered division of household chores (Meiksins and Whalley, 2002).

12.2.4 Job Sharing

Definition. Job sharing is defined as a work arrangement in which two people agree to share the workload of a full-time job (see Barnett, 2003). It is similar to regular part-time in that it involves reduced hours and reduced work load for the individual. However, it is different because two employees are involved and arrange to divide the hours and workload of one position in a way that is mutually beneficial. By thus using temporal workplace flexibility, the employees involved make choices about when they will work to enhance work-life integration. The pay and benefits for each job sharing employee are pro-rated according to their relative contribution to the full-time job.

Usage. There appears to be a decline in the proportion of employers allowing job sharing in the United States. According to the 2008 National Study of Employers (Galinsky et al., 2008), only 31% allow employees to share jobs, which is down from 38% in 1998. Using their multi-organization sample comprised mostly of professional employees, Hill et al. (2008b) found differences by gender and life stage in the usage of job sharing. Neither men (0%) nor women (0%) used job sharing in Life Stage 1 (age less than 35 and no children). However, in Life Stage 2 (oldest child less than age 6) some women (6%) and very few men (1%) used job sharing. The difference was even more striking in Life Stage 3 (at least one child under age 6 and at least one child age 6 or older). In this stage, 16% of the women, but none of the men, used job sharing. Using a sample of professionals from a financial services company in Ontario, Canada, Frone and Yardley (1996) found that women rated job sharing as being of higher importance than men did.

Findings. There have been relatively few scholarly studies examining job sharing. One qualitative study, completed in Canada during the 1990s, found that mothers, some of whom were professionals, reported experiencing a richer quality of life by using job sharing (Turner, 1996). Another study in Canada found that nurses who used job sharing reported greater job satisfaction and job retention (Kane, 1999). A more recent study examined the benefits and disadvantages of job sharing, by interviewing managers and employees (Branine, 2003). Managers gave varied responses regarding job sharing, with some feeling that it was effective and offered a good flexibility option, while others reported that it was not effective and created problems with continuity and effectiveness of some jobs. This same study also stated that many managers felt that providing job sharing allowed them to keep qualified and trained individuals in the workplace. The employees in the study appreciated job sharing because it provided them with a way to integrate work and family commitments and avoid career breaks that would otherwise be necessary to meet family needs. Similarly, another recent study using medical professionals showed that job sharing facilitated the combination of professional occupation and family life, as well as prevented job burnout (Als and Brautigam, 2006). Australian workers reported that job sharing allows companies to retain and attract staff (The Hudson Report, 2006). A separate study conducted in Australia examined how various work/life balance strategies being offered by employers have changed in frequency between 1997 and 2000 (De Cieri et al., 2005). The study found the availability of job sharing went from 75 to 82% in 3 years.

Job sharing has become more acceptable among those in academic positions. Those professors engaged in job sharing arrangements often report that the extra time they can devote to family and other personal pursuits is worth the corresponding decrease in salary (Schneider, 2000). Department chairs and deans of universities reported that there were potential benefits from job sharing that would extend to both employees and employers in academia (Freeman and Coll, 2009). For example, they felt that job sharing employees would benefit from more family time and less stress due to faculty member work overload, while the number one benefit to the department or university would stem from having "two different viewpoints and academic strengths to complement the [research/faculty] team" (p. 69).

While job sharing is generally accepted as a valuable form of temporal workplace flexibility, it is not without criticism in certain professional and managerial positions. In the study previously mentioned, Freeman and Coll (2009) noted job sharing may introduce complications in the tenure process in academia, especially if one partner is clearly more competent than the other. Many corporate executives believe that job sharing is inappropriate for managers because positions at that level ought to be fulltime. Although job sharing may not be the best temporal flexibility option for some fields, other positions seem to provide a much better fit for this option. For example, nursing is a professional position which is frequently cited as being an excellent context for job sharing (Pershun, 1992).

12.2.5 Transition from Leave

Definition. A leave of absence is "any authorized time away from work" (FlexPaths, n.d.). Parental leave (also called maternal, paternal, or family leave) is a common leave of absence. After a leave of absence, a temporal workplace flexibility option is to work part-time during the transition period of returning to work. During this transition, the workload may gradually increase incrementally toward the full workload.

Usage. According to the 2008 National Study of Employers (Galinsky et al., 2008), 82% of employers in the United States allow employees to return to work gradually after childbirth or adoption. Around the world, many countries have enacted laws providing salary subsidies to facilitate a part-time transition period after birth. Sweden was the first country to support transition from parental leave. In 1978, parents were given the option to reduce daily work hours by 25% until the youngest child was 8 years old (Hegewisch and Gornick, 2008). Today Sweden offers 16 months of paid leave (2 months must be taken by the father) after the birth of a child. This leave may be spread out over 3 years providing for a transition period of part-time employment returning from parental leave. Portugal's policy provides 3 months full-time parental leave or 12 months of part-time leave (Ray et al., 2008). Twelve countries have policies specific for a gradual return to work after birth or adoption; Austria, Belgium, Denmark, France, Germany, Greece, Luxembourg, Netherlands, Norway, Portugal, Spain, and Sweden (Hegewisch and Gornick, 2008).

Partly due to the variety in parental leave policies in different countries, usage of such policies differs substantially from country to country. For example in Germany, approximately 7,000 men and 388,000 women used parental leave in 1995. Yet in the Netherlands, 9,000 men and 12,000 women used parental leave in the same year (Bruning and Plantenga, 1999).

Findings. Hill (1992) examined employees using IBM's flexible work leave of absence program. These professional employees were transitioning from full-time leave and working for 20–32 h per week. Within 3 years they were required to work full-time again. The program participants identified the following personal benefits: less personal stress, better personal health, better parenting, ability to care for elderly parents, and greater job satisfaction. They also reported workplace benefits, including: greater productivity, less expense for the company, reduced attrition, less retraining, higher morale, and greater loyalty. In the US, longer maternity and paternity leaves had positive effects on self-esteem and marital satisfaction, as well as an increased interest in infant developmental milestones; while shorter leaves had a negative impact on self-esteem, marriage, support from husbands, and parental interest and connection with the infant (Feldman et al., 2004). In Europe, extended parental leave was found to be associated with lower infant mortality rates (Tanaka, 2005). The Australian Institute of Family Studies reported that mothers returning from maternity leave transitioned back to work at different rates depending on several factors (Baxter and Gray, 2008). For example, mothers who used paid leave returned to work faster than those with unpaid leave, with 83 and 77% (respectively) returning to work within 18 months after the birth. Also, mothers who worked as permanent employees transitioned back to the workplace sooner than contract workers and casual employees. The most common reason for returning to work within the child's first year was that mothers reported needing to work because of financial needs.

12.2.6 Phased Retirement

Definition. The aging of the baby-boomer generation toward retirement, in combination with plummeting fertility rates, has elicited the specter of massive labor force shortages. This is particularly true for employees in demanding professional positions, who would choose retirement over continued full-time employment. Oftentimes these "mature workers" have a wealth of knowledge and experience highly valued by the organization. Phased retirement is a type of temporal workplace flexibility in which mature employees work "a part-time schedule or [step] down to less responsibility in preparation for retirement" (Hewitt Associates, 2008, p. 1). This may be a gradual or incremental process of reduced hours or reduced workload, anticipating the time of complete retirement. Phased retirement is now being requested as an alternative to full-time hours and full retirement (Drago et al., 2009).

Usage. Reports of the prevalence of phased retirement are varied, but indicate that use and interest in phased retirement is growing. In 1992, 6% of older workers in the United States reported that they were partially retired, which more than doubled in 6 years, with 13% partially retired in 1998 (Gustman and Steinmeier, 2000). Defining partial retirement as working less than 1,750 h annually (or less than 35 h a week 50 weeks of the year), 22% of working 50–58 year olds were partially retired. For those over age 80 who were still employed, 72% were partially retired (Haider and Loughran, 2001). Of the employers who would allow an employee to reduce hours before retirement, 36% have actually had an employee use this option in the last 3 years, with those in professional positions having the most opportunity for phased retirement (Hutchens, 2003). One indicator of phased retirement is examining the prevalence of part-time work among older adults. Older workers participating in part-time employment are not necessarily participating in a phased retirement program, but it does indicate an overall trend in reduction of hours. Compared with 13 other European countries, the Netherlands have substantially higher rates of part-time work among older men and women (Kantarci and Van Soest, 2008). In the United States, those who do continue to work in their elderly years are among the most educated, the wealthiest, and the healthiest of the elderly (Haider and Loughran, 2001).

Older employees have expressed their interest and expectations in some work during their retirement. Nearly 80% of baby boomers report that they expect to work at least part-time in retirement (Roper ASW, 2004). Most professional employers are willing to offer phased retirement programs (73%), but mostly as an informal arrangement (Hutchens, 2003; Hutchens and Grace-Martin, 2006). Interest and willingness, however, has not always equated with successful programs. European countries have had varying levels of success with phased retirement programs. Sweden, France and Germany have implemented successful phased retirement programs, while Denmark and Finland have had less success in launching such programs (Delsen, 1996; Reday-Mulvey, 2000).

Findings. Review of the scholarly literature indicates benefits associated with phased retirement, though few of the studies focus specifically on professional employees. In a study of 601 retired professionals and nonprofessionals from Australia, those who retired abruptly reported worse health than those who retired gradually (de Vaus et al., 2007). In a longitudinal US sample of 2,389 retired employees (not limited to professionals), there was no difference in the levels of happiness between those who retired abruptly and those who retired gradually. The difference came in the sense of control that workers felt about their retirement (Calvo et al., 2009). This emphasizes the importance of the perceived control or the availability of flexible work options (see Eaton, 2003 and Jones et al., 2008).

In the United States, there are various legal barriers to adopting a phased retirement program including confusion about current laws and policies of the IRS (Kelly et al., 2008). This supports the findings of Hutchens and Grace-Martin (2006) that employers are willing to permit the option of phased retirement, but as an informal arrangement. Also, employers who are willing to offer phased retirement usually have a flexible work environment of part-time schedules, job sharing options, and flexible schedules.

12.3 Conclusion

In summary, the preponderance of scholarly literature supports temporal workplace flexibility as a positive factor in work-life integration for professionals. A large body of literature documents that schedule flexibility is associated with less work-family conflict and enhanced job satisfaction. Being overworked is often a major cause of stress for professionals. Reduced-hours options, in the form of regular part-time employment, job sharing, transition from leaves, and phased retirement; all offer alternatives for professionals to continue their careers, albeit at a less stressful pace, while simultaneously having time and energy for their families and personal endeavors. In sum, temporal workplace flexibility is an essential component in a strategy to enhance work-life integration for professional employees as they attempt to provide for, and nurture their families in harmony.

Acknowledgements The authors thank the Family Studies Center of the BYU School of Family Life for its support of this project. Address correspondence to: E. Jeffrey Hill, School of Family Life, Brigham Young University, 2052 JFSB, Provo, UT 84602 USA (jeff_hill@byu.edu) (801) 422-9091.

References

Allen TD, Shockley K (2009) Flexible work arrangements: help or hype?. In Crane DR, Hill EJ (eds) Handbook of families and work: interdisciplinary perspectives. University Press of America, Lanham, MD, pp 265–284

Als C, Brautigam P (2006) Job-sharing in nuclear medicine: an 8-year experience (1998–2006). Nucl Med Rev 9:82–83

Armstrong-Stassen M (1998) Alternative work arrangements: meeting the challenges. Can Psychol 39:108–123

Aryee S (1992) Antecedents and outcomes of work-family conflict among married professional women: evidence from Singapore. Hum Relat 45:813–837

Baltes BB, Briggs TE, Huff JW, Wright JA, Neuman GA (1999) Flexible and compressed workweek schedules: a meta-analysis of their effects on work-related criteria. J Appl Psychol 84:496–513

Barnett RC (1998) Toward a review and reconceptualization of the work/family literature. Genet Soc Gen Psychol Monogr 124:125–182

Barnett RC (2003) Reduced hours work/part-time work. In Kossek EE, Pitt-Catsouphes M (eds) Work and family encyclopedia. Sloan Work and Family Research Network, Chestnut Hill, GA. http://wfnetwork.bc.edu/encyclopedia_entry.php?id=252&area=All. Accessed on 16 Feb 2010

Barnett. RC, Gareis KC (2000) Reduced-hours employment: the relationship between difficulty of trade-offs and quality of life. Work Occup 27:168–187

Barnett RC, Gareis KC (2002) Full-time and reduced-hours work schedules and marital quality. Work Occup 29:364–379

Barnett RC, Hall DT (2001) How to use reduced hours to win the war for talent. Organ Dyn 29:192–210

Baxter J, Gray M (2008) Paid maternity, paternity and parental leave. Submission from the Australian Institute of Family Studies to the Productivity Commission Inquiry. http://www.aifs.gov.au/institute/pubs/parental-leave.pdf. Accessed on 9 Feb 2010

Branine M (2003) Part-time work and job sharing in health care: is the NHS a family-friendly employer?. J Health Organ Manage 17:53–68

Browning DS, Miller-McLemore BJ, Couture PD, Lyon KB, Franklin RM (2000) From culture wars to common ground. Westminster John Knox Press, Louisville, KY

Brunning G, Plantenga J (1999) Parental leave and equal opportunities: experiences in eight European countries. J Eur Soc Pol 9:195–209

Budd JW, Mumford KA (2006) Family-friendly work practices in Britain: availability and perceived accessibility. Hum Resour Manage 45:23–42

Calvo E, Haverstick K, Sass SA (2009) Gradual retirement, sense of control, and retirees' happiness. Res Aging 31:112–135

De Cieri H, Holmes B, Abbott J, Pettit T (2005) Achievements and challenges for work/life balance strategies in Australian organizations. Int J Hum Resour Manage 16:90–103

de Vaus D, Wells Y, Kendig H, Quine S (2007) Does gradual retirement have better outcomes than abrupt retirement? Results from an Australian panel study. Ageing Soc 27:667–682

Delsen L (1996) Gradual retirement: lessons from the Nordic countries and the Netherlands. Eur J Ind Relat 2:55–67

Drago R, Wooden M, Black D (2009) Who wants and gets flexibility?: changing work hours preferences and life events. Ind Relat Rev 62:394–414

Eaton SC (2003) If you can use them: flexible policies, organizational commitment, and perceived performances. Ind Relat 42:145–167

Edwards JR (1996) An examination of competing versions of the person-environment fit approach to stress. Acad Manage J 39:292–339

Facer RL, Wadsworth L (2008) Alternative work schedules and work-family balance: a research note. Rev Publ Pers Adm 28:166–177

Fast JE, Frederick JA (1996) Working arrangements and time stress. Can Soc Trends Winter:14–19

Feldman R, Sussman AL, Zigler E (2004) Parental leave and work adaptation at the transition to parenthood: individual, marital, and social correlates. J Appl Dev Psychol 25:459–479

FlexPaths (n.d.) Glossary. http://www.flexpaths.com/resources/glossary. Accessed on 9 Feb 2010

Freeman BJ, Coll KM (2009) Solutions to faculty work overload: a study of job sharing. Career Dev Q 58:65–70

Frone MR, Yardley JK (1996) Workplace family-supportive programmes: predictors of employed parents' importance ratings. J Occup Organ Psychol 69:351–366

Galinsky E, Bond JT, Sakai K (2008) 2008 National study of employers. Families and Work Institute, New York, NY

Grzywacz JG (2009) Workplace flexibility: implications for worker health and families. In Crane DR, Hill EJ (eds) Handbook of families and work: interdisciplinary perspectives. University Press of America, Lanham, MD, pp 219–240

Grzywacz JG, Carlson DS, Shulkin S (2008) Schedule flexibility and stress: linking formal flexible arrangements and perceived flexibility to employee health. Commun Work Fam 11:199–214

Gustman AL, Steinmeier TL (2000) Retirement outcomes in the health and retirement Study. Working Paper No. 7588. National Bureau of Economic Research, Cambridge

Haider S, Loughran D (2001) Elderly labor supply: work or play? RAND Working Paper, DRU-2582. http://www.rand.org/pubs/drafts/2008/DRU2582.pdf. Accessed on 9 Feb 2010

Hayman J (2009) Flexible work arrangements: exploring the linkages between perceived usability of flexible work schedules and work/life balance. Commun Work Fam 12:327–338

Hegewisch A, Cornick JC (2008) Statutory routes to workplace flexibility in cross-national perspective. Institute for Women's Policy Research, Washington, DC. http://www.iwpr.org/pdf/B258workplaceflex.pdf. Accessed on 9 Feb 2010

Hewitt Associates (2008) Timely topics: about flexible work arrangements. http://www.hewittassociates.com/_MetaBasicCMAssetCache_/Assets/Articles/2008/TimelyTopics_FlexibleWorkArrangements.pdf. Accessed on 9 Feb 2010

Hill EJ (1992 August) Creating a more flexible and a more productive workplace. American Psychological Association, 100th Annual Convention, Washington, DC, pp 14–18

Hill EJ, Grzywacz JG, Allen S, Blanchard VL, Matz-Costa C, Shulkin S, Pitt-Catsouphes M (2008a) Defining and conceptualizing workplace flexibility. Commun Work Fam 11:149–163

Hill EJ, Hawkins AJ, Ferris M, Weitzman M (2001) Finding an extra day a week: the positive effect of job flexibility on work and family life balance. Fam Relat 50:49–58

Hill EJ, Jacob JI, Shannon LL, Brennan RT, Blanchard VL, Martinengo G (2008b) Exploring the relationship of workplace flexibility, gender, and life stage to family-to-work conflict, and stress and burnout. Commun Work Fam 11:165–181

Hill EJ, Jacob JI, Ferris M (2009) Workplace flexibility and the relationship of work hours to work-life difficulty: comparisons in four regions representing 75 countries., 3rd International Community, Work and Family Conference, Utrecht/Netherlands, 16–18 April 2009

Hill EJ, Martinson V, Ferris M (2004a) New-concept part-time employment: a work-family adaptive strategy for women professionals with small children. Fam Relat 53:282–292

Hill EJ, Martinson V, Ferris M (2004b) Beyond the mommy track: the influence of new-concept part-time work for professional women on work and family. J Fam Econ Issues 25:121–136

Hill EJ, Mead NT, Dean LR, Hafen DM, Gadd R, Palmer AA, Ferris M (2006) Researching the 60-h dual-earner work week: an alternative to the "opt out revolution." Am Behav Sci 49: 1184–1203

Human Resources and Skills Development Canada (n.d.) Flexibility in hours of work: work and family provisions in Canadian collective agreements. http://www.hrsdc.gc.ca/eng/lp/spila/wlb/wfp/07Flexibility_in_Hours_of_Work.shtml. Accessed on 9 Feb 2010

Hutchens R (2003) The Cornell study of employer phased retirement policies: a report on key findings. http://digitalcommons.ilr.cornell.edu/lepubs/1/. Accessed on 9 Feb 2010

Hutchens R, Grace-Martin R (2006) Employer willingness to permit phased retirement: why are some more willing than others?. Ind Lab Relat Rev 59:525–546

Jones BL, Scoville DP, Hill EJ, Childs G, Leishman JM, Nally KS (2008) Perceived versus used workplace flexibility in Singapore: predicting work-family fit. J Fam Psychol 22:774–783

Kane D (1999) Job sharing: a retention strategy for nurses. Can J Nurs Adm 12:16–22

Kantarci T, Van Soest A (2008) Gradual retirement: preferences and limitations. De Econ 156: 113–144

Kelly EL, Dahlin EC, Spencer D, Moen P (2008) Making sense of a mess: phased retirement policies and practices in the United States. J Workplace Behav Health 23:147–164

Latack JC, Foster LW (1985) Implementation of compressed work schedules- participation and job redesign as critical factors for employee acceptance. Pers Psychol 38:75–92

Lee MD, MacDermid SM (1998) Improvising new careers: accommodation, elaboration, transformation. The Center for Families at Purdue University, West Lafayette

Lo S (2003) Perceptions of work-family conflict among married female professionals in Hong Kong. Pers Rev 32:376–390

Martens MFJ, Nijhuis FJN, van Boxtel MPJ, Knottnerus JA (1999) Flexible work schedules and mental and physical health: a study of a working population with non-traditional working hours. J Organ Behav 20:35–46

Meiksins P, Whalley P (2002) Putting work in its place: a quiet revolution. Cornell University Press, Ithaca, NY

OECD.Stat Extracts (2009) Incidence of FTPT employment – common definition. Online OECD employment database. http://www.stats.oecd.org/Index.aspx?DatasetCode=FTPTC_I. Accessed on 27 Nov 2009

Pershun PG (1992) Job sharing – two who made it work. Am J Nurs 92:75–80

Raabe PH (1998) Being a part-time manager: one way to combine family and career. In Vannoy D, Dubeck PJ (eds) Challenges for work and family in the twenty-first century. Aldine de Gruyter, New York, NY, pp 81–92

Ray R, Gornick JC, Schmitt J (2008) Parental leave policies in 21 countries: assessing generosity and gender equality. http://www.cepr.net/documents/publications/parental_2008_09.pdf. Accessed on 9 Feb 2010

Reday-Mulvey G (2000) Gradual Retirement in Europe. J Aging Soc Pol 11:49–60

Ronen S (1981) Flexible working hours: an innovation in the quality of work life. McGraw-Hill, New York, NY

Roper ASW (2004) Baby Boomers envision their retirement II- key findings: survey of Baby Boomers' expectations for retirement. American Association of Retired Persons, Washington, DC

Rosa RR (1991) Performance, alertness and sleep after 3–5 years of 12 h shifts: a follow-up study. Work Stress 5:107–116

Schneider A (2000) Job-sharing rises as professors seek flexible schedules. Chron High Educ 47:10–12

Tanaka S (2005) Parental leave and child health across OECD countries. Econ J 115:F7–F28

The Hudson Report (2006) Jobsharing: a fresh look at flexible working. Part three. HR Insights, Australia

Turner L (1996) Time out with half-time: job sharing in the Nineties. Can J Couns 30:104–113

Voydanoff P (2007) Work, family, and community: exploring interconnections. Lawrence Erlbaum, Mahwah, NJ

Wadsworth L (2009) Compressed workweek schedules. In Kossek EE, Pitt-Catsouphes M (eds) Work and family encyclopedia. Sloan Work and Family Research Network, Chestnut Hill, GA, http://wfnetwork.bc.edu/encyclopedia_entry.php?id=16389&area=All. Accessed 9 February 2010

Chapter 13
Causes and Consequences of the Utilization of Work-Life Policies by Professionals: "Unconditional Supervisor Support Required"

Bram Peper, Josje Dikkers, Claartje Vinkenburg, and Marloes van Engen

13.1 Introduction

The European workplace has changed. Employees increasingly ask for organizational policies that allow them to combine their work and their private lives (Lewis et al., 2009). In the Netherlands it is estimated that no less than 40% of employees face troubles in combining their work and private lives (Geurts et al., 2003), which brings high costs, both for individuals and for organizations (Allen et al., 2000). Organizations also increasingly have become aware of the work-life conflicts of their employees. Work-life conflict is often defined as a form of inter role conflict in which the role pressures from the work and family domains are mutually incompatible in some respect. That is, "participation in the work (family) role is made more difficult by virtue of participation in the family (work) role" (Greenhaus and Beutell, 1985, p. 77). In addition to these negative influences between work and home, researchers nowadays also differentiate positive influences between the two domains: work-life enrichment or facilitation. Greenhaus and Powell (2006) recently defined work-life enrichment as "the extent to which experiences in one role improve performance or the quality of life in the other role" (p. 73).

According to recent work-life studies, the utilization of work-life policies does not unequivocally lead to lower levels of conflict or to higher levels of enrichment between work and life. Kossek et al. (2006), for example, found that formal access to telework may not necessarily reduce work-to-life conflict for professionals who typically already have some informal job autonomy. Formal use of telework was, however, significantly related to higher performance, although other work-life policies were not. It is, therefore, vital to distinguish between different types of policies; policies that enable employees to work while they can hire others for their caring tasks, and policies that give employees more flexibility and control over their working time which enables them to fulfill the caring tasks themselves (cf. Appelbaum et al., 2005). In this study, we examine the association of the utilization of flexible

B. Peper (✉)
Erasmus University Rotterdam, Rotterdam, Netherlands
e-mail: peper@fsw.eur.nl

S. Kaiser et al. (eds.), *Creating Balance?*, DOI 10.1007/978-3-642-16199-5_13,

policies (targeted at increasing employees' temporal and spatial flexibility) and care-related policies (aimed at assisting employees in combining work with care giving) with work-life enrichment of professionals.

Research further suggests that managerial support is critical when it comes to the utilization and effectiveness of work-life programs (e.g. Allen et al., 2000; Fried, 1999; Maxwell, 2005; Perlow, 1997; Thompson et al., 1999; Veenis, 2000), as it is up to managers or supervisors to communicate, implement and manage work-life policies in organizations (Lewis, 2003). In addition, research consistently shows that the level of support that employees receive from their supervisor is crucial in alleviating conflicts between work and life (e.g., Mesmer-Magnus and Viswesvaran, 2006). Supervisor support is a core aspect of work-life culture, or "the shared assumptions, beliefs, and values regarding the extent to which an organization supports and values the integration of employees' work and private lives" (Thompson et al., 1999, p. 394). Case study research emphasizes that managerial attitudes and practices are shaped by the organizational culture in which managers operate. A manager's response to a request is influenced not only by official policy, but also by the "unwritten rules" of an employing organization. Such "unwritten rules" (the set of shared assumptions, opinions and values, also referred to as "organizational culture") are a decisive factor in managers' attitudes (Lewis and Taylor, 1996; Perlow, 1995). Fried (1999) shows that in a company with a typical "overtime" culture – one in which working long hours is regarded as a sign of productivity and commitment – managers regard requests to take parental leave or the actual utilization of such leave as contrary to the prevailing standards. In their view, taking parental leave is a sign of a "negative" attitude, with all that that implies for the relevant employee's career. On the one hand, managers are influenced by the organizational culture in which they operate (Kossek and Friede, 2006); on the other hand, the way managers respond to requests by employees can change – or maintain – an organizational culture. Consequently, this study also relates work-life culture in general and supervisor support in particular to the uptake of (flexible and care-related) work-life policies and work-life enrichment.

Finally, few previous studies have examined the associations of work-life integration with professionals' career progression. In their review of work-life studies, Eby et al. (2005) showed that career attitudes (e.g., career mobility and career satisfaction) were not frequently studied as criteria (2.9% of the studies included in the review). Objective career success in particular has rarely been related to work-life balance. Here, we study the associations of utilization of work-life policies, work-life culture and work-life enrichment with professionals' career advancement.

In this chapter, the utilization and management of work-life policies by professionals is the central focus. In Sect. 13.2 we will analyze the utilization of work-life policies in association with work-life culture, work-life enrichment, and career advancement among Dutch professionals. This analysis is based on survey research in two samples.

In Sect. 13.3 we elaborate on the role of supervisors in the (non-)utilization of policies by professionals employed by a Dutch financial organization. This analysis is based on qualitative interviews. Supervisors were interviewed on the use of work-life policies by their employees, and their attitude toward granting work-life policy requests. Section Sect. 13.4 is devoted to the conclusion and discussion.

13.2 Utilization of Work-Life Policies by Dutch Professionals in Association with Work-Life Culture, Work-Life Enrichment and Career Advancement

This section aims to answer four research questions:

1. What is the utilization of work-life policies among Dutch professionals?
2. What is the association of work-life policies with work-life culture?
3. What is the association of work-life policies and work-life culture with work-life enrichment?
4. What is the association of utilization of work-life policies with career advancement?

These questions will be answered by examining two Dutch samples: (i) a Dutch subsidiary of a financial consultancy with headquarters in the United States of America ($N = 638$) that will be referred to as Company X, and (ii) a more heterogeneous sample with Dutch professionals from different sectors ($N = 131$) which we will refer to as Sample Y. In both samples, the same questionnaire has been administered, in order to collect data on professionals' utilization of work-life policies, their work-to-life (and life-to-work) balance or enrichment (measured with the SWING; Geurts et al., 2005), and their perceptions of their company's work-life culture (see Dikkers et al., 2007).

In this section, four hypotheses will be developed on the basis of the four research questions described above. A graphical depiction of the relationships between the main variables is given in Fig. 13.1.

13.2.1 Utilization of Work-Life Policies by Professionals

In order to answer our first research question – What is the utilization of work-life policies among Dutch professionals? –, we selected those employees with a job at

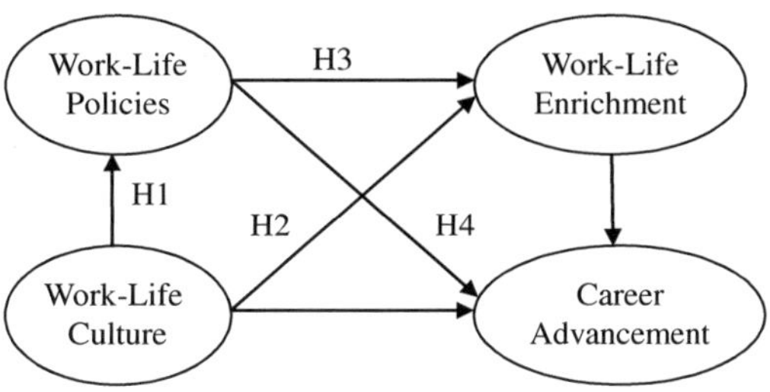

Fig. 13.1 Research model reflecting the hypotheses tested in Sect. 13.2.1

the (lower, intermediate, or higher/top) management level. Thus we ensured that all the employees in the sample were professionals with executive responsibilities. The work-life policies selected can be divided into *flexible* (i.e., targeted at increasing employees' spatial or temporal flexibility) and *care-related* (i.e., aimed at assisting employees in combining work with care giving responsibilities) arrangements. We examined professionals' utilization of four flexible work-life policies (flextime, working from home, part-time work, and compressed workweek) and two care-related work-life policies (parental leave, and childcare subsidies by the employer). Two of these policies are the subject of legislation in the Netherlands (i.e., part-time work, and parental leave) and therefore can be used by all employees who are eligible, whereas the other arrangements are part of the selected companies' employment contracts. In Company X, the compressed workweek was not offered.

In Table 13.1, the percentages of professionals (parents versus non-parents, and men versus women) using these work-life policies within the two samples is given. In Company X, flexible working times (flextime; 51%) and working from home (40%) were the most frequently used policies in the total sample, followed by part-time work (29%), childcare arrangements (12%), and parental leave (5%). We found two significant gender differences in utilization of work-life policies, with men making more use of working from home ($t_{(1,592)} = 4.35, p < 0.001$) and women working part time more frequently ($t_{(1,451)} = -9.97$, $p < 0.001$). Furthermore, parents made significantly more use of flextime ($t_{(1,451)} = 3.01$, $p < 0.01$), working from home ($t_{(1,414)} = 5.23, p < 0.001$), and part-time work ($t_{(1,393)} = 4.26, p < 0.001$) compared with professionals without children in Company X.

In the more heterogeneous Sample Y, flextime (63%), working from home (35%), and part-time work (25%) also constituted the most popular policies. The arrangements that were used least frequently were the compressed workweek (8%), childcare subsidy (6%), and parental leave (3%). Again, women worked part time more often than men ($t_{(1,129)} = 5.81$, $p < 0.001$). In the Netherlands, more than 60% of women work part time, working on average 24 h per week (OECD, 2007). Consequently, the Netherlands has the highest rate of female part-time work throughout the OECD countries. Parents also used this particular policy more often compared with professionals without children in this sample ($t_{(1,126)} = 3.10$, $p < 0.01$). This renders additional support to the idea that Dutch parents (in particular mothers) use part-time work as a means to combine work with care giving responsibilities.

13.2.2 Work-Life Policies in Relation to Work-Life Culture

The second research question guiding this section – What is the association of work-life policies with work-life culture? – focuses on the uptake of work-life policies in association with work-life culture (Thompson et al., 1999, p. 394). According to a review of work-life culture studies (Kinnunen et al., 2005), employees experiencing a supportive culture toward the integration of work and private life make more frequent use of work-life policies. Here we distinguish between two central dimensions

Table 13.1 Professionals' utilization of work-life policies (in percentages)

Policies	Company X					Sample Y				
	Total ($N = 638$)	Men Parents ($N = 138$)	($N = 354$) Non-parents ($N = 216$)	Women Parents ($N = 88$)	($N = 284$) Non-parents ($N = 196$)	Total ($N = 131$)	Men Parents ($N = 51$)	($N = 74$) Non-parents ($N = 23$)	Women Parents ($N = 30$)	($N = 57$) Non-parents ($N = 24$)[a]
Flextime	51.1	59.0	49.5	59.5	43.3	63.4	66.7	52.2	66.7	58.3
Working from home	39.9	63.2	37.4	39.0	26.5	35.1	35.3	30.4	43.3	29.2
Part time	29.1	10.4	15.5	86.9	31.9	25.2	7.8	4.3	73.3	16.7
Compressed workweek	–	–	–	–	–	8.4	5.9	8.7	13.3	8.3
Parental leave	5.0	6.1	–	25.6	–	3.1	2.0	–	10.0	–
Childcare subsidy	11.6	22.1	–	46.4	–	6.1	11.8	–	6.7	–

[a]Due to missing values not all subgroups add up to $N = 131$ in Sample Y; Significant subgroup differences are reflected by: $** = p < 0.01$, $*** = p < 0.001$

of work-life culture (see Dikkers et al., 2004, 2007): (i) *support* from the organization, supervisor and colleagues regarding the integration of employees' work and family life, and (ii) *barriers* reflected by negative career consequences associated with the utilization of work-life arrangements and organizational demands to make long work days.

We expect that employees who perceive their company's culture toward work-life balance and the uptake of work-life policies to be supportive will feel more comfortable in using these arrangements than employees perceiving a less supportive or even hindering work-life culture. Therefore, it is expected that employees perceiving a supportive work-life culture will make more use of work-life policies compared with those perceiving an unsupportive culture.

> Hypothesis 1: High levels of work-life culture support and low levels of barriers to it are related to high utilization of work-life policies

In order to test this hypothesis, we performed a series of hierarchical regression analyses. The work-life policies were entered separately as dependent variables, and in a first step, the covariates (i.e., gender, and parental status) were entered as the independent variables. In a second step, work-life culture (support and barriers) was added to the covariates. The results of these regression analyses for those policies that are significantly related to perceptions of work-life culture are represented in Table 13.2 for both samples.

In Sample Y, the utilization of two policies was significantly related to work-life culture. High levels of culture supportive of work-life integration were associated with high utilization of working from home ($\beta = 0.29$, $p < 0.001$). In other words, employees who perceive their organization's culture as supportive toward the integration of work and private life work from home more frequently compared with those who perceive the culture to be less supportive. Furthermore, barriers to a culture supportive of work-life integration were positively related to utilization of childcare subsidies ($\beta = 0.22$, $p < 0.05$). Possibly, the causal path linking culture to this specific policy is "reversed" in nature: Only when employees start using this policy will they be confronted with barriers to the integration of work and private life.

In Company X, positive associations were found between cultural barriers to work-life integration and the utilization of three policies; working from home ($\beta = 0.14$, $p < 0.001$), childcare subsidy ($\beta = 0.10$, $p < 0.01$), and parental leave ($\beta = 0.09$, $p < 0.05$). Again, it may be possible that as soon as employees start using these particular policies they will be confronted with barriers to the integration of work and private life. The work-life culture within this particular company may be predominantly "contradictory" or "obstructing" in nature. This means that employees perceive high levels of both support and barriers, respectively, or low levels of support and high barriers in balancing work with private life.

In sum, Hypothesis 1 (High levels of work culture supportive of work-life integration and low levels of barriers to it are related to high utilization of work-life policies) is only supported for the utilization of working from home in Sample Y.

Table 13.2 Covariates (Model 1) and work-life culture (Model 2) in relation to work-life policy uptake in Sample Y and Company X

	Sample Y ($N = 131$)				Company X ($N = 638$)			
	Childcare subsidy		Working from home		Childcare subsidy		Working from home	
	Model 1	Model 2	Model 1	Model 2	Model 1	Model 2	Model 1	Model 2
Gender[a]	–0.06	−0.02	0.05	0.04	0.15***	0.16***	−0.16***	−0.15***
Parental status[b]	19*	0.21*	0.09	0.07	0.48***	0.48***	0.20***	0.19***
Culture – support	–	−0.04	–	0.29**	–	−0.01	–	0.03
Culture – barriers	–	0.22*	–	0.16	–	–	–	0.14***
F	2.79	3.43*	0.60	2.63*	95.45***	50.51***	22.08***	14.02***
ΔR^2	0.04	0.10*	0.01	0.08*	0.24***	0.25*	0.07***	0.08**

[a]1 = men, 2 = women
[b]0 = non-parents, 1 = parents
$* = p < 0.05$, $** = p < 0.01$, $*** = p < 0.001$

13.2.3 Work-Life Policies and Work-Life Culture in Relation to Work-Life Enrichment

As part of the third research question – What is the association of work-life policies and work-life culture with work-life enrichment? – we wanted to examine the utilization of work-life policies and work-life culture in relation to work-life enrichment. In their recent meta-analysis, Mesmer-Magnus and Viswesvaran (2006) examined the effectiveness of five facets of family-friendly work environments in reducing work-life conflict. The 38 studies (total $N = 13{,}605$) included in the meta-analysis showed that a family-friendly work culture was the most influential factor in reducing work-life conflict.

This suggests that perceiving a culture supportive of the integration of work and private life is favorable in terms of reduced levels of work-life conflict. In a similar vein, high levels of support and low levels of barriers are expected to be related to increased levels of work-life enrichment (Ten Brummelhuis, 2009; Van Steenbergen, 2007). Greenhaus and Powell (2006) recently defined work-life enrichment as "the extent to which experiences in one role improve performance or the quality of life in the other role" (p. 73). This definition is inherently bi-directional, and comprises both positive spillover from work to private life (work-life enrichment), and positive spillover from private life to work (life-work enrichment). We expect that employees who believe that their supervisor – or the organization in general – supports them in their integration of work and private life will perceive their work in a positive light, which will improve their quality of life or even their performance of tasks at home (i.e., work-life enrichment).

Hypothesis 2: High levels of work-life culture support and low levels of barriers are related to high levels of work-life enrichment.

Moreover, we examined the association of the utilization of work-life policies with work-life enrichment. When employees use policies aimed at facilitating them in integrating work and private life, they are expected to experience higher levels of work-life enrichment. More specifically, those using flexible arrangements (e.g., flextime) will have more freedom in adjusting their work to their responsibilities at home. This may increase the extent to which experiences in one role (either at work or at home) improve their performance or quality of life in the other role (i.e., work-life and life-work enrichment). Additionally, employees using care-related policies (e.g., parental leave) are assisted in fulfilling their care-related responsibilities, which may also increase positive spillover between work and private life. Therefore, we expect the following:

Hypothesis 3: Employees using work-life policies will report high levels of work-life and life-work enrichment.

To test these hypotheses, work-to-life and life-to-work enrichment were entered as the dependent variables in a series of regression analyses. The covariates were entered as the independent variables in a first step, and utilization of the work-life policies was added in a second step; in a final step, work-life culture was entered as an independent variable. The results of these regression analyses are given in Table 13.3 for both samples.

Table 13.3 Covariates (Model 1), work-life policy uptake and work-life culture (Model 2/3) in relation to work-life enrichment in Company X and Sample Y

	Company X (N = 638)				Sample Y (N = 131)					
	Work-life enrichment		Life-work enrichment		Work-life enrichment			Life-work enrichment		
	Model 1	Model 2	Model 1	Model 2	Model 1	Model 2	Model 3	Model 1	Model 2	Model 3
Gender[a]	0.00	–0.02	0.10*	0.09*	0.13	0.12	0.11	0.01	–0.10	–0.10
Parental status[b]	0.00	0.01	0.08*	0.08*	–0.07	–0.09	–0.10	–0.08	–0.15	–0.15
Working from home		–		–	–	0.21*	0.16	–	–	–
Part time work		–		–	–	–	–	–	0.22*	0.22*
Culture – support		0.15***		0.15***	–	–	0.21*	–	–	0.05
Culture – barriers		–0.04		0.00	–	–	0.09	–	–	0.06
F	0.00	4.65***	4.98**	6.01***	1.52	2.93*	2.67*	0.40	1.71	1.10
ΔR^2	0.00	0.02***	0.01**	0.03***	0.02	0.07*	0.10	0.01	0.04*	0.04

[a] 1 = men, 2 = women
[b] 0 = non-parents, 1 = parents
$* = p < 0.05$, $** = p < 0.01$, $*** = p < 0.001$

In Company X, utilization of the work-life policies was not significantly related to work-life enrichment or life-work enrichment. Work-life culture support was significantly and positively associated with both types of enrichment ($\beta = 0.15$, $p < 0.001$); professionals perceiving high levels of support in integrating work and private life reported high levels of positive spillover between both domains. However, only small amounts of variance in enrichment were explained by the covariates, work-life policies, and work-life culture in this company (varying from $R^2 = 0.00$, ns to $R^2 = 0.03$, $p < 0.001$).

In Sample Y, only utilization of working from home was related significantly to work-life enrichment ($\beta = 0.21$, $p < 0.05$). After adding work-life culture support to the equation in the third step ($\beta = 0.21$, $p < 0.05$), this association disappeared, however ($\beta = 0.16$, ns). Since utilization of this policy is significantly related to work-life culture support ($\beta = 0.23$, $p < 0.01$), this finding may hint at full mediation of the association of working from home with work-life enrichment via support. In other words, employees working from home perceive their organization as supportive of the integration of work and private life, and therefore report high levels of work-to-life enrichment. Furthermore, employees working part time reported high levels of life-to-work enrichment ($\beta = 0.22$, $p < 0.05$). Work-life culture was not related significantly to life-work enrichment in this sample.

In sum, Hypothesis 2 (High levels of work-life culture support and low levels of barriers are related to high levels of work-life enrichment) is supported for the association of work-life culture support and work-life enrichment in Sample Y, and for the association of support with both types of enrichment in Company X. Hypothesis 3 (Employees using work-life policies will report high levels of work-life and life-work enrichment) is supported for two policies (i.e., working from home and part time work) in Sample Y.

13.2.4 Work-Life Policies in Relation to Career Advancement

A career is often defined as "the evolving sequence of a person's work experiences over time" (Arthur et al., 1989, p. 8). Previous research conceptualizes career success as objective (extrinsic) and subjective (intrinsic) success (e.g., Judge et al., 1995; Ng et al., 2005). Extrinsic career success refers to objectively observable career accomplishments (e.g., salary) whereas intrinsic success relates to the individual's own subjective feelings regarding his/her career accomplishment (e.g., career or job satisfaction). In the current study, we focus on job level as an indicator of objective or extrinsic career success.

In order to answer the final research question of this section – What is the association of utilization of work-life policies with career advancement? –, the utilization of flexible and care-related work-life policies (see Sect. 13.2.2 for a description of the selected policies) by Dutch professionals was related to job level among the heterogeneous sample of professionals from different sectors ($N = 131$) which was previously referred to as Sample Y.

Previous research has found that availing themselves of work-life policies may harm employees' career progress (Anderson et al., 2002; Thompson et al., 1999). Judiesch and Lyness (1999), for example, found that managers who took leave of absence were promoted less often and received smaller salary increases. Using these arrangements may signal lower commitment and perhaps even lower performance to the employer (Almer et al., 2004; Glass, 2004). In some organizational cultures, the amount of time spent at work is seen as the best indication of investment and career dedication by working parents (Perlow, 1995; Starrels, 1992). It is therefore not surprising that prior empirical research has indicated that transition to part-time work often has a negative influence on employees' career options (Higgins et al., 2000; Tomlinson, 2006). Consequently, we expect that work-life policies that reduce professionals' work hours (i.e., part time work, and parental leave) will result in negative career consequences. Utilization of the other flexible or care-related arrangements, however, does not significantly reduce the amount of work-related time, thereby not negatively affecting the professionals' career prospects.

Hypothesis 4: Employees working part time or taking up parental leave have lower job levels compared to those working full time or not using parental leave.

In a series of regression analyses, career advancement (i.e., job level) was entered as the dependent variable. The covariates (i.e., gender and parental status) were entered as the independent variables in a first step, and utilization of the work-life policies was added in a second step. Because we are also interested in the associations of work-life culture and enrichment with career advancement, work-life culture and work-to-life and life-to-work enrichment were added to the independents in a third and fourth step respectively. The final results of these regression analyses are represented in Table 13.4 for Sample Y.

Table 13.4 Covariates (Model 1), policy uptake (Model 2), work-life culture (Model 3), and enrichment (Model 4) in relation to job level in Sample Y ($N = 131$)

	Job level			
	Model 1	Model 2	Model 3	Model 4
Gender[a]	−0.10	0.05	0.02	0.00
Parental status[b]	0.02	0.13	0.11	0.10
Part-time work[c]		−0.25*	−0.25*	−0.24*
Parental leave		−0.21*	−0.20*	−0.19*
Compressed week		0.15#	0.14	0.14
Culture – support			0.09	0.08
Culture – barriers			−0.12	−0.12
Work-life enrichment				0.06
Life-work enrichment				−0.07
F	0.64	3.09*	2.85**	2.22*
ΔR^2	0.01	0.11**	0.14	0.15

[a]1 = men, 2 = women
[b]0 = non-parents, 1 = parents; c 0 = no usage, 1 = usage
$\# = p < 0.10$, $* = p < 0.05$, $** = p < 0.01$, $*** = p < 0.001$

In Sample Y, employees working part time ($\beta = -0.25, p < 0.05$) or using parental leave ($\beta = -0.21, p < 0.05$) apparently have lower job levels compared with those working full time or not taking up parental leave. Usage of compressed workweek was marginally related to job level ($\beta = 0.15, p < 0.10$); employees working on a 4 × 9 schedule (compressed workweek) reported higher job levels than those working on a regular schedule. Work-life culture and enrichment were not related significantly to job level in this sample.

In sum, Hypothesis 4 (Employees working part time or taking up parental leave have lower job levels compared with those working full time or not using parental leave) is supported in Sample Y. The results presented in this section all point to the influence that companies – direct supervisors in particular – can have on professionals' uptake of work-life policies, their work-life enrichment or integration, and their career advancement. In order to gain a more in-depth understanding of this influence, we will now examine the role that supervisors' attitudes and behavior play in professionals' utilization of work-life policies with the aid of a qualitative case study.

13.3 Role of Supervisors in Utilization of Work-Life Policies

13.3.1 The Role of Supervisors for the Utilization of Work-Life Policies

Despite the key role of supervisors in the practical implementation of work-life policies, few researchers have examined the factors that shape managerial attitudes and behavior (Den Dulk and De Ruijter, 2008; Lewis, 2003; Poelmans and Beham, 2005). Supervisors can influence the utilization of work-life policies in several ways. To start with, their influence lies in their response to work-life requests. Secondly, they influence the utilization of work-life policies by the way they supervise employees who make these requests on a day-to-day basis. Finally, their own use of work-life policies and/or the way they manage their own work-life issues can have an influence (Lewis, 2003).

Supervisors' behavior toward work-life policies may vary considerably. Yeandle et al. (2003) found four types of supervisors in relation to their knowledge and awareness of work-life policies. In order of decreasing interest in work-life policies, they distinguish: supervisors who take a "progressive" approach to work-life issues; supervisors who have a "vague" understanding of family-friendly policies; supervisors who displayed ignorance of family-friendly policies; and supervisors who were "resistant" to the family-friendly approach.

In this section we analyze the experiences and attitudes of supervisors concerning the utilization of work-life policies by their subordinates, based on interviews with managers in Company X. First we focus on the vision of the supervisors regarding the extent to which Company X is a caring organization. We then address the way supervisors deal with the two types of work-life policies, e.g., flexible and care-related policies. Therefore, our fifth research question is: What are the experiences with and attitudes of supervisors regarding work-life policies? We conclude by discussing the behavior and attitudes on the allowance decisions of supervisors with regard to granting work-life policy requests of subordinates. Our final research question is: How can we understand the allowance decisions made by supervisors on work-life policy request by subordinates?

13.3.2 Supervisory Experiences with and Attitudes on Flexible and Care-Related Policies

In the fall of 2002 we interviewed eleven supervisors at their work place. These interviews were part of a larger research project regarding the utilization of work-life policies at Company X. All the supervisors were partners in Company X, and they were responsible for approving the work-life requests of their subordinates. The organizational work-life culture is very important for the utilization of work-life policies by employees, and it is also an important factor for their work-life balance. Company X is characterized by a contradictory work-life culture, with much support for work-life issues but at the same time barriers like time and career constraints (Den Dulk and Peper, 2007). Most of the supervisors experience Company X as an organization that takes good care of its employees and their work-life balance.

> For people to be able to deal with the amount and pace of the work, as [an] organization you need to facilitate employees in such a way they can handle their work fitting their personal situation. (M007, female, 1964)

The contradictory aspect of the work-life culture is also clearly mentioned by the supervisors. To be a caring organization in the highly competitive financial sector in which Company X operates can be walking a tight rope.

> As [an] organization we have to watch out not to proclaim two contradictory messages; because on the one hand we want to do a lot in relation to the work-life balance, and on the other hand we have productivity targets on which we put emphasis. These are messages which are at right angles to each other. (M007, female, 1964)

The contradictory aspect of the work-life culture at Company X was a constant factor in the interviews with the supervisors when we discussed their experiences and opinions on flexible and care-related policies.

13.3.2.1 Flexible Policies

We elaborate on three types of flexible policies: part-time work, working a day at home and telework. Most supervisors had subordinates using the part-time policy, which is a statutory right in the Netherlands.[1] The experiences with part-time work were mixed; good examples as well as bad examples were mentioned. However,

[1]Employees are allowed to adjust their working hours, unless it conflicts heavily with organizational interests (Den Dulk, 2001).

when we asked the supervisors for their opinion of this policy, they were much more outspoken. Almost all supervisors mentioned a four-day week as the absolute minimum if you want to make a career within this company. On top of that, the employee is expected to be flexible.

> I am not positive [about part time work]. I have nothing against it. Some part timers work for quite a while at our organization, and I am glad they keep working with us. The problem, however, is when people start working part time they lose their flexibility. Often as a result of having children. If the children have to be picked up from kindergarten, an employee cannot do a job which requires flexibility. [...] Four days a week will just do, but three is really not enough. (M003, male, 1960)

Although part-time work is a statutory right, and supervisors are not unsupportive towards its use, part-time work can interfere with managerial control. Most subordinates will be evaluated on their output, and visibility at the work place still is an important aspect for supervisors.

> Sometimes it drives me crazy because I don't know who is working when, but I do have good experiences [with part time work]. (M007, female, 1964)

Company X is an organization where clients are of utmost importance, and some of the supervisors relate their opinion of the use of part-time work explicitly to the expectations of the clients.

> Four days is all right, that's the law and nobody bothers. However, for us it becomes problematic when it becomes less then four days, in relation to our clients. You are not accessible for the client. If the client accepts that, it's ok. We have two 'girls' working for three days, and they have less clients and clients who accept it [working three days]. (M005, male, 1948)

> It is manageable, but difficult in the organization. Sometimes clients think it is annoying, and it is difficult to realize because of the high work demands and the extensive amount of clients. (M009, male, 1964)

Next to part-time work, working a day at home and telework are both policies that provide employees with the flexibility to perform work-related tasks at home. There is an important difference between working a day at home and telework. The first is usually not a formal policy, and is often granted when subordinates have to finish work which requires undisturbed attention.

> Occasionally it's handy. When you have to finish a report which needs much concentration. I do not promote it, but it is possible. (M008, male, 1960)

> I think it's good for the employees who perform well. I forbid working at home to the people who don't perform well, otherwise it is not visible what they do. In any case I am unwilling towards a structural day working at home, I can't sell this to my client. (M011, female, 1967)

Telework on the other hand is a formal policy and means working from home on a structural basis. Therefore, telework is seen as more problematic by supervisors. Most supervisors mentioned the importance of working in a team, which requires visibility on the work floor. Next to this, supervisors especially mentioned the relation with the clients of Company X as a reason to restrict telework possibilities.

> Negative. We don't allow it for employees in this team. It's not possible for me to tell a client one of our team cannot meet the client on for example Tuesday, because it is the team members' regular working at home day. (M011, female, 1967)

> People who put in a request for telework to me, don't have a chance. I think they don't have much chance at Company X at all. Here not many have a positive attitude towards telework. (M005, male, 1948)

13.3.2.2 Care-Related Policies

Care-related policies are a more specific type of work-life policies. Whereas flexible policies can also be used to schedule leisure time, care-related policies are explicitly formulated for employees' caring tasks. Working people have to find a way to deal with the balance between work and caring duties. For instance, it is not possible to postpone the daily care of children because of heavy work demands at a certain time. We discussed supervisors' experiences with parental leave, child care, and request for short-term leave of absence as a consequence of a sudden family crisis, such as a sick child.

Parental leave is a statutory policy in the Netherlands, and the supervisors had no real problems with employees taking parental leave. However, most supervisors viewed parental leave as a policy mainly used by female employees.

> Sometimes, the ladies attach parental leave to their maternity leave. The gentlemen don't take up parental leave. (M004, male, 1952)

Not all supervisors had experience with subordinates taking parental leave, but some feared the eventuality that several employees would use this policy at the same moment.

> I don't want to think of the situation they take it up all at the same time. Then I would have an empty department. (M011, female, 1967)

Using parental leave does have career consequences, and considering the gendered aspect, this means that women are trading off their career to caring.

> In fact it means you break off your career. After such a leave you have to start all over to build your portfolio. (M003, male, 1960)

We asked whether the supervisors thought child care should be a task of the organization. This question provided a wide range of answers, from supervisors who thought the organization should do more on this subject to managers who thought child care as being primarily a task of the parents.

> No, I think it is the responsibility of the parents. How they wish to arrange it is their business. (M001, male, 1944)

> Yes and no. No, because it is primarily one's own responsibility. On the other hand, you can facilitate your employees for a large part. And if you want to do something for your employees, then this is something you can take good care of. Also because, politically, it is not very well organized. (M012, male, 1965)

Parental leave and child care are long-term policies, whereas short-term leave is meant to help employees incidentally with unexpected care tasks. These events, especially those concerning one's relatives, can happen at any time.

> That always is a priority. You need to help people out if they encounter sudden unexpected problems. If you don't do that, the employee will not work. Only, when you pay attention to this request for a while, you will notice that some employees never ask for help and others always. In that case you have to address this with that person. (M005, male, 1948)

However unfortunate this may be for the employee, supervisors are deal with this problem in the context of the need of the organization, and their responsibility towards the rest of the team.

> You can't refuse, of course, but it is not comfortable and can cause problems. It puts an enormous burden on the colleagues when they have to take over these tasks. (M008, male, 1960)

For some supervisors it is a clear cut question, private life prevails above the organization, also because it can harm the organization if it does not take the employees' private life into account.

> If someone can't handle the combination of work and care, that person can better go home and first deal with the problems at home. (M007, female, 1964)

13.3.2.3 Concluding Remarks

The interviews with the supervisors in Company X show their ambivalent or contradictory attitudes towards the utilization of work-life policies by their subordinates. They support the utilization of certain policies, like part-time work, parental leave and short-term care leave, but they are less supportive regarding policies like telework, working a day at home, or child care. All supervisors acknowledge the "trade off" that follows when employees use work-life policies; utilization implies (negative) career consequences. Especially long-term work-life policies, like working part time or parental leave, are considered to damage one's career in Company X. Next to the personal consequences, the utilization of work-life policies can also harm the team or the organization, because of the increase of work demands when a team member is on leave. Even the supervisors who are most supportive mention these negative consequences. Clearly, supervisor support is not unconditional. This leads us to the question in what way the attitudes of supervisors are influencing their decisions when subordinates request the use of work-life policies.

13.3.3 Supervisor Allowance Decisions[2]

In this section we discuss how supervisors grant requests by their subordinates, i.e., the "allowance decision" concerning work-life policies in order to answer our

[2]The arguments developed in this section are based on Den Dulk and Peper (2009).

final research question (How can we understand the allowance decisions made by supervisors on work-life policy request by subordinates?). First, we discuss several theoretical expectations concerning the allowance discussion. Subsequently, we will present the empirical evidence for these expectations as found by Den Dulk and De Ruijter (2008). In their study they use the dataset of the supervisors of the Dutch Company X, and data from supervisors of several other organizations in the financial sector. Our main aim here is to unravel allowance decisions made by supervisors.

Powell and Mainiero (1999) assume that supervisors take the impact of utilization of work-life policies on work outcomes into account when considering an employee request. They developed the disruptiveness hypothesis to explain managerial decision-making on work-life requests for subordinates. Their hypothesis states that supervisors take the degree to which the utilization of a work-life arrangement will disrupt the organization's work into account when considering such a request. Work-life policies can make supervisors' jobs more complex and difficult because they then have to deal with various schedules and arrange replacements when employees are on leave, while still making sure that the necessary work gets done. Powell and Mainiero (1999) distinguish several factors that influence the degree to which a supervisor will view a subordinate's request to use work-life policies as disruptive: (1) the type of work-life policy requested and the reason offered by the subordinate for making the request, and (2) the nature of the tasks, skills and responsibilities of the subordinate making the request.

Another study on the allowance decision examines dependency theory (Klein et al., 2000). The main assumption of dependency theory is that supervisors depend – to varying extents – on their subordinates. Supervisors are responsible for the results and performance of the department they manage. Their subordinates contribute – in varying degrees – to this performance. This makes supervisors dependent on their employees and the greater the employee's contribution, the more dependent the supervisor is. This may give such employees more power, not only in salary negotiations but also regarding the use of work-life policies.

Den Dulk and De Ruijter (2008) argue that both disruptiveness and dependency considerations play a role in allowance decisions and can be seen as complementary rather than contradictory. In understanding the attitudes of supervisors, it is important to consider the dilemma they face when deciding whether or not to grant employees' requests to utilize work-life policies: should they give priority to short-term departmental and organizational goals (i.e., ensure that the necessary work gets done and prevent disruption to the conduct of work), or should they give priority to long-term goals (i.e., retain valuable employees by responding to their personal and family needs)? This dilemma is particularly striking when the two sets of goals are perceived as conflicting. Supervisors cope with the dual agenda of caring for employee needs and pursuing organizational goals by utilizing different strategies, for instance, by being supportive during crises but very demanding in day-to-day working life (Das Dores Guerreiro et al., 2004)

Den Dulk and De Ruijter (2008) tested the two theories in a vignette study among a sample of 46 Dutch and British financial sector supervisors working in four different firms. Instead of focusing on whether or not supervisors grant a request, this study examined the attitudes of supervisors toward requests by hypothetical employees. Evidence from this study generally supported the disruptiveness theory. The findings indicated that the person making the request (female versus male, supervisory position or not) and the nature of the request itself do matter. Requests by women are judged more positively than requests by men, in particular when they concern taking up leave, indicating that care duties are seen mainly as a woman's responsibility. Requests made by supervisors are judged more negatively than requests by employees who do not supervise others. In sum, the study found considerable support for disruptiveness theory. Regarding less disruptive requests, for example short-term leave, the study found that dependency arguments were also important. When the labor market is tight and it is difficult to find new employees, supervisors were (even) more positive about short-term care leave. The conclusion is that supervisors consider both the degree of disruptiveness and employee needs, as well as the risk of losing valuable personnel in allowance decisions.

However, the study did not produce the same results with respect to requests to work from home occasionally, another arrangement that can be considered as relatively non-disruptive. This finding might be explained by the fact that working from home gives rise to new coordination and control problems (e.g., Peters and Den Dulk, 2003), which supervisors may perceive as making their tasks more complex and difficult. We also have to consider the possibility that supervisors may combine a negative attitude with positive decision-making. Even though a supervisor might be negative about requests by employees on whom he/she depends most, as granting the request will complicate the work in the short term, the risk of the employee leaving the department might lead the supervisor to grant the request anyway. This is an issue that should be taken into account in future research.

13.4 Conclusions and Discussion

13.4.1 Conclusions

The central findings of Sect. 13.2 are linked to the four research questions and hypotheses guiding this section in Table 13.5. With regard to the first research question (What is the utilization of work-life policies among Dutch professionals?), we can conclude that flextime, working from home, and part-time work were the most frequently used arrangements in both samples. Every Dutch employee has the right to adjust his or her working hours and this particular arrangement is used frequently in the Netherlands, particularly by women (OECD, 2007). Working flexible hours and occasionally working from home are both flexible work-life policies, which may greatly enhance professionals' flexibility regarding the time and place they work. However, in contrast to part-time work – which has to be formally arranged in one's employment contract – utilization of these two policies may be based upon an *informal* agreement between employee and supervisor. Moreover, both flexible work-life policies do not reduce professionals' work hours and, therefore, may not impede their career advancement (see the fourth research question). This may explain the high utilization of these policies, in particular within the "up-or-out" culture characterizing Company X.

Table 13.5 Overview of research questions, hypotheses, findings, and practical implications

Research question	Hypothesis	Findings	Practical implications
(1) What is the utilization of work-life policies among Dutch professionals?	–	Flextime, working from home, and part time work are the most frequently used arrangements in both samples	Make professionals (and supervisors) familiar with the whole spectrum of work-life policies offered
(2) What is the association of work-life policies with work-life culture?	High levels of work-life culture support and low levels of barriers are related to high utilization of work-life policies (H1)	Support is positively related to utilization of working from home in Sample Y; barriers are positively associated with the uptake of childcare subsidies (both samples), working from home, and parental leave (Company X)	Supervisors should explicitly support their employees' work-life integration when they want to increase the uptake of work-life policies or positive spillover between their employees' work and private life by, for example, clearly defining expectations, goals and resources at work and at home in order to make a "road map" for professional and personal success
(3) What is the association of work-life policies and work-life culture with work-life enrichment?	High levels of work-life culture support and low levels of barriers are related to high levels of work-life enrichment (H2) Employees using work-life policies will report high levels of work-life and life-work enrichment (H3)	Support is positively related to enrichment in both samples; working from home is positively related to work-life enrichment (via support), and part-time work is positively associated with life-work enrichment in Sample Y	

Table 13.5 (continued)

Research question	Hypothesis	Findings	Practical implications
(4) What is the association of utilization of work-life policies with career advancement?	Employees working part time or taking up parental leave have lower job levels compared to those working full time (H4)	Working part time and taking up parental leave are unfavorable in terms of Dutch professionals' career advancement in Sample Y	An open discussion between supervisors and professionals on utilization of work-life policies and its impact on career advancement is important, as well as the introduction of incentives or rewards for implementing work-life policies and supervisor training in the tools that would allow them to successfully implement work-life policies
(5) What are the experiences with and attitudes of supervisors regarding work-life policies?	–	Supervisors show a contradictory attitude towards the uptake of work-life policies by their subordinates by only supporting certain policies (e.g., part-time work and parental leave), and by acknowledging the "trade off" when employees use work-life policies for their career and work team	
(6) How can we understand the allowance decisions made by supervisors on work-life policy request by subordinates?	–	Supervisors consider both the degree of disruptiveness and employee needs as well as the risk of losing valuable personnel in allowance decisions; inconsistency between attitudes and allowance behavior is possible	

The uptake of the compressed (4 × 9) workweek and the care-related policies (i.e., parental leave, and childcare subsidies) was low. This is not surprising given the fact that parental leave (26% in 2007) is not used frequently in general by Dutch parents who are eligible to use this arrangement (Merens and Hermans, 2009). In addition, childcare subsidies are nowadays granted by a standardized union of the Dutch government and employers. Therefore, additional subsidies or arrangements are not frequently offered by Dutch employers. The compressed workweek is a relatively new and innovative work-life policy in the Netherlands, which is not yet present in many companies.

Turning to the second research question (What is the association of work-life policies with work-life culture?), we can conclude that – as expected – work-life culture support was positively related to utilization of working from home in Sample Y. However, we also found that barriers were positively associated with the uptake of childcare subsidies in both samples, and that professionals perceiving high levels of barriers worked from home and used parental leave frequently in Company X. A possible explanation for this counterintuitive finding may be that these associations are reversed in nature. Because we have employed a cross-sectional design to collect data from both samples, we cannot test the causal direction of the associations found. From a hypothetical point of view it is plausible, however, that employees who do not use work-life policies are not aware of the negative career consequences or organizational time demands associated with their uptake. In contrast, employees who start using work-life policies may be confronted with these barriers. In a recent longitudinal study by Dikkers and Demerouti (under review), support was found for a reversed causal path between utilization of work-life policies and work-life culture: utilization of working from home predicted employees' perceptions of work-life culture 1 year later.

With regard to the third research question (What is the association of work-life policies and work-life culture with work-life enrichment?), we can conclude that support was positively related to work-life enrichment in both samples and to life-work enrichment in Company X. Furthermore, working from home was positively associated with work-life enrichment via support, and part- time work was positively associated with life-work enrichment in Sample Y.

With regard to the fifth and sixth research questions, the present study clearly shows that the attitudes of supervisors are very important in understanding why many employees are not taking advantage of the wide range of work-life policies on offer nowadays. The attitudes of supervisors are not only important in understanding managerial decision-making with respect to requests to utilize work-life policies, but are also crucial in understanding the requesting behavior of employees. Supervisor support is by no means unconditional. The evidence found in several studies suggests that managers generally take a short-term view of work-life policies rather than a long-term view that cherishes human capital. They do not (yet) consider that employees' work-life balance contributes to organizational goals (see also the "Dual Agenda" concept, Rapoport, 2002). If, however, organizations wish to retain valuable human capital, future policies should offer supervisors additional incentives or rewards for implementing work-life policies, for example, by introducing facilities to manage the disruption in work. Moreover, several studies note that managers are generally unaware of existing policies and lack training in the tools that would allow them to implement policies successfully. Not surprisingly, many studies find inconsistencies in policy implementation and variations in management attitudes and behavior toward work-life policies.

13.4.2 Practical Implications

We would like to distinguish between practical implications intended for professionals and those targeted at their supervisors or employers. One important implication of the conclusions drawn above for the professional employees is that they should become well acquainted with the work-life policies offered by their employer if they have not yet done so. In both samples studied in this section, professional employees did not frequently use most policies. It may be possible that professionals are only aware of the existence of several well-known policies such as flextime, and therefore only use these arrangements, while their employer offers a larger variety of policies such as the compressed workweek or calamity leave. Being familiar with the whole spectrum of work-life policies available to them may supply professionals with the tools needed to balance work with responsibilities at home. This is supported by our finding that working from home was positively associated with work-life enrichment and that part-time work was positively associated with life-work enrichment.

Simultaneously, employers should explicitly communicate the work-life policies available to their professional employees. Hopkins (2005) suggests that supervisors can support their employees in (better) balancing their work with their responsibilities at home by being knowledgeable about work-life programs and policies and by disseminating information about these policies. In addition, she offers several other proposals for supervisors to demonstrate concrete support for work-life integration. Case studies of supervisors who are supportive show that they clearly define work expectations in terms of results and simultaneously "ask employees to identify the important goals, concerns, and demands outside the office hat require time and energy" (Friedman et al., 1998, p. 121). Supervisors subsequently use this information to draw a "road map" toward professional and personal success.

Another practical implication is that supervisors should take work-life culture into account when considering utilization of work-life policies. We found that professionals perceiving high levels of support in integrating work and private life reported high utilization of working from home, as well as high levels of work-life and life-work enrichment. However, those working from home or using parental leave also reported negative career consequences and time demands (barriers). Therefore, it is imperative that employers consider their company's (and their own) attitude towards their employees' work-life integration when they want to increase the uptake of work-life policies or positive spillover between their employees' work and private life.

Finally, we would like to advise professionals to seriously consider the possible impact that utilization of different types of work-life policies may have on their (extrinsic) career success. If one primarily aims at getting ahead in one's career within a short period of time, the uptake of arrangements such as part-time work and parental leave may not be advisable. However, if one prefers a career that varies from the traditional linear career altogether, and in which other priorities (such as gaining expertise or raising a family) are considered to be important, the impact of utilization of work-life arrangements on one's extrinsic career success may be less relevant than, for example, its impact on intrinsic career success (e.g., job satisfaction).

In a similar vein, we would like to advise supervisors to support their professional employees in using the work-life policies of their choice. Supervisors are the ones making decisions regarding employees' career advancement. If they are opposed to their subordinates using policies which will (structurally or temporarily) reduce their work hours, they may feel less inclined to consider them for promotion. It is, therefore, imperative, that supervisors realize their role in the impact of work-life policies on their subordinates' extrinsic career success. By openly discussing perspectives on work-life policies and their possible effect on (different types of) career advancement with their subordinates, supervisors can reduce the uncertainty or even barriers withholding employees from using (certain) policies. This discussion could be part of regular work meetings in which the "road map for professional and personal success" as mentioned above is jointly developed and refined by supervisors and their employees.

13.4.3 Theoretical Implications for Future Research

Research has shown it is often very difficult to change an organization, and even more difficult to change an organization's culture (see, e.g., Haas et al., 2000). An organization's work-life culture may range from very positive to very negative. The work-life culture is positive when the organization considers work-life balance important, when utilizing work-life arrangements has few consequences for employees' careers, and when the standards set for working hours allow scope for family duties. However, the work-life culture of an organization can also contain contradictory elements (Den Dulk and Peper, 2007; Kirby, 2000; Lewis, 2003). An organization may show concern for its employees' work-life balance, for example, but at the same time associate employee commitment with attendance and working long hours. In other words, the organization may support employees in their efforts to achieve a good work-life balance – for example, by introducing work-life arrangements – but simultaneously place time demands on the employees that conflict with the actual use of such measures. Future research would profit from examining this contradictory work-life culture in more detail, instead of crudely distinguishing between cultures that are either supportive or unsupportive towards the utilization of work-life policies and work-life balance in general.

References

Allen TD, Herst DEL, Bruck CS, Sutton M (2000) Consequences related to work-to-family conflict: a review and agenda to future research. J Occup Health Psychol 5:278–308

Almer ED, Cohen JR, Single LE (2004) Is it the kids or the schedule?: the incremental effect of families and flexible scheduling on perceived career success. J Bus Ethics 54:51–65

Anderson SE, Coffey BS, Byerly RT (2002) Formal organizational initiatives and informal workplace practices: links to work–family conflict and job-related outcomes. J Manage 28:787–810

Appelbaum E, Bailey T, Berg P, Kalleberg A (2005) Organizations and the intersection of work and family: a comparative perspective. In Ackroyd S, Batt R, Thompson P, Tolbert PS (eds) The Oxford handbook of work and organization. Oxford University Press, Oxford and New York, pp 52–73

Arthur MB, Hall DT, Lawrence BS (1989) Generating new directions in career theory: the case for a transdisciplinary approach. In Arthur MB, Hall DT, Lawrence BS (eds) Handbook of career theory. Cambridge University Press, Cambridge, pp 7–25

Das Dores Guerreiro M, Abrantes P, Pereira I (2004) Transitions case studies summary report for the EU Framework 5 study 'Gender, parenthood and the changing European workplace'. RIHSC, Manchester

Den Dulk L (2001) Work-family arrangements in organisations: a cross-national study in the Netherlands, Italy, the United Kingdom and Sweden. Rozenberg Publishers, Rotterdam

Den Dulk L, De Ruijter J (2008) Managing work-life policies: disruptive versus dependency arguments. Explaining managerial attitudes towards employee utilization of work-life policies. Int J Hum Resour Manage 19:1224–1238

Den Dulk L, Peper B (2007) Working parents' use of work-life policies. Sociol Probl Prát 53:51–70

Den Dulk L, Peper B (2009) Managing work-life policies in the European workplace: explorations for future research. Working Papers on the Reconciliation of Work and Welfare in Europe No. 04-2009. Available at SSRN: http://ssrn.com/abstract=1489879

Dikkers JSE, Demerouti E (under review) Mapping terra incognita: work-home culture and its associations with work-related antecedents over time. J Vocat Behav

Dikkers JSE, Geurts SAE, Den Dulk L, Peper B, Kompier MAJ (2004) Relations among work-home culture, the utilization of work-home arrangements, and work-home interference. Int J Stress Manage 11:323–345

Dikkers JSE, Geurts SAE, Den Dulk L, Peper B, Taris TW, Kompier MAJ (2007) Dimensions of work-home culture and their relations with the use of work-home arrangements and work-home interaction. WorkStress 21:155–172

Eby LT, Casper WJ, Lockwood A, Bordeaux C, Brinley A (2005) Work and family research in IO/OB: content analysis and review of the literature (1980–2002). J Vocat Behav 66:124–197

Fried M (1999) Parental leave policy and corporate culture. Women and Work 1:11–26

Friedman S, Christensen P, De Groot J (1998) Work and life: the end of the zero-sum gamer. Harv Bus Rev 119–127 volume 76 No 6 p119–29 Nov-Dec 1998

Geurts SAE, Kompier MAJ, Rosburgh S, Houtman ILD (2003) Does work-home interference mediate the relationship between workload and well-being?. J Vocat Behav 63:532–559

Geurts SAE, Taris TW, Kompier MAJ, Dikkers JSE, Van Hooff MLM, Kinnunen UM (2005) Work home interaction from a work psychological perspective: development and validation of a new questionnaire, the SWING. Work Stress 19:319–339

Glass J (2004) Blessing or curse? Work-family policies and mother's wage growth over time. Work Occup 31:367–394

Greenhaus JH, Beutell NJ (1985) Sources of conflict between work and family roles. Acad Manage Rev 10:76–88

Greenhaus JH, Powell GN (2006) When work and family are allies: a theory of work-family enrichment. Acad Manage Rev 31:72–92

Haas LL, Hwang P, Russell G, (Eds) (2000) Organizational change and gender equity: international perspectives on fathers and mothers at the workplace. Sage, Thousand Oaks, CA

Higgins G, Duxbury L, Johnson KL (2000) Part-time work for women: does it really help balance work and family?. Hum Resour Manage 39:17–32

Hopkins K (2005) Supervisor support and work-life integration: a social identity perspective. In Kossek EE, Lambert S (eds) Work and life integration: organizational, cultural, and individual perspectives. Lawrence Erlbaum, Mahwah, NJ, pp 445–467

Judge TA, Cable DM, Boudreau JW, Bretz RD (1995) An empirical investigation of the predictors of executive career success. Pers Psychol 48:485–519

Judiesch MK, Lyness KS (1999) Left behind? The impact of leaves of absence on managers' career success. Acad Manage J 42:641–651

Kinnunen UM, Mauno S, Geurts SAE, Dikkers JSE (2005) Work-family culture in organizations: theoretical and empirical approaches. In Poelmans SAY (ed) Work and family: an international perspective. Lawrence Erlbaum, Mahwah, NJ, pp 87–120

Kirby EL (2000) Should I do as you say or do as you do? Mixed messages about work and family. The Electronic Journal of Communication/La Revue de Electronique de Communication, 10. http://www.cios.org/EJCPUBLIC/010/3/010313.html. Accessed on 10 Feb 2010

Klein KJ, Berman LM, Dickson MW (2000) May I work part-time? An exploration of predicted employer responses to employee request for part-time work. J Vocat Behav 57:85–101

Kossek EE, Friede A (2006) The business case: managerial perspectives on work and the family. In Pitt-Catsouphes M, Kossek EE, Sweet S (eds) The work and family handbook: multi-disciplinary perspectives, methods, and approaches. Lawrence Erlbaum, Mahwah, NJ, pp 611–626

Kossek EE, Lautsch BA, Eaton SC (2006) Telecommuting, control, and boundary management: correlates of policy use and practice, job control, and work-family effectiveness. J Vocat Behav 68:347–367

Lewis S (2003) Flexible work arrangements: implementation, outcomes and management. In Cooper CL, Robertson IT (eds) International review of industrial and organizational psychology, vol 18. Wiley, New York, NY, pp 1–28

Lewis S, Brannen J, Nilsen A, (Eds) (2009) Work, families and organisations in transition. European perspectives. The Policy Press, Bristol

Lewis S, Taylor K (1996) Evaluating the impact of family-friendly employment policies: a case study. In Lewis S, Lewis J (eds) The work-family challenge: rethinking employment. Sage, London, pp 112–127

Maxwell GA (2005) Checks and balances: the role of managers in work-life balance policies and practices. J Retailing Consum Serv 12:179–189

Merens A, Hermans B (2009) Emancipatiemonitor 2008 (SCP-publicatie 2009/2. Sociaal en Cultureel Planbureau, The Hague

Mesmer-Magnus JR, Viswesvaran C (2006) How family-friendly work environments affect work/family conflict: a meta-analytic examination. J Lab Res 27:555–574

Ng TWH, Eby LT, Sorensen KL, Feldman DC (2005) Predictors of objective and subjective career success: a meta-analysis. Pers Psychol 58:367–408

OECD (2007) Employment outlook 2007 (Statistical Annex). OECD Publishing, Paris. http://www.oecd.org/dataoecd/29/27/38749309.pdf. Accessed on 1 Oct 2007

Perlow LA (1995) Putting the work back into work/family. Group Organ Manage 20:227–239

Perlow LA (1997) Finding time: how corporations, individuals and families can benefit from new work practices. ILR Press, Ithaca, NY

Peters P, Den Dulk L (2003) Cross cultural differences in managers' support for home-based telework: a theoretical elaboration. Int J Cross Cult Manage 3:329–346

Poelmans S, Beham B (2005) A conceptual model of antecedents and consequences of managerial work-life policy allowance decisions. International Center of Work and Family, Inaugural Conference, Barcelona/Spain, 7–9 July 2005

Powell GN, Mainiero LA (1999) Managerial decision making regarding alternative work arrangements. J Occup Organ Psychol 72:41–56

Rapoport R (2002) Beyond work-family balance: advancing gender equity and workplace performance. Jossey-Bass, San Francisco, CA

Starrels ME (1992) The evolution of workplace family policy research. J Fam Issues 13:259–278

Ten Brummelhuis L (2009) Family matters at work. Depleting and enriching effects of employees' family lives on work outcomes. ICS, Utrecht

Thompson CA, Beauvais LL, Lyness KS (1999) When work-family benefits are not enough: the influence of work-family culture on benefit utilization, organizational attachment, and work family-conflict. J Vocat Behav 54:392–415

Tomlinson J (2006) Routes to part-time management in service sector organizations: implications for women's skills, flexibility and progression. Gen Work Organ 13:585–605

Van Steenbergen EF (2007) Work-family facilitation: a positive psychological perspective on role combination. Universiteit Leiden, Leiden

Veenis E (2000) Als het bedrijf er maar niet teveel last van heeft: De invloed van het ouderschapsbeleid en de ouderschapscultuur op de combinatieproblemen van mannelijk en vrouwelijke werknemers met jonge kinderen in supermarkten en boekhandels. Proefschrift, Universiteit Utrecht. Pasmans, The Hague

Yeandle S, Phillips J, Scheibl F, Wigfield A, Wise S (2003) Line managers and family-friendly employment. Roles and perspectives. The Policy Press, Bristol

Part V
Boundary Issues of Professionals' Work-Life Integration

Chapter 14
The Pressures of Commitment: Taking Software Home

Jeff Hyman and Chris Baldry

14.1 Introduction

The current era has seen a number of academic and policy debates over the claimed increase in the porosity of the boundary between the work and domestic spheres of social activity (Baldry et al., 2007; Houston, 2005; Warhurst et al., 2008). The causes of this have been identified variously as the shift to more "flexible" forms of employment (Department of Trade and Industry, 2004), an increasing rate of female participation in the labour market with consequent demands on child-care resources (Cousins and Tang, 2004; Crompton, 2002), the intensification and extensification of the labour process in time of heightened competition (Cousins and Tang, 2004) or economic crisis and the enabling qualities of IT which have made possible an increase in both mobile work (Hislop and Axtell, 2009) and teleworking at home (Haddon and Brynin, 2005).

We can identify several contrasting perspectives in the way these debates have been conducted. What we could call the established or managerial view assumes that the prime reason life and work might be out of "balance" is because of the demands made by domestic caring responsibilities (mainly if not entirely focused on women workers): here the origin of the problem lies outside the workplace but the state and progressive employers can encourage or offer remediating "family-friendly" policies.

Those discussions which see the tension arising more from the changing nature of work tend to gravitate towards two opposing poles. A somewhat dystopian view sees the blurring of the boundary as an encroachment of capitalist control structures into domestic life, and as a colonising of the worker's private world by the public world of employment. Examples here might be the incidence of extended working hours or teleworking at home (Limburg and Jackson, 2007).

Such criticisms have extended to the way the issue has been conceptualised. Fleetwood (2007), drawing on the work of Harvey, has argued that the neoliberal project has seen the responsibility for wellbeing relocated onto the individual under

---

J. Hyman (✉)
University of Aberdeen, Aberdeen, UK
e-mail: j.hyman@abdn.ac.uk

S. Kaiser et al. (eds.), *Creating Balance?*, DOI 10.1007/978-3-642-16199-5_14,

the rhetoric of "freedom of choice" and rationalised by the concept of flexibility (see also Hakim, 2000). Similarly, Webster (2004) points out that the increased attention given in policy debates to family-friendly working has displaced calls for universal child-care provision and, in so doing, has pushed the responsibility for balancing work and domestic responsibilities onto the individual (usually female) employee. Equally significantly, the focus on families, almost exclusively defined in terms of caring responsibilities, ignores the very real possibility of any worker, carer or not, experiencing a significant disjunction between the demands of work and non-work life. Hence, suggest Lewis et al. (2007), the concept of work life-balance (WLB) was intended to be more inclusive, although still very much located within the neo-liberal discourse of personal choice and continuing the lack of attention given to structural and institutional constraints.

The opposite, more utopian view, has hailed the developments as representing a partial liberation from the tyranny of conventional work and a move closer to non-alienated work, where the worker puts him or herself into the product or service so that work *is* life, and their separation becomes less relevant (Leadbeter, 1999; Reeve, 2000). An example here might be the so-called creative industries (Florida, 2004; Townley et al., 2009). Even in this case, though, there are fine distinctions between what is willingly offered and that which is manipulated through management expectations.

Clearly these various perspectives on the experiential nature of WLB are often incompatible with each other while, in addition, some observers argue that systematic imbalance between work and life has yet to be satisfactorily demonstrated empirically (see e.g., Hooker et al., 2007). In a parallel development, the frequency and amplitude of occupational stress, has been seen by many observers as a manifestation of incompatibility between growing work demands and domestic lives (European Agency for Safety and Health at Work, 2007).

If there is a blurring of boundaries between work and life (Ransome, 2008), we have to ask what exactly is meant by the "life" part of the concept of WLB. Time away from work (re-creation) is not necessarily leisure: for women it may include a large element of unpaid domestic work (including but not exclusively care work); this is of course also possible, but less likely, for men. In this chapter therefore we argue that we cannot separate the concept and attainment of WLB from an analysis of the gendered politics of the workplace and labour market.

This chapter takes a group of "new professionals" – software workers – and shows how the work often permeates their non-work lives both quantitatively (through the length and pattern of working time) and qualitatively (ranging from exhaustion and stress to continuously thinking and talking about work when not in the workplace). Survey and interview data from five case study organisations are used to show that the work-life mix for many software workers is composed of both a voluntary extension of work into non-work time through professional commitment and membership of the software occupational community, but is also partly imposed through managerial expectations and the intensification of project targets. These processes are further examined for evidence of differential impacts and consequent behaviours between men and women software workers.

14.2 The Changing History of the Work-Home Boundary

In whatever way WLB is interpreted, we can say that the economic transitions of the late twentieth century, which have seen the shift in employment from manufacturing to services, increased female labour market participation and the almost universal adoption of the goal of flexibility in work organisation, have contributed to make the delineation of the work-life boundary and the way it is experienced subject to a process of redefinition.

To put these developments, and their contending interpretations, into perspective we need to recognise that the boundary between work and home has seldom been as rigidly defined as the contemporary stereotype suggests. Much pre-industrial work was either performed in the domestic dwelling (for example spinning and weaving) and fitted around the agricultural cycle, or involved the artisan (the smith, the miller, the shoemaker) living where he worked. What was revolutionary about the industrial revolution was not the technology but the change in social relationships: the labour in these tasks was now to be sold on the market and the purchaser of labour power could put it to work under a single roof where the transformation of labour power into value could be controlled and maximised. This has conventionally been seen as the origin of the functional and spatial separation between work and home (Marglin, 1974).

Even under industrial capitalism, however, the boundary between the two spheres was never finite. Marx saw them as two sides of the same coin: domestic labour embodied the function of re-production, in which the waged worker was to be re-fuelled and rested sufficiently for the next day's shift. Even if we take this statement to be objectively true, it has been accepted that, subjectively, we treat work and domestic regimes as being experientially different because they encompass different relationships, demands and responsibilities.

It has been suggested by Webster that this perceived qualitative difference between work time and domestic time was most objectified during the mid-twentieth century when the Fordist work paradigm was dominant (Webster, 2004). In most of the advanced industrial societies, the rhythms of standard regular employment influenced the wider patterns of social and institutional life and temporal periods outside employment were clearly defined. This was however also the period of the "male breadwinner" model of employment where many women were excluded from the spatial and temporal world of employment and relegated to the domestic sphere.

This having been said, a century and more of sociological study has demonstrated various instances where the work-life boundary has been experienced as porous by workers in specific occupational, socio-economic and labour market contexts. We can identify several examples:

(a) Where the nature of work created geographical communities around a single industry, such as mining (Dennis et al., 1956) or shipbuilding (Brown and Brannen, 1970). Here workers lived next to fellow workers in the same industry and often with the same employer, and recreational activity in the community often overlapped with work-based groupings, either unofficially (in the pub or club where talk revolved around work issues) or officially (in the form of company-sponsored brass bands or choirs).

(b) Where the efforts of paternalist employers were directed at improving the social and living conditions of their employees. The earliest of these in the UK were probably Robert Owen's mills at New Lanark (1800), although most of the later nineteenth and early twentieth century examples were the works of members of non-conformist churches (such Methodists and Quakers) who saw a concern for the conditions of their workers as a moral duty. In addition to the provision of high quality workers' housing, often in the form of company communities such as New Lanark, Saltaire, Bourneville (Cadbury) or Port Sunlight (Lever), organised social activities were encouraged (often with a strong teetotal theme, as an alternative to more alcoholic pursuits); at Cadbury's community of Bourneville for example there were company sports days and gardening and camera clubs.

(c) Where the nature of the job creates clear occupational communities, the members of which feel and see themselves as "different" from the wider society; examples are police (Salaman, 1974) or railways workers. While geographically more diffuse than example (a), members of occupational communities tend to socialise outside their work with fellow members, even if they may work for different organisations, because they "speak the same language".

(d) Where the work is characterised by a high degree of autonomy and personal commitment or calling or is performed in a self-employed capacity. Here the work can often seem a part of "life" and barriers to where and when it is performed may seem irrelevant.

Where example (a) was mainly characteristic of a limited number of manual working class occupations, characteristics (c) and (d) are shared by many professions and quasi-professions (Marks and Scholarios, 2007). Additionally, the above examples are not mutually exclusive. One can see close affinity between example (c)'s occupational community and the professional emphasis evident in example (d). Recent studies reveal increasing recognition of other attitudinal, behavioural and structural influences on work-life activity which can impact on the actual and desired porosity of work and life. Hence, Warhurst and his colleagues (2008) focus on different labour markets, the impact of work intensification and, in particular, growing dissatisfaction among professionals, (represented by group (d) in our typology) at their perceived loss of occupational autonomy and authority. Similarly, there is recent pan-European evidence for more variety of (enforced) working time patterns, reduced employee controls over their working time and "internalisation of working demands (especially in high qualified occupations)" (Krings et al., 2009). Other writers draw upon a synthesis between work and life-style to explain patterns and experiences of adaptation (Warhurst et al., 2008, p. 14). A similar holistic framework to examining work-life dynamics has been developed by McKie and her colleagues, under their organisation carescapes conceptualisation (McKie et al., 2008). The important development from earlier WLB characterisations common to all these approaches is that work and life are interconnected, and that neither can be examined in isolation without reference to broader external and internal influences, an approach which the present authors adopt in their analysis. Another important dimension is the setting: as indicated above, work-life characteristics and behaviour can be expected to link to different occupations, commercial pressures and sectors. The changing perspectives of professional workers have already been touched upon: the intention of this chapter is to examine a sub-set of professionalised or "knowledge" workers, with particular reference to examining the extent of their voluntary or imposed commitment to work as a contributory factor in work-life dynamics.

14.3 The Growth and Nature of Software Work

The ubiquity of software and its manifold impacts on consumer goods, technology, service and industrial processes are integral to modern life. For example, software underpins the financial services sector, which is estimated to account for nearly a fifth of UK business software revenue (Developing the Future, 2008, p. 8). The underlying strength of the sector is confirmed with the UK accounting for a quarter of Europe's biggest software companies and having the largest share of software venture capital investment in Europe (Developing the Future, 2008, p. 16). Nevertheless, the sector has not been immune to the current economic downturn, with many software companies suffering weak profits and, in 2008, "the lowest level of software company Initial Public Offerings on the London Stock Exchange for more than a decade" (Developing the Future, 2008, p. 9). Despite this slowdown, the sector has an annual turnover of some £50bn and was estimated to employ about 430,000 in the UK in 2006 (Developing the Future, 2008, p. 17). While most of the small and medium-sized enterprises (SMEs) are based in London and South-East England, there are a number of regional clusters, including Edinburgh, with strong links to the Scottish capital's formerly dominant finance sector. In Scotland generally, the software sector remains relatively buoyant, with 100,000 employed in software, telecommunications and IT services. 60,000 of these are employed directly on the supply side with the rest providing ICT expertise in-house (ScotlandIS, 2009).

Software has been regarded as at the forefront of the "knowledge economy". Nevertheless, whilst technology specialists such as software workers have been labelled as knowledge workers (Frenkel et al., 1999), within this field there can exist wide skill differences. Even so, a substantial proportion of software and allied workers are graduates, as many as 60% according to recent UK Labour Force Survey (LFS) estimates (Heckley, 2005). Knowledge work is defined by key features including extended prior education; continuing professional development; a measure of task autonomy; team working and interaction with information technology (Castells, 2000) and there is no question that in developed economies, access to work has become contingent on higher levels of educational attainment and specialist expertise. Moreover, many software workers work in "young" industries and companies and are themselves young. In the present study, only 6% of 320 survey respondents were over 50 years of age, and three-quarters were aged 40 or less. The comparative youthfulness of the software respondents is demonstrated by estimates of 27% of employees aged over 50 for the working population as a whole (Kersley et al., 2006, p. 32).

14.4 The Studies

Five Scottish case study organisations were chosen for the study: an independent software house employing 248 people (Omega), a software division of 275 people which was part of a large national telecommunications company (Beta), and three medium-sized to smaller independent firms (Gamma, Pi, and Lambda) employing 150, 50 and 20, respectively.[1] Within each case study employee questionnaires and semi-structured interviews were conducted. Background data on company history, operating procedures, employment policies and staff characteristics were gathered as part of an intensive process of case study analysis and observation involving four researchers in each of the five companies for approximately 4 months.

Beta was part of an ex-public sector utility and can be distinguished in nearly all respects from the other four relatively new Scottish companies, but particularly in terms of its size, more conventional bureaucratic structure, the formalisation of its HRM policies such as training, performance appraisal and communication mechanisms, and in its recognition of a union and harmonised practices. HR policies in the owner-managed companies tended to be informal, unwritten, haphazard and rudimentary.

A majority of males (72% overall) were found in the survey, although Omega, Pi and Lambda employed between 32 and 41% females. Forty six percent of men had first degrees compared with 34% of women but more women 26% (cf. 23% men) had postgraduate qualifications, probably linked to the growing number of women taking conversion information technology courses. There were qualitative differences in the work undertaken: the higher-end of software work, such as programming and design, tended to be performed by men. Omega, which had a woman owner, had 15 in its technical services team, 13 of whom were men.

14.4.1 A Porous Boundary

Our evidence showed that the work life of software workers intruded into their domestic sphere in two major ways, which we have terms quantitative and qualitative intrusions.

Quantitative intrusions consisted of frequently working long hours and irregular patterns of working time. Although all the companies had specified contractual hours, there was evidence of considerable temporal variability, reflecting both heavy demand for software services and the nature of the work. The degree to which

[1] Full details of the methodology employed in the study can be found in Baldry et al. (2007).

working beyond contractual time was formalised varied between companies; while neither Omega nor Pi paid for overtime, Beta operated a standard 42-h week, with shift and unsocial hours allowances, operated through a flexi-time agreement. When combined with the software workers' commitment to their job (see below) the result was that 59% of respondents reported working up to 10 h unpaid overtime and 13% worked in excess of 10 h weekly unpaid overtime. Software work is typified by the necessity to meet project deadlines and providing effective responses to client needs and these were cited by our software workers as the primary reasons for working long hours, while a smaller proportion cited not wanting to let down clients or colleagues.

> If I have to, I'll work late. I'll work late for every night for a month to get a release out but I'd rather estimate properly. because I don't think I can do my job properly if I'm going (flat) out every night... (Beta: female software engineer).

Apart from Beta, formal shift-working was rare but all the companies had call-out systems for at least some grades of their software staff. Thus Omega worked an informal "peaks and troughs" system, whilst Pi employed a flexi-time system (at staff request) with added flexibility expected when required.

These responses indicate a high level of occupational time commitment which was further demonstrated by the two thirds of respondents who reported taking work home at least occasionally.

> At the beginning of the project I would say I was working maybe 50–52 h a week...Weekends, not both days, normally a Saturday or something or I would take work home and do some work at home. (Omega: female software engineer and team leader).

The qualitative aspects of work intrusion have been less examined by previous WLB researchers. Essentially these consist of the degree to which the work experience subjectively, or physically, gets in the way of or obstructs normal domestic social patterns, even when the employee is not actually engaged in work. For example, over half of our survey respondents reported that they thought about work away from the workplace either "quite often" or "all the time". 41% claimed to feel exhausted after work and 13% were kept awake by work problems. Interestingly, nearly one-fifth claimed that their work prevents them from spending enough time with family and friends, whilst one third claimed to feel stressed in their job, again "quite often" or "all the time".

However, the existence of intrusive patterns of work-related activity into what would otherwise be non-work or domestic time/space does not in itself mean that these are subjectively experienced as an imbalance by our knowledge workers. Such a perception will depend on whether such patterns are felt to be imposed or are willingly undertaken. Within the context of the HRM debate for example, it might be concluded that, if we see evidence for discretionary effort, such as working long hours or taking work home, among a specific group of employees, this could be an indicator of a high or at least satisfactory level of organisational commitment.

14.4.2 The Contours of Commitment of Knowledge Workers

Organizational psychology's treatment of commitment rests on a componential account of organisational commitment encompassing at least three psychological states: affective commitment (an emotional identification with the organisation), normative commitment (a sense of obligation towards the organisation), and continuance commitment (an exchange based concept based on a perceived need to stay with the organisation due to the high costs of leaving) (Meyer and Allen, 1991). While Castells (2000) argues that knowledge workers should possess at least continuance commitment if they are to direct discretionary effort towards organizational goals, the "high commitment" approach centres on affective organisational commitment as the main vehicle through which discretionary effort can be cultivated.

In contrast to these ideals, the survey data from our software cases indicated that in all cases, software workers' commitment to the occupation received higher ratings than affective commitment to the organization; for example while 50% of the total sample agreed with the statement "I feel a strong sense of belonging to my company", 63% agreed that "I am proud to tell others of my occupation". Statements measuring continuance commitment generally received the lowest rating: 63%, for example, disagreeing with the statement "I have too few options to consider leaving the company" (Baldry et al., 2007, p. 102). While the image of the knowledge worker identifying with the goals of their organization found more support in the smaller independent companies than in Beta, it was clear that, in all companies, commitment to the occupation was greater than to the organization.

From the discussion at the start of this chapter it might be supposed that software firms, and particularly recent start-ups, as knowledge organizations, would have adopted high commitment management practices to fit with the highly autonomous and intrinsically motivated nature of software jobs. This proved not to be the case, with only Beta showing evidence of formal HRM practices, such as a formal training programme, appraisal linked to pay and promotion and profit-sharing via a share-save scheme. However, if the reason for lower affective commitment across our cases was due to this as yet underdeveloped nature of commitment-raising practices, we would expect that affective commitment would be higher in Beta than the independents. In fact the reverse was the case: in the one company with formalized HRM practices, there was lower affective commitment than in the smaller companies which utilized a variety of informal paternalist and owner-manager initiatives.

The interview responses throw some light on these patterns. Although the amount of unpaid overtime and flexible inputs might suggest a willingness to expend discretionary rather than prescribed effort (Fox, 1974, p. 16), the software workers themselves clearly differentiated between any extra effort which they expend for the company and the long hours, working nights and weekends which are seen to be part of the job of software – you do the hours to get the project delivered because that is part of the identity of being a software professional.

For many of the developers we interviewed it was the job that drove their effort, rather than the company:

> ....I think in development most of us are committed more to the job than to the company because we are all in it because we enjoy programming and that's the first thing, the second thing is what company you work for and what sort of work you get to do . . .so it is probably first the job, then work and probably every one of us would say the same. (Pi: female programmer)

> So, if you like, the commitment's to Beta inasmuch as they are paying me to do what I like and I like to do the job to the best of my ability (Beta: male applications project analyst)

It is thus clear that the discretionary "work effort" evidenced by the willingness to work was distinctly removed from the fact that they were employed by a particular company:

> Again, as with many engineers, I'm highly motivated by what I do for myself and the fact that it carries the company is a secondary but not insignificant part of that. So that's a kind of inherited commitment to the company, the fact that I'm finding the work interesting and challenging as well. (Gamma: Principal Software Engineer)

> To the project yes. . .(since the changes) we don't actually see much about where our place is in the whole company. So I've probably got more commitment to the project than I have to the Centre or the company, if that makes any sense. . .because I know what's happening with the project more than I do about anything else that is happening outside the project. (Beta, female software engineer)

When questioned about willingness to stay with the company, the interviews suggested that the software developers would stay with an interesting project, or if the company continued to offer them opportunities to update or enhance their skills (and thus their leverage in the labour market), but they were well aware (not least though their social networks) what developments and conditions were available elsewhere and that they were in a strong labour market position:

> I like working for Beta and I would like to continue working for Beta but Beta aren't the only company. So if I don't like what I'm doing, I'm not going to hang around and do something I don't like, I'll just move on until I find something I do.(Beta software developer)

> Q: Would you leave for a pay increase elsewhere.?

> No, I would leave. If I stayed I guess it wouldn't be through commitment to the company, it would be because I enjoyed the work, which isn't the same thing. No I don't think I would (stay). I'll defend them but not to that extent, not where it's causing me personal injury. (Beta software developer)

Thus it would seem that software developers' commitment to their organisation was markedly indirect and was sustained only in so far as the organisation offered the prospect of enhancing personal development and labour market leverage.

Their market position depended on them staying up to date and knowledgeable about the latest software developments: they thus had a keen interest in not only what companies were paying but in which companies were doing the latest projects. What became clear during these interviews was the role that social contacts outside the workplace, and outside working time, played in relaying this knowledge and, in so doing, supporting the software workers' sense of occupational identity; this is examined in more detail in the following section.

14.4.3 Evidence for a Software Community

Despite frequent statements by our interviewees that they liked to keep work and home life separate, what emerged time and time again was a high degree of inter-professional socialising, extending in several cases to partners also working in software, sometimes even in the same company. A husband and wife who both worked at Beta, reviewed their inter-related social lives in the following terms:

> (Wife) Both of us have friends at work. One of my friends at work I actually went to University with plus I've made some other friends, there is a sort of group of us. So I mean I go out with them outside of work and see them, plus we've had parties and things when we've had folk from work to our house. We do have like team nights out with the people that we work with that we go on every now and then
>
> (Husband) I've got one really good friend at Beta and I've got other friends that we go to the football and go out for lunch and things like that and there are other people that have left Beta and now working for different companies. We meet up with them regularly, go for lunch and things like this... (Beta)

These knowledge workers expressed some of the characteristic attitudes and social patterns of an occupational community (Salaman, 1974) in which a clear sense of collective identity extended across workplace and social contexts:

> Q: Do you feel you identify more with being a software engineer, for example?
>
> I mean I guess I can talk shop with other software engineers and my boyfriend's a software engineer, that is just by chance we ended up doing the same thing. Our closest friends that are another couple are both software engineers (Beta software developer)
>
> ... all the people I know are somehow connected with this company. So it's very much like that yes. I mean a lot of my old friends are people who used to work here as well as people who do work here....Glasgow's a small village in the first place – a little software industry. When I go out with my Glaswegian friends in the evening they also run into someone they know. (Gamma: IT support engineer)

However it was clear that these networks of social contacts were also importantly functional in serving to disseminate information about developments in the industry and the state of the labour market.

> Q: Do you socialise with people in the industry?
>
> I try to keep in touch with my peer group from University days, in fact we are still all in pretty regular contact and we generally are in fairly good knowledge of the positions that everybody else is in... and it's good for me because it means I can keep in touch with what is happening in the industry and where I sit in terms of what the industry average is and that kind of stuff. (Beta software developer)

These responses seem on the face of it to offer a clear example of "work centrality": where the challenges and the autonomy of the job are an intrinsic part of the employee's life:

> (Work) is actually quite central. Possibly more than a lot of people would like it to be but I don't see it as a chore .... and every day is different then I do quite enjoy it to be honest .... So I would say it is quite a major part – don't get me wrong – I do obviously go out and relax and forget about work for a while. I think you have got to, let your mind wind down from day to day...(Lambda software engineer).

However, to assume that this quantitative and qualitative overspill of software work into non-work life simply indicates that software workers fit into the type (d) stereotype we indicated at the start of the chapter would be to ignore other key determining factors in their situation and, in particular, the role of management.

14.4.4 Capturing the Work-Life Boundary

There was some evidence that management (particularly the owner-managers) recognised that supporting the norms of this occupational group could be the basis for an attempted "organizational embrace". The values of an internalised work ethic were commonly expressed in interviews by our software respondents:

> The fact is that doing a good job is not finishing... at the end of the day and walking away and that's it. (Beta, male developer).
> I don't mind doing the long hours because I know that ultimately it's going to be worth it ... just the sort of type of person I am – I like to see, at the end of the day, well we did a good job and we got it done. (Gamma, female software engineer).

However, there was little doubt that the owner-managers also helped to give substance, direction and form to these values through developing organisational cultures emphasising both professional norms and competitive realities. At Lambda, the owner aimed to recruit people "who want to join the club, not with a 9–5 mentality". Moreover, socialisation with both current and previous work colleagues was common and generally welcomed, and at social gatherings work usually represented, in the words of a male software engineer, "the main topic of conversation". At Pi, the owner-manager commented on the flat and informal structure he had established in which "the doors are kept open so that people can come and talk all the time ... I go to the pub every Friday night with them". Similar policies were in place at Lambda:

> I thought would be quite difficult, and in some cases it is, but we do get on very well and we all do socialise and I certainly try to create a team spirit, whether it's in simple things like on Friday afternoons bringing in sort of Pizza or something or going out after Fridays work for a couple of drinks or something. I try to create a team there as well. (Software Manager, Lambda)

These organisational cultures emphasised the relative youthfulness and often single status of the employees and were also heavily gendered. For example, in the case of Pi, the owner-manager described how, to maintain a "family" culture:

> We've created traditions such as a boys' (sic) golf outing ... we've gone to the same place every year and very few people from outside the company are invited (Pi, owner-manager).

It was the newer software companies in particular that displayed this somewhat "laddish" culture in which after-work drinks and social events were commonly mentioned. These male respondents rarely mentioned any domestic restrictions to their working or social lives.

For female and older staff however, and those with family responsibilities, such cultural expectations could be experienced as both excluding and coercive. This experience was certainly expressed by one female developer at Pi:

> I'm the only married person in development. Most of them are single men so in terms of things in common, I think they have – there's a lot of sporting interests but I don't really get involved in that sort of stuff . . . because most of them are guys, sometimes I don't think I have anything in common with them at all – the Sun newspaper, looking at page 3 when you come in the morning, discussing Kylie Minogue on MTV the night before, it gets a bit silly at times.

At Omega, a high-ranking manager recalled that when present at a prolonged senior management meeting he had had to leave at 6 pm to pick up his children from child-care. As he got up to leave the meeting, his director's cynical parting query of "leaving already?" still lingered with him a year following the event.

Further examples from these companies indicated the extent to which employee commitment to their occupation could be manipulated to the benefit of the company. Thus, entry to elite projects was contingent upon demonstrating appropriate behaviours, including willingness to work long and often unsociable hours. Not surprisingly, perhaps, team leaders' decisions helped to maintain and reinforce the divisions between employees working on prestigious and intellectually demanding work and those allocated to more routine, standard procedures. As we consider below, it is equally unsurprising that men tended to occupy the former group.

14.4.5 Role of Domestic Responsibilities

A further perspective on the links between commitment, professionalism and work-life pressures concerns the assumed and actual allocation of domestic responsibilities and their relationships with working lives. We first asked survey respondents whether they felt that there had been changes in men and women's attitudes toward paid and domestic labour. Although there was a general feeling that men and women's attitudes towards sharing domestic responsibilities were converging, men in our survey still hold more defined views of women's responsibilities and view them as having prime responsibility for domestic duties. When we enquired about actual behaviour, software women, whatever their attitudes, did take prime responsibility for domestic roles whilst men tended to participate at their discretion. As a female business manager at Omega remarked:

> I seem to have to do most of the washing up and ironing, all these household chores. . .I think the onus is always back on the women if the man isn't going to do it then the woman is going to do it. The buck stops here if you like.

Only about a third of women in our software companies had dependent family responsibilities (29/90), leading us to expect initially that both men and women would have equal access to software employment opportunities. This proved not to be the case. Whilst there was little difference between men and women in terms of the salience of software work, the dual responsibilities of work and home for women resulted in rather different job and career profiles.

Women were more likely to work part-time and were also less likely to work unpaid overtime. In terms of a career with the existing company, women with children rated a career as less important than women without children, whose orientations (even if in practice not realised) were not significantly different from those of men. However, in dual-earner households, software women contributed a smaller proportion to household income, irrespective of the presence of dependents. The women respondents reported they made sacrifices in paid work in order to sustain domestic work and in dual-career households, women, even without children, tended to give priority to their male partners' careers.

> In a relationship, you have got to decide, it's very difficult for two people to have full on careers, quite often one person has got to give a bit and in our relationship, it happens to be me and I don't mind doing that. (Gamma: female software developer)

A further aspect of the subordination of women's careers emerges from migration patterns, where women software workers sacrificed their own opportunities to accommodate partners' relocation and promotion prospects.

These gender based life-choices resulted in differences in work content. The higher-end activities (programming, systems analysis, testing, design, and user/application support) tended to be dominated by men. Pure development work was masculinised: the 11 developers in Pi were mostly male, young and single. Only two out of 15 of Omega's technical services team were women. A male senior software developer at Gamma summed up the distinctions:

> This is sexist. Sorry, I apologise for this but you tend to find more women in the support roles, which is what we've actually got. Both girls we have do primarily support. So I guess if anything, that's putting women down again. Sorry.

So the better qualified, younger and generally male employees tended to be offered or pushed for the more prestigious work whilst women, with their combinations of more fragmented career history, attributed domestic allegiances, lower technology qualifications and less training and development, have tended to remain on lower status single projects and customer-facing functions. A male business development manager at Gamma summarised the consequences:

> There is a certain amount of freedom within the organisation to actually develop your skills which I think is good – but . . . if people are not on a project or not visible, they tend to get left behind.

Unsurprisingly, a substantial majority of software women believed that advancement depended on putting job before family; only one-fifth of men shared this view.

14.4.6 Work Preferences?

From the willingness of women to relocate, to work part-time and to subordinate career advancement to family, the question emerges as to whether these choices reflect informed preferences by women in order to better accommodate a balanced role between home and work. The findings might indicate that, in the case of parents, choices were being made by both partners but that the choices were different.

Mothers were more likely to sacrifice work for family, whilst fathers inclined to the opposite: actual working hours, including unpaid overtime, were highest for fathers. As many of the married men claimed that they would like to spend more time with their families, it would appear that by willingly working long irregular hours there was, in addition to professional commitment, possibly a more long term calculation of the positive effects that this would have on pay, prospects and security.

As the more prestigious jobs were occupied by men, it can be speculated that this strategy was enjoying some success. The male finance director at Pi was moved to comment:

> You've only got to look at this company at the Board level, there are seven of us and we are all men.

This could however be a fairly high-risk strategy. An engineer at Beta reminisced that he was married for a brief period and that nightly call-outs "had a fairly notable effect on the marriage at the time".

14.5 Conclusions

It is clear from the above that software workers conform to many of the claimed characteristics of professional knowledge workers: they possess a strong professional identity; they are predominantly young and highly mobile; they have a commitment to the job which extends to a desire to push the boundaries of their own skills and competencies. These characteristics contribute to a freely admitted pattern of quantitative and qualitative spillovers from work to non-work life.

We have shown that when we examine the degree to which these patterns are experienced either as an inevitable and non-problematic part of the occupational culture or as an unwanted *im*balance between work and non-work, we have to build into the picture a deliberate manipulation by management of aspects of the shared occupational value system and their incorporation into corporate culture, with explicit expectations with regard to overt behavioural manifestations of job commitment. Arguably, this element of management expectation or even pressure has been underplayed by some revisionist accounts of work-life relationships which have focused on employee articulations of commitment without sufficiently teasing out the underlying and potentially challenging factors which contribute to these apparently positive expressions (see e.g., Kanter, 1989 and for a critique of "high-commitment" work practices, Legge, 2001).

Secondly, we have to be aware of the dimension of gender politics within the profession which results in these professional and managerial demands being experienced in quite different ways by male and female software professionals. The examples discussed in this chapter demonstrate that, where managerial expectations about patterns of working time and work intensity are based on what is acceptable to a workforce with a particular profile (young, male, single), this actually creates discriminatory structures for employees whose biographies lie outside that profile. A form of professionalised occupational and work-orientated community appears to be emerging which may also consolidate masculine solidarity in the same way that working class communities such as coal-mining did in the past and that professions such as law, continue to do today.

Perhaps male informal behaviour may often be experienced by female software professionals as "silly" but behind the observation lies a host of potentially exclusionary attitudes and practices which, in combination with structural factors such as fewer opportunities for continuing development, can serve to restrict women's occupational opportunities. Women software workers, especially those with dependents, may have attained some balance between their working and domestic lives, but this has come at a significant cost to their own career trajectories.

These conclusions suggest that we need a far more holistic grasp of the pressures, constraints and choices that combine to define contemporary work-life boundaries, in order more fully to understand the experiences of different segments of the contemporary workforce.

References

Baldry C, Bain O, Taylor P, Hyman J, Scholarios D, Marks A, Watson A, Gilbert K, Gall G, Bunzel D (2007) The meaning of work in the new economy. Palgrave Macmillan, Basingstoke

Brown R, Brannen P (1970) Social relations and social perspectives amongst shipbuilding workers – a preliminary statement. Sociology 4:71–84

Castells M (2000) The rise of the network society. Blackwell, Oxford

Cousins C, Tang N (2004) Working time and work and family conflict in the Netherlands, Sweden and the UK. Work Empl Soc 18:531–549

Crompton R (2002) Employment, flexible working, and the family. Br J Sociol 53:537–558

Dennis N, Henriques F, Slaughter C (1956) Coal is our life. Tavistock, London

Department of Trade and Industry (2004) Flexible working in the IT industry: long-hours cultures and work-life balance at the margins?http://www.diversityatwork.com.au/.../Flexible%20Working%20in%20the%20IT%20industry.pdf. Accessesd on 11 Feb 2010

Developing the Future (2008) Challenges and opportunities facing the UK software development industry

European Agency for Safety and Health at Work (2007) Expert forecast on emerging psychosocial risks related to occupational safety and health. Office for Official Publications of the European Communities, Luxembourg

Fleetwood S (2007) Why work-life balance now?. Int J Hum Resour Manage 188:387–400

Florida R (2004) The rise of the creative class. Basic Books, New York, NY

Fox A (1974) Beyond contract: work, power and trust relations. Faber & Faber, London

Frenkel, S. Korczynski, M. Shire, K. Tam, M (1999) On the front-line: organization of work in the information economy. Cornell University Press, Ithaca, NY

Haddon L, Brynin M (2005) The character of telework and the characteristics of teleworkers. New Technol Work Empl 20:34–46

Hakim C (2000) Work-lifestyle choices in the 21st century. Oxford University Press, Oxford

Heckley, G (2005) Offshoring and the labour market: the IT and call centre occupations considered. Office for National Statistics – Labour Market Trends – September, 373–385

Hislop D, Axtell C (2009) To infinity and beyond? The spatial implications of mobile teleworking. New Technol Work Empl 24:60–75

Hooker, H, Neathey, F, Casebourne, J & Munro, M (2007) The third work-life balance employee survey: main findings. Employment Relations Research Series No 58. Department of Trade and Industry, London

Houston D (2005) Work-life balance in the 21st century. Palgrave Macmillan, Basingstoke
Kanter RM (1989) When giants learn to dance: mastering the challenge of strategy, management and careers in the 1990s. Simon & Schuster, New York, NY
Kersley B, Alpin C, Forth J, Bryson A, Bewley H, Dix G, Oxenbridge S (2006) Inside the workplace: findings from the 2004 workplace employment relations survey. Routledge, London
Krings BJ, Nierling L, Pedaci M, Piersanti M (2009) Working time, gender and work-life balance, WORKS project. Higher Institute of Labour Studies, Leuven
Leadbeter C (1999) Living on thin air. Penguin, London
Legge K (2001) Silver bullet or spent round: assessing the meaning of the "high commitment"/performance relationship. In Storey J (ed) Human resource management. Thomson Learning, London, pp 21–36
Lewis S, Gambles R, Rapoport R (2007) The constraints of work-life balance approach: an international perspective. Int J Hum Resour Manage 188:360–373
Limburg D, Jackson PJ (2007) Teleworkflow: supporting remote control with workflow management Ssystems. New Technol Work Empl 22:146–167
Marglin S (1974) What do bosses do? The origins and functions of hierarchy in capitalist production. Rev Radic Pol Econ 6:60–112
Marks A, Scholarios D (2007) Revisiting technical workers: professional and organizational identities in the software industry. New Technol Work Empl 22:98–117
McKie, L, Hearn, J, Bowlby, S, Smith, A & Hogg, G (2008) Organisation carescapes: researching organisations, work and care, Hanken School of Economics Working Papers No. 538. Helsinki: Department of Management and Organisation
Meyer JP, Allen NJ (1991) A three-component conceptualization of organizational commitment. Hum Resour Manage Rev 1:61–89
Ransome P (2008) The boundary problem in work-life balance studies: theorising the total responsibility burden. In Warhurst C, Eikhof D, Haunschild A (eds) Work less, live more? Critical analysis of the work-life boundary. Palgrave Macmillan, Basingstoke, pp 62–79
Reeve R (2000) Happy mondays. Momentum, London
Salaman G (1974) Community and occupation: an exploration of work/leisure relationships. Cambridge University Press, Cambridge
Scotland IS (2009) Scottish technology industry survey: measuring up to economic challenges. Livingstone, Scotland
Townley B, Beech N, McKinley A (2009) Managing the creative industries: managing the motley crew. Hum Relat 62:939–962
Warhurst C, Eikhof D, Haunschild A (2008) Out of balance or just out of bounds? Analysing the relationship between work and life. In Warhurst C, Eikhof D, Haunschild A (eds) Work less, live more? Critical analysis of the work-life boundary. Palgrave Macmillan, Basingstoke, pp 1–21
Webster J (2004) Working and living in the European knowledge society: the policy implications of developments in working life and their effects on social relations. Report for the project Infowork. Employment Research Centre, Trinity College, Dublin

Chapter 15
The "Myth of Separate Worlds": An Exploration of How Mobile Technology has Redefined Work-Life Balance

Linda Duxbury and Rob Smart

15.1 Introduction

Technology has changed the business practices of millions of professional workers worldwide. The hardware and software that revolutionized how people worked were initially found only in the workplace and were used only during traditional 9-to-5 working hours. The typewriter stayed on the desk when its user went home, the office telephone was physically tied to its spot by a cable and computers were too heavy to be carried anywhere. Nowadays, however, the act of performing work is not limited to specific hours at a specific location. The latest incarnations of work related technology for professional employees, such as laptops, cell phones and the Blackberry, support work outside the confines of the office, at almost any time of the day or night (Towers et al., 2006).

These tools, referred to as mobile technologies, have been defined to include "handheld IT artefacts that encompass hardware (devices), software (interface and applications) and communication (network services)" (Jarvenpaa and Lang, 2005, p. 8). In this chapter mobile devices are operationally defined to include mobile phones, portable digital assistants, laptop computers with e-mail, and integrated wireless devices such as the Blackberry. An important characteristic of each of these mobile technologies is fact that they are not bound to a specific place but remain present regardless of place (Pica and Kakihara, 2003). In the language of boundary role theory, mobile technology can be considered "a topology whereby neither boundaries nor relations mark the difference between one place or another. Instead boundaries come and go, allow leakage or disappear altogether" (Pica and Kakihara, 2003, p. 6).

Mobile technologies are of special interest to researchers seeking to understand the balance between work and personal-life as they have been linked to a fundamental shift in how the boundaries between work and home are constructed in Western societies (Golden and Geisler, 2007; Shumate and Fulk, 2004). When Kahn et al. (1964) published their seminal work on role conflict and ambiguity, the separation

L. Duxbury (✉)
Carleton University, Ottawa, ON, Canada
e-mail: linda_duxbury@carleton.ca

S. Kaiser et al. (eds.), *Creating Balance?*, DOI 10.1007/978-3-642-16199-5_15,

of work and family roles in both time and space was common. Work occurred during designated hours at a location away from the home and transitions between work and home-based roles were well scripted in societal expectations. Today, advances in computing and communication technologies have made it possible to conduct more work at home, to conduct personal business on the job and to do both in the same time frame, swapping in and out of roles as needed (Shumate and Fulk, 2004, p. 59). As technology transcends many aspects of users' lives, organizational expectations with respect to how these technologies should be used have the potential to clash with personal desires and family expectations. Mobile technology means that many employees are not able to escape workplace controls at the end of the workday. Indeed, as pointed out by Shumate and Fulk (2004, p. 56), mobile technology has shifted our view of the family domain from a "place of refuge ... leisure and entertainment" where the worker is "free from outside expectations and surveillance" to a location where "e-lancers" (Pica and Kakihara, 2003) perform legitimate work.

Claims about how mobile technology influences work and family have become widespread and debates concerning the impacts of these devices are proliferating (Gephart, 2002; Pica and Kakihara, 2003). An extensive review of the literature indicates, however, that very little is known about how these technologies are used by employees and what impact their use is having on key individual, organizational and family outcomes. Furthermore, there is little consensus within the available literature on how these devices affect work-life balance. While some researchers (Bolino, 1999; Towers et al., 2006) talk about the benefits that technology can bring to individuals and their families, others talk about the potential such devices have to disrupt work-life balance (Duxbury et al., 1992; Salazar, 2001). Little empirical evidence is available to support or dispute either of these points of view. This chapter sets the stage for further research in this area by summarizing extant research addressing the following issues:

- How can mobile technologies impact boundaries in time and space?
- Why do professional employees use mobile technologies?
- How much time do professional employees spend using mobile technologies to perform work outside office hours in a location other than their office?
- What impact does the use of mobile technology have on work-life balance?

The chapter is structured as follows. Section 15.2 outlines the approach taken in this chapter to explore the relationship between mobile technology use and work-life balance. Sections 15.3, 15.4, 15.5, and 15.6 address each of the issues noted above. The paper ends, in Sect. 15.7, with a number of suggestions for future research in this area.

15.2 Terminology and Empirical Background

The chapter examines the relationship between the use of mobile technology and work-life balance. Very few authors explicitly define work-life balance (Carlson and Grzywacz, 2008; Greenhaus and Allen, 2006). While the most widely held meaning of work-family balance is a lack of conflict or interference between work and family roles (Frone, 2003), recent research suggests that work-life balance is more than a lack of inter-role conflict (Eikhof et al., 2007). A more holistic definition is offered by Greenhaus and Singh (2003) who define work-life balance as "the extent to which individuals are equally involved in – and equally satisfied with – their work role and their family role."

One of the first tasks we had as researchers was to define a term that aptly described the type of work facilitated by mobile technology (i.e., work done in the evening, on weekends and during vacations by employees who spend a "normal" working day in the office). Our goal was to find an expression that reflected the fact that mobile technology makes it possible for professional employees – managers and highly qualified employees working on complex knowledge intensive tasks – to be as productive working outside the office and outside conventional working hours as they are working inside the office, during "regular" working hours. In other words, the term chosen needed to reflect the fact that employment need no longer be limited to specific hours at a specific location. The ability to be productive outside traditional office hours means that work can be extended in the dimension of time. The fact that it is possible to work outside the walls of the office means that work can also be extended in the dimension of space. Accordingly we coined the term work extension to describe those circumstances where employees performed work outside office hours in any location other than their office. We also chose to refer to the mobile technology which makes this type of work possible as *work extending technology* or WET (Towers et al., 2006).

This chapter focuses on professional workers who toil (with other colleagues) in the office during office hours and use a range of technologies to perform supplemental work outside office hours and/or when they are no longer in the office. While such employees are generally expected to be physically present in the office 5 days a week for the whole workday, they may sometimes work at home during the day when there is a pressing reason (e.g., the need to be able to work undisturbed or to be able to deal with a non-work issue, like a sick child or the visit of a workman). They may travel occasionally for business, but they are not mobile workers (such as consultants or sales people) who spend the majority of their time at client sites or on the road.

Empirical data presented in this chapter is taken from two different studies into the relationship between mobile technology use and work-life balance done by Linda Duxbury, Chris Higgins and their colleagues:

- The 2001 *National Study on Work-life Conflict* ($N = 31{,}571$ Canadian employees who work for public, private and not-for-profit sector organizations)
- The 2005 *Mobile Technology Users Study* (845 Canadian federal civil servants who used mobile technology were surveyed, 61 participated in in-depth follow up interviews and 18 took part in focus groups).

A complete description of the 2001 study can be found in Duxbury et al. (2007) while details on the 2005 study are given in Towers et al. (2006).

15.3 Mobile Technology and Boundaries in Time and Space

Boundary theory "describes how employees transition between home and work roles, and is based on the assumption that individuals seek to minimize difficult role transitions and the frequency of role transitions" (Thompson et al., 2006, p. 295). Boundaries are conceptualized as the lines of demarcation between domains and are used to define the point at which domain relevant behaviour begins or ends (Clark, 2000, p. 756). Ashforth (2001, p. 262) defines boundaries as "mental fences . . . used to simplify and order the environment." The concept of boundaries has been used in many disciplines "to refer to the physical, emotional, temporal, emotional, cognitive and/or relational limits that define entities as separate from one another" (Ashforth, Kreiner, and Fugate, 2000, p. 474). The idea of a boundary between work and life is shaped by the notion of "separate spheres" – the existence of two distinctly different domains of experience and activity, the workplace and the home (Gephart, 2002; Golden and Geisler, 2007, p. 520).

A review of the relevant literature suggests that mobility technology has blurred the boundaries between work and non-work (Eikhof et al., 2007; Fleming and Spicer, 2004; Golden and Geisler, 2007; Greiner, 2006; Middleton et al., 2005; Pica and Kakihara, 2003; Prasopoulou and Pouloudi, 2006; Rakow and Navarro, 1993; Shumate and Fulk, 2004). Employees who use mobile technology often find themselves inhabiting multiple worlds and multiple roles simultaneously (Shumate and Fulk, 2004; Warhurst et al., 2008) and existing synchronously in two environments, a physical one and a virtual one (Pica and Kakihara, 2003). For these individuals, boundary transitions are both more frequent and more challenging (Shumate and Fulk, 2004) as the "various boundaries that have traditionally separated the workplace from the more private domain of life" become porous (Flemming and Spicer, 2004, p. 74).

Time and space are important organizing mechanisms that help to organize an individual's day into perceived regions and offer cues to appropriate role behaviour (Shumate and Fulk, 2004). The fact that mobile technologies have the ability to compress space and time and erode and reconstitute the spatial division between work and non-work that characterizes post-industrial societies gives them the potential to disrupt the fragile boundaries between work and non-work domains (Arnold, 2003; Fleming and Spicer, 2004; Prasopoulou and Pouloudi, 2006). Gergen (2002) has coined the phrase "absent presence" to describe the situation in which an employee is physically present in one domain but intellectually transported to another (e.g., sitting at home doing e-mail for work).

Any debate about the relationship between work and non-work life usually involves a discussion of the importance of time allocation (Thompson and Bunderson, 2001). Through the establishment and preservation of specific temporal boundaries people organize their activities and attach meaning to their experiences (Prasopoulou and Pouloudi, 2006). The premise in much of the work-life literature is that "because the quantity of available time is finite, managing work and non-work domains is a balancing act that seeks the magical appropriate distribution of hours among important domains" (Thompson and Bunderson, 2001, p. 20). Most scholars have adopted a "depletion" argument in discussing the relationship between work and non-work – assuming that work and non-work are inherently conflicting because they both make claims to the same resources in general, and time in particular (Thompson and Bunderson, 2001). The ability of time to facilitate the establishment of order in daily activities is gradually subsiding as temporal boundaries that were once distinct (e.g., work time versus private time) have become more fluid (Prasopoulou and Pouloudi, 2006). The speed of the change can be realized by considering the following. In 2001, Thompson and Bunderson (p. 19) noted that "work organizations directly compete with non-work domains such as family, friends and leisure for the employee's personal stock of time....because workers *cannot be in two place at once*, work time necessarily subtracts from available non-work time and vice versa"(italics added). In 2009, however, this statement no longer holds as mobile technology enables employees to be in one place physically and another virtually at the same time. This connectivity can be both intrusive and disrupting (Middleton et al., 2005).

The ability of mobile technology to erode time seems to have three sources: the enabling characteristics of the technology, the culture of the organization, and the users' own behaviour (Prasopoulou and Pouloudi, 2006). Technically, devices like mobile phones and the Blackberry function as "electronic leashes" making it easier for employees to transcend social distance which used to function as a protective shield against disturbances (Prasopoulou and Pouloudi, 2006). Work by Prasopoulou and Pouloudi (2006) suggests that the "any time" communication allowed by mobile technology presents more work-life challenges for individuals who work for organizations whose culture dictates a 24/7 expectation. While mobile technology allows for immediate response to organizational issues, it simultaneously requires the user to disengage from whatever else they are currently doing (Middleton et al., 2005). Finally, Prasopoulou and Pouloudi (2006) point out that mobile technology is more problematic for professionals who are unable to draw a definite temporal boundary between their work and non-work activities by turning off their mobile technology or not answering business calls/e-mails.

15.4 Why Use Mobile Technologies

Many professional employees make use of mobile technology. This section examines why this should be the case by first summarizing current speculation with respect to the advantages and disadvantages of mobile technology use to employers, employees and families and then presenting a summary of our recent findings with respect to this issue.

15.4.1 Advantages of Mobile Technology

Our literature review suggested that organizations that supply their professional employees with mobile technology may realize three main benefits. First, their employees may work longer hours. As Ipsos-Reid (2001) note, by accepting the technology, the employee consents to being in a position to do work away from the traditional office location and consents to a (potentially) longer workday. The Gartner Group (Dulaney et al., 2000) reported that professional WET users gain

5 h of productivity each week by having a notebook computer for use at home. Second, their employees will be more accessible. An employee with mobile technology can be reached at any time, anywhere. In theory, the employee is able to regulate the pace at which work is performed and can do work at home and in his or her off-hours, thereby making more time for family etc. This would be an ideal situation if workload and productivity expectations were held constant. Unfortunately our research indicates that this is not the case as many of the mobile technology users in our samples reported that work expectations increase with the adoption of mobile technology (Towers et al., 2006). Third, WET provides opportunities for the employer to monitor and control the employee. Green (2002) notes that that it is not necessary that the monitoring actually takes place for it to be an effective source of control; the individual merely needs to feel that they may be monitored for them to behave as if they are being monitored.

The above list of benefits raises an interesting question. If using mobile technology is biased in favour of the employer, as it appears to be, why do employees accept the technology? Our review of the literature suggested that employees who use mobile devices might realize two primary benefits. First, the technology gives employees more control over where, how and when they do their work (Towers et al., 2006). The ability to do work at their own pace in locations of their choosing often serves as a powerful inducement to agreeing to adopt mobile technology. Second, mobile technology facilitates impression management which is "the process by which people attempt to influence the image others have of them" (Bolino, 1999, p. 84). More specifically mobile technology allows individuals to outwardly project their image as an "in demand," hard-working, committed employee. The importance of impression management as a possible benefit for mobile technology users emerges when we consider their willingness to perform work during weekends and vacations, a strategy that exemplifies their dedicated to the firm. Sending an e-mail to one's manager at 10:30 pm can be as effective a method of impression management as physically sitting in the office. The act of working after hours can also be seen to be part of a strategy of intimidation; by sending e-mails at 11 pm, a worker may be attempting to make colleagues appear less dedicated – the colleagues are then forced to (or must appear to) work equally long hours (Towers et al., 2006).

15.4.2 Disadvantages of Mobile Technology

Two potential disadvantages of mobile technology use for the employee were identified from our review of the literature. First, the fact that most work extension is done in ones' residence (Towers et al., 2006) means that the home, traditionally a place of sanctuary, becomes an extension of the workplace. The obvious consequence is that time is taken away from family and other personal activities. The fact that the employee can be reached at any time by cell phone or e-mail contributes to the blurring of the boundary between work and home. Salazar (2001) reviewed the literature on work and boundaries at home and found that people working at home need to define boundaries in time and space. She found that this process is becoming increasingly difficult to manage and notes that for career professionals the "relationship between home and work is more like a continuum."

Second, WET may also increase the potential for family conflict. As Bolan (2001) noted, trying to balance the needs of employer and family at the same time and more importantly, in the same place, can cause an increase in conflict at home. Family members may resent the amount of time that the work extender is actually spending on work, and feel that this is time taken from them. Furthermore, as the work extender works at home, the family members may feel that their space is being taken away from them. Thus the two dimensions of work extension may contribute to conflict. Finally, technologies that facilitate work outside the office often act as an agent to legitimize work extension at home (Duxbury et al., 1992; Venkatesh and Vitalari, 1992) thereby increasing the extent to which work interferes with family (Duxbury and Higgins, 2009) and the level of role interference for professional men and women.

15.4.3 Advantages and Disadvantages: A View from the Trenches

The literature abounds with lists of the advantages and disadvantages of mobile technology use. One of the objectives of our in-depth study of mobile technology use was to empirically validate these lists (Towers et al., 2006). We did so by asking WET users what they had personally found to be the most important work and non-work related benefits and drawbacks of using mobile technology. We asked them to answer these questions from two perspectives: benefits and drawbacks that they personally had experienced and benefits and drawbacks that they perceived were realized by their employer. Findings with respect to these issues are summarized in Table 15.1 and discussed below.

Five work related benefits of WET use were indentified by a substantive number (10% or more) of the WET users in our sample. Key benefits mentioned by respondents included the fact that they perceived that the technology had increased their efficiency and productivity (34%), given them more flexibility with respect to when and where they performed their work (32%), facilitated work from home (25%) and increased accessibility to colleagues and clients when they were not in their office. This list of benefits is very similar to their perception of how their organization had benefited from their adoption of WET: "they get more work out of us" (63%), "it is easier for the organization to contact me" (30%). Respondents were also able to identify several ways in which the technology had benefited them outside of work. Key benefits noted by these users included increased accessibility to family (26%), the ability to work from home in the evening (22%) and the ability to manage unexpected circumstances at home without serious disruptions at work.

Several conclusions can be drawn by scanning this list of benefits. First, employees value WET because it makes it easier for them to engage in work extension activities and manage their time more efficiently. Second, employees perceive that WET enhances their productivity, a by-product of the fact that they put more hours into work and are more able to manage their time. Third, the vast majority of

Table 15.1 Perceived advantages and disadvantages of mobile technology

Benefits and drawbacks of mobile technology	%
Work related advantages for the employee	
I am more efficient and productive; I can manage my time more effectively	34
I have more flexibility and freedom with respect to when and where I work – "I am not tied to my desk"	32
I can now work from home	25
I am able to reach people/people can reach me when I am in transit/travelling/off site	24
It is easier to get work done outside of "normal" working hours	12
Non-work related advantages for the employee	
My family can get in touch with me	26
I can be at home (don't have to put in overtime hours in office)	22
I can work from home in an emergency (i.e., sick kids)	19
It helps me with my work life balance	14
Advantages for the employer	
Higher productivity – "they get more work out of us"	63
Higher efficiently – "employees work smarter"	39
More able to get in touch with their eemployees	30
Higher employee satisfaction	16
Work related disadvantages for the employee	
I am never off duty – "I can be reached at any time", "they expect ne to always be accessible to work"	40
It increases work expectations with respect to workload and turnaround time	20
It has lengthened my work day	20
There are no work related drawbacks	16
Non-work related disadvantages for the employee	
It takes away from my time with my family/ my personal time – this time is now given to work	48
There are no non-work related drawbacks	38
It is intrusive – it is hard to get away from work	19
Disadvantages for the employer	
Cost of devices, updating and training	39
Loss of control – managers don't know where employees are, what they are doing	20
Employees make personal use of the technology during work hours	20
Employee burnout, stress due to the high volume of e-mail	10
There are no disadvantages to the employer	10

the benefits mentioned by users involve employees giving preference to work over non-work activities. In fact, many of the behaviours they cite when describing the advantages of mobile technology (i.e., can work from home, people can reach me all the time, can put in overtime hours at home, can work outside regular hours) give priority to work at the expense of family and are likely to make work-life balance more challenging.

Relatively few individuals mentioned work-life balance as an advantage of mobile technology – a finding that is consistent with the findings with respect to the drawbacks of using WET mentioned by the respondents. What are the work-related drawbacks of using WET? More work – less life! The users that we interviewed talked about the fact that for them, mobile technology meant that they were "never off duty" (40%), that the organization had increased expectations with respect to workload and turn around time (20%) and that it had resulted in a lengthening of their work day. From a non-work perspective, respondents noted that the main disadvantage of the technology was the fact that it facilitated work during family time and that as a result they now had less time for their family, and that the technology was intrusive and hard to get away from. The following quotes from our participants sum up general tendencies with respect to the disadvantages of the technology.

> E-mail is widely used as the primary method of communication with employees at all levels, however there is a universal expectation that everyone reads, clears and responds to all email on a daily basis, actually a constant basis. This is unrealistic. Memos sent late afternoon are expected to be read by morning for discussion. This is unreasonable. Everyone needs to understand that communications need to be managed; time must be allowed to read, understand and respond. (Male manager)

> The amount of work has increased dramatically in the last decade, mainly due to the increased use of technology. With today's technology we can have someone in our office, an incoming phone call, voice-mails and e-mails – all of which we are supposed to respond to at once. Technology has added the expectation of immediate response and solution to the workplace. There is no balance. (Female professional)

15.5 Time Spent Using Mobile Technology

How much time do professional employees spend using mobile technology to extend when or where they work? Half of the professionals who responded to our 2006 survey indicated that they used technology to work at home while 34% used it to work remotely but outside the home. These professionals spent a substantial amount of time per week working remotely (3.2 h per week at home and 2.1 h per week in other locations). This finding is consistent with the responses we obtained with respect to the advantages and disadvantages of WET – they increase expectations from the organization with respect to availability, response time and workloads.

To further understand how much time the professionals in our sample used WET to extend their work hours and to work remotely we subdivided our sample into three groups of users based on the amount of time per week they spent using WET: low users (1–20% of total work week), moderate users (21–60% of total work week) and high users (>60%). While the plurality of users fell into the low (28%) and moderate (48%) user groups, one in four were high users. Data on how much each of these groups use WET are shown in Table 15.2 and discussed below.

Table 15.2 Use of mobile technology to support mobile work

	Work at home		Work activities all locations	
User group	Total hours	Hours using WET	Total hours	Hours using WET
Low	5.1	2.1	43.3	26.7
Medium	11.8	9.3	52.4	28.9
High	28.7	25.2	68.8	53.6
Total sample	5.84	3.0	44.2	27.97

Examination of the data in Table 15.2 show that employees who were classified as high users of mobile technology spend substantially more hours per week in work than those in the low or medium user groups. These hours are not, however, spent at the office. Nor are they spent in alternative work locations; they are spent working from home. It is not possible from these data to determine the direction of causality – do people who have access to more tools spend time working because the technology allows them to do so or do they find it hard with the technology on hand to separate work from non-work? Are they workaholics and use mobile technology to support overwork or do they have very heavy work demands and so acquired mobile technology to help them meet these demands by working from home? Future research is needed to disentangle these findings.

15.5.1 Time Spent Using Mobile Technology: The View from the Home Front

While it makes intuitive sense that work extension will impact the family, little research in this area currently exists. To close this gap our 2005 study looked at how the family of the WET user felt about their use of this technology. Unfortunately, since we did not have access to the family members themselves, we had to ask the employee themselves to speak on behalf of their family. More specifically we asked respondents how their family felt about their use of technology to work during family time (i.e., evenings, weekends, vacation). We also asked them to explain how they managed the work-life boundary.

Sixty percent of respondents indicated that they tried to keep a strict separation between their work and their home life. The data would, however, suggest that most are not successful in this regard: 44% of respondents confessed that they used WET to work when they were on holidays, 60% engaged in supplementary work at home on week nights and 75% used it to work from home on the weekends. Respondents who indicated that they did not work evening/weekends stated that this had been a conscious decision on their part – that "they needed time with their family – they are important to me" (male systems analyst) and that "it is just not healthy – I work hard and I need the downtime" (female professional). It is interesting to note, however, that all of the respondents who chose not to work weeknights and weekends indicated that they understood where their colleagues were coming from with respect to their need to engage in work extension activities during family time. They all noted that one consequence of keeping work and family separate was that they often felt overwhelmed with work (especially with e-mail) when they began their workweek or got back from holidays. The reaction of this group to the idea of working when on holidays was also interesting and worthy of note: they felt sorry for people who had to engage in such activities (45% gave this response).

Finally, it is interesting to note that while the majority of WET user respondents perceived that their family understood their need to work on "family time" one in three stated that their family was not happy about such activities and was "pushing back."

15.6 Mobile Technology and the Work-Life Boundary

Just under half of the 31,571 respondents to our 2001 national survey on work-life balance used mobile technology (e-mail, computers, and internet) to support work outside the office (Duxbury and Higgins, 2001). These people spent an average of 5 h per week (or 20 h per month) in work extension activities. Additional analysis indicated that virtually all of this extra work was undertaken as unpaid overtime at home. The tendency to use mobile technology to extend the workday was strongly linked to job type but, within job type, not to gender. Whereas the majority of managers (73%) and professionals (56%) in our national survey engaged work extension activities, only 32% of those in the clerical and administrative sample did so.

We asked respondents "to what extent their use of mobile technology had increased, decreased or had no impact on their workloads, their amount of stress they are under and their ability to balance work and family." Responses to this question are given in Fig. 15.1.

Examination of the findings for the total 2001 National Work, Family, and Lifestyle sample indicate that while the majority of the professional employees in our sample felt that mobile technology had increased their workloads (70%) and the amount of stress they are under (50%), they did not feel that they it had affected their ability to balance work and family (49%). Also worthy of note is the fact that approximately the same percent of respondents felt that WET had enhanced their ability to balance work and family as those who stated that it had made balance more difficult. These conflicting results suggest that there are two distinct vantage points when looking to understand the impact of mobile technology on the work-life balance on employees. Specifically, our findings support the idea that mobile technology can both support and hinder work-life balance at the same time.

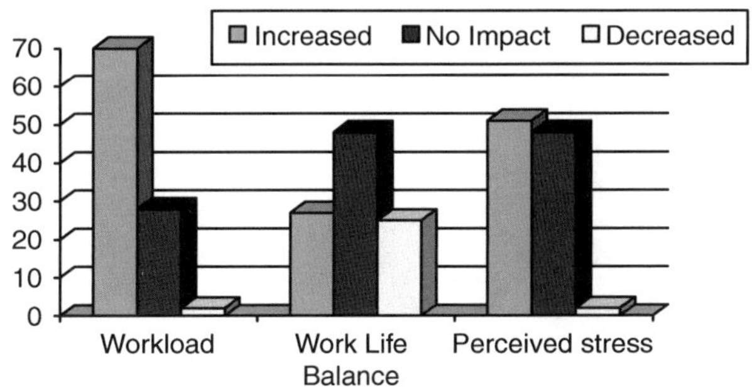

Fig. 15.1 Impact of mobile technology on workload, work-life balance and perceived stress

15.6.1 The Janus Face of Technology: Why the Contradictions?

The paradoxical nature of mobile technology observed in our research reflects the idea put forth by Arnold (2003) that technology may often "have consequences of one kind, while at the same time be implicated in a contrary set of consequences or implications" (p. 231). Arnold used the Roman deity Janus as a metaphor for technology in order to highlight the contradictory outcomes of increasing technology use. The metaphor emphasises the ambiguity of mobile technology uncovered by our research and may help to explain our findings with respect to mobile technology and work-life balance.

A review of the literature suggests that the paradoxical impact of mobile technology might be due to the fact that the potential gains in productivity offered by WET are offset by the human costs (workload, stress) associated with an inappropriate or extensive use of these work-extending technologies. Furthermore, they suggest that the ability to complete tasks more quickly through the use of technology typically does not lead to an increase in the amount of free time available to an individual, but rather an increase in the number of tasks to be completed. Researchers who hold this view argue that the effect on the individual of 24/7 access to work is a feeling that there is "no time to waste" and that employees should be doing something "productive" at all times (Hassard, 1999).

Not all researchers paint a bleak picture with respect to the impact of mobile technology on work-life conflict. In fact, a substantive body of work can be identified which contends that the increased sense of control and flexibility offered by wireless devices helps employees integrate work and family domains (e.g., Darrah et al., 2007; Golden and Geisler, 2007; Middleton et al., 2005; Rakow and Navarro, 1993). Golden and Geisler (2007), for example, concluded that the "spirit of the PDA device is control" and that employees used the PDA to manage the work-life boundary through both integration and segmentation of work and personal-life. Similarly, Rakow and Navarro (1993) concluded that cellular phones were liberating for professional workers, giving them increased flexibility and time as well as an edge over their competition. These authors also noted that mobile technology facilitates remote mothering (i.e., the ability to accomplish mothering responsibilities across time and space) thereby allowing women to work "parallel" shifts rather than the double shift described by Hochschild (1989). Middleton et al. (2005) concluded their study on the Blackberry by noting that users have embraced the anytime, anywhere connectivity that mobile devices enable, valuing the ability to better control the heavy demands of their jobs. They also reported that access to a Blackberry provided employees with greater control over both their work and non-work environments and enabled them to manage the heavy demands of their work in ways that best suited them. Prasopoulou and Pouloudi (2006) observed that technology gives users the ability to move between different locations and arrange more flexible working conditions – to "be distant as well as close." Katz and Aakhus (2002, p. 2) note that mobile technology can "increase the pace and efficiency of life, allowing more flexibility at business and professional levels as well as in family and personal life" by making it easier for employees to multitask work with activities that require minimal attention.

15.7 Conclusions and Recommendations

This paper has provided some new information about how and why work extension technologies are being used and how such use affects the work life balance of professional employees. The evidence from the literature and from our own work in the area suggests that mobile technologies are shifting temporal boundaries and that the walls of the office have moved outward to include the home and the myriad of other locations where work can now be done. Furthermore, while work extenders need to deal with the impact of work extension on their work and non-work spheres, it would appear that the mechanisms they currently use to control their use of technology to perform work extension are only partially successful in helping to keep a clear boundary between work and non-work spheres. They are provided with little guidance from the employer about how to put limits on the encroachment of their space and their time by work. Indeed, the culture of the organisation encourages work extension and promotes those professionals who are available for work 24/7.

Employers benefit by providing WET in terms of work output and responsiveness. There also appear to be clear benefits for employees who adopt mobile technologies as WET increases flexibility with respect to the timing and location of work, which makes it easier to accommodate both work and family. At the same time, of course, WET increases expectations: managers and colleagues alike expect work extenders to be almost always available to do work, which makes it easier for work to encroach on family time and also leads to a greater workload. The ability to perform work extension is a dual-edged sword. One of our interviewees captured this point very clearly:

> I believe that while technology has increased the ability to work from the home and outside regular business hours, it has also increased the expectation that you do so. So while it has enhanced the ability to balance work and family, it also has complicated it. (Male professional).

At this point in time we have only a preliminary understanding of the role mobile technology plays in management of the work-life boundary. While there is widespread perception that these technologies complicate work-life boundaries there is little empirical documentation of the ways in which individuals and organizations use these technologies to either segment and/or integrate work and personal life. Given the proliferation of such devices more research is needed to understand how mobile technology disrupts and challenges not only time and identity boundaries, but also the spatial dimensions of what is considered work and non-work. The paucity of scholarly research on this phenomenon is particularly striking in light of the fact that the popular press abounds with anecdotal descriptions of the use of these technologies.

Unfortunately, many of the empirical studies that are available have limitations. For one, most studies are cross sectional and correlational in nature which limits our ability to determine whether or not mobile technology increases or decreases work-life conflict over time. For another, much of the work on mobile technology has not been device specific, focusing instead on categories of devices like mobile phones. This focus has limited our ability to generalize findings and may be responsible for many of the inconsistencies in findings observed in the literature. Furthermore, much of the research in the area has focused on the spatial aspects of this mobility. The temporal aspects have received limited attention. Another problem is that many of the studies look at how mobile technologies affect the organization (the assumption is that such tools are intended for organizational use only) or focus on the functionality of the technology, and give few insights into how professionals in business organizations manage the impact of these technologies on the temporal boundaries separating work from non-work activities.

Our review of the literature has identified a number of issues that remain to be addressed in this area. First, the fact that much of the research in this area has been cross sectional means that the direction of causality for increased mobile technology usage is difficult to ascertain – do people who have access to more tools spend time working because the technology allows them to do so or do they find it hard with the technology on hand to separate work from non-work? Are they workaholics and use mobile technology to support overwork or, alternatively, are individuals with very heavy work demands more likely to acquire WET to help them meet these demands by working from home? In this case, respondents with a high dependence on mobile technology would still work these long hours if they did not have mobile technology, but the location where the work was performed would be quite different – they would work in the office. Future studies with a longitudinal design should be undertaken to address this issue.

Second, it is evident that the use of mobile technology has been beneficial to both employers and employees. Despite this apparent symbiosis, the benefits are often at logical odds to each other. For instance, a positive for the employer may be viewed as a negative for the employee and vice versa. Future research should seek ways to reconcile these issues.

Third, the advantages of using mobile technology, regardless of whether they pertained to work or non-work, were all associated with the fact that mobile technology gives employees greater control over when and where they worked and increases their ability to communicate easily and asynchronously with others. These two characteristics of mobile technology can be seen as the reasons for the other advantages namely increased flexibility and greater work-life balance. Unfortunately, the same features of mobile technology that increase perceived control and facilitate communication also appear to be the source of many of the disadvantages of mobile technology such as the feeling of being "on call" all the time (for work and family) and the lack of a clear boundary between work and home. Disadvantages unique to the work domain include increased workloads and increased work expectations (speed of response, how much work can be done in a given time period). The main disadvantage relating to the family was that less time was available for spending with the family. This appears to be a direct consequence of these increased workloads and expectations as well as the increased ease with which one can work from home and elsewhere. Again, there is a need to determine how best to deal with these competing issues.

Acknowledgements The authors would like to acknowledge Dr. Christopher Higgins and Dr. Ian Towers – who worked with us to collect much of the data reviewed in this chapter.

References

Arnold M (2003) On the phenomenology of technology: the "Janus-Faces" of mobile phones. Inf Organ 13:231–256

Asforth B, Kreiner G, Fugate M (2000) All in a day's work: boundaries and micro role transitions. Acad Manage Rev 25:472–491

Ashforth B (2001) Role transitions in organizational life: an identify based perspective. Lawrence Erlbaum, Mahwah, NJ

Bolan S (2001) New retention route for IT shops. Comput Dealer News 17:40

Bolino M (1999) Citizenship and impression management, good soldiers or good actors. Acad Manage Rev 24:82–98

Carlson D, Grzywacz J (2008) Reflections and future directions on measurement in work-family research. In Korabik K, Lero D, Whitehead D (eds) Handbook of work-family integration. Elsevier, New York, NY, pp 57–74

Clark S (2000) Work/family border theory: a new theory of work/family balance. Hum Relat 53:747–770

Darrah C, English-Lueck J, Freeman J (2007) Living in the eye of the storm: controlling the Maelstrom in Silicon Valley. Working Paper, San Jose State University. http://www.sjsu.edu/depts/anthropology/svcp/pdfs/svcpmael.pdf. Accessed on 4 Jul 2007

Dulaney K, Egan R, Purchase E (2000) How to build a wireless office: the next wireless revolution. A Garnter Group Research Note, 4 February 2000

Duxbury L, Higgins C (2001) Work-life balance in the new millennium: where are we? Where do we need to go?. Canadian Policy Research Networks, Ottawa, ON

Duxbury L, Higgins C (2009) Key findings and recommendations from the 2001 national work-life conflict study – Report Six. Ottawa: health Canada.
http://www.hc-sc.gc.ca/ewh-semt/pubs/occup-travail/balancing_six-equilibre_six/index-eng.php. Accessed on 16 Feb 2010

Duxbury L, Higgins C, Mills S (1992) After hours telecommuting and work-family conflict: a comparative analysis. Inf Syst Res 3:173–190

Duxbury L, Towers I, Higgins C, Thomas A (2007) From "9 to 5" to "24/7": how technology has redefined the workday. In Law W (ed) Information resources management: global challenges. Idea Publishing, Lodon, pp 305–331

Eikhof DR, Warhurst C, Haunschild A (2007) What work? What life? What balance? Critical reflections on the work-life balance debate. Empl Relat 29:325–333

Fleming P, Spicer A (2004) You can checkout anytime, but you can never leave: spatial boundaries in high commitment organizations. Hum Relat 57:75–94

Frone M (2003) Work-life balance. In Quick J, Tetrick L (eds) Handbook of occupational health psychology. American Psychological Association, Washington, DC, pp 143–162

Gephart R (2002) Introduction to the brave new workplace: organizational behaviour in the electronic age. J Organ Behav 23:327–344

Gergen K (2002) The challenge of absent presence. In Katz JE, Aakhus M (eds) Perpetual contact: mobile communication, private talk, public performance. Cambridge University Press, Cambridge, pp 227–241

Golden A, Geisler C (2007) Work-life boundary management and the personal digital assistant. Hum Relat 60:519–551

Green N (2002) Who's watching whom?. In Brown B, Green N, Harper R (eds) Wireless world: social and interactional aspects of the mobile age. Springer, London, pp 32–45

Greenhaus JH, Allen TD (2006) Work–family balance: exploration of a concept. Family Studies Center – School of Family Life at Brigham Young University, Families and Work Research Conference, Provo, 20–22 March 2006

Greenhaus JH, Singh R (2003) Work-family linkages. In Kossek EE, Pitt-Catsouphes M (eds) Work and family encyclopedia. Sloan Work and Family Research Network, Chestnut Hill, GA. http://wfnetwork.bc.edu/encyclopedia_entry.php?id=263&area=All. Accessed on 16 Feb 2010

Greiner L (2006) Technology is throwing our work life balance out of whack. Computing Canada, 32. http://www.itbusiness.ca/it/client/en/ComputerCanada/News.asp?id=39062&bSearch=True. Accessed on 16 Feb 2010

Hassard J (1999) Images of time in work and organization. In Clegg SR, Hardy C (eds) Studying Organization. Sage, London, pp 327–344

Hochschild A (1989) The second shift: working parents and the revolution at home. Viking, New York, NY

Jarvenpaa S, Lang K (2005) Managing the paradoxes of mobile technology. Inf Syst Manage Fall:7–23

Kahn R, Wolfe D, Quinn R, Snoek J, Rosenthal R (1964) Organizational stress: studies in role conflict and ambiguity. Wiley, New York, NY

Katz JE, Aakhus MA, (Eds) (2002) Perpetual contact: mobile communication, private talk, public performance. Cambridge University Press, Cambridge, MA

Middleton C, Scheepers H, Cukier W (2005) Exploring the contradictions of mobility: a case study of BlackBerry users in Canada. In Thong J, Tam K (eds) Proceedings of the Hong Kong mobility roundtable.HKUST, Hong Kong, pp 299–309

Pica D, Kakihara M (2003) The duality of mobility: designing fluid organizations through stable interaction. Association for Information Systems, 11th European Conference on Information Systems, Naples, 19–21 June 2003

Prasopoulou E, Pouloudi A (2006) Enacting new temporal boundaries: the role of mobile phones. Eur J Inf Syst 15:277–284

Rakow L, Navarro V (1993) Remote mothering and the parallel shift: women meet the cellular telephone. Crit Stud Mass Commun 10:144–157

Reid I (2001) Analyzing the return on investment of a BlackBerry deployment. Research Study prepared for Research in Motion: RIM, Waterloo, ON

Salazar C (2001) Building boundaries and negotiating work at home. J Appl Behav Sci 43: 162–170

Shumate M, Fulk J (2004) Boundaries and role conflict: when work and family are co-located: a communication network and symbolic interaction approach. Hum Relat 57:55–74

Thompson CA, Beauvais LL, Allen TD (2006) Work and family from an industrial/organizational psychology perspective. In Pitt-Catsouphes M, Kossek EE, Sweet S (eds) The work and family handbook. Multi-disciplinary Perspective and Approaches. Lawrence Erlbaum, London, pp 283–307

Thompson J, Bunderson JS (2001) Work-nonwork conflict and the phenomenology of time: beyond the balance metaphor. Work Occup 28:17–39

Towers I, Duxbury L, Higgins C, Thomas A (2006) Time thieves and space invaders: technology, work and the organization. Organ Change Manage 19:593–618

Venkatesh A, Vitalari NP (1992) An emerging distributed work arrangement: an investigation of computer-based supplemental work at home. Manage Sci 38:1687–1706

Warhurst C, Eikhof DR, Haunschild A (2008) Out of balance or just out of bounds? Analysing the relationship between work and life. In Warhurst C, Eikhof DR, Haunschild A (eds) Work less, live more? Critical Analysis of the work-life boundary. Palgrave Macmillan, Basingstoke, pp 1–21

Chapter 16
Working in Polycontextual Environments: An Empirical Analysis of Flexpatriates' Lifestyles

Helene Mayerhofer, Barbara Müller, and Angelika Schmidt

16.1 Introduction

As international organizations enter new markets and exploit the advantages of location to increase their global scope they face the challenge of transferring resources. Very often this has to be done on a short-term basis and cannot be met by classical expatriation or local adaptation. In this chapter we focus on flexpatriate assignments, short-term, unaccompanied business travel assignments of workers who hold a job in a home office and carry a workload in other countries while their family remains in their home location (Mayerhofer et al., 2004a, b).

In contrast to expatriates, Human Resource Management (HRM) has so far had little involvement with flexpatriates, leaving them to forge new work patterns and relationships and new lifestyles with peer support and at their own initiative. Thus, academic research in International Human Resource Management (IHRM) has largely focused on questions regarding expatriates. Also, the IHRM literature has yet to address the challenge of balancing work-life. Research and practical management show work-life and family balance issues are mostly considered a domestic-based matter and seem to be developed and implemented from a domestic point of view (Harzing and Ruysseveldt, 2004).

Flexpatriates blur the boundaries between work and non-work domains (Ilies et al., 2009) while they function as key facilitators of cross-unit interaction and exchange. An increased international workload for professionals means opportunities to develop skills and competences in different cultural contexts but also new physical and psychological demands and new types of work-family tensions (Harvey et al., 2010; Suutari, 2003). Work-life issues; especially family separation, health and travel stress are amongst the most often cited problems in the context of international working patterns for employees with children, for those in relationships without children and for single employees (DeFrank et al., 2000; Kappelhoff et al., 2006; Welch and Worm, 2006).

H. Mayerhofer (✉)
WU Vienna University of Economics and Business, Vienna, Austria
e-mail: helene.mayerhofer@wu.ac.at

S. Kaiser et al. (eds.), *Creating Balance?*, DOI 10.1007/978-3-642-16199-5_16,

The literature on International Human Resource Management has yet to address the challenge of balancing work-life. Research and practical management shows that work-life and family balance issues are mostly considered a domestic-based notion and developed and implemented from a domestic point of view (Harzing and Ruysseveldt, 2004). In this chapter we consider balancing, but we will also explore whether working in polycontextual environments creates a flexpatriates' lifestyle. We understand lifestyle as a set of elements of behaviour that can be observed by others. This understanding differs in two ways from the understanding of lifestyle used by Welch and Worm (2006) as a single variable and as a positive outcome of these working conditions. We aim to analyse and describe various aspects of lifestyle in order to address major issues about the work-life integration of flexpatriates.

We review the nature of flexpatriate-lifestyle influenced by international work and explore how employees face a range of issues in their work, personal and family lives. We define lifestyle as a set of observable behaviours (based on further developing Richter's model, 2006), which reflect the polycontextual working environment and characterize it by drawing upon empirical data from qualitative semi-structured interviews with 99 male and female managers working as flexpatriates for multinational enterprises located in Austria. Flexpatriate assignments are employed in various industries and can be found among all hierarchical levels. In exploring the dimensions of lifestyle four empirically grounded types emerged: Tough Travellers, Enjoyers, Cosmopolitans and Contactors. Our analysis of flexpatriates' everyday lived behaviour shows multiple facets of their lifestyles. From the perspective of the organization and especially from HRM it is good management practice to develop and implement strategies that fit the needs of this important type of employees in order to enhance flexpatriates' commitment to stay with the company and continue doing their job.

16.2 Polycontextual Environments – Nature of Flexpatriates

The importance of acquiring and diffusing knowledge across the global organization is recognized as critical to global success. By leveraging, learning and utilizing different global management competencies organizations can gain global competitive advantages (Bartlett and Ghoshal, 1989; Prahalad and Hamel, 1990). The most common way of filling positions, transferring knowledge and developing cultural competencies among a management cadre is through long term assignments of expatriates (e.g., Collings et al., 2007; Edström and Galbraith, 1977; Harris et al., 2005; Harvey et al., 1999; Johnson et al., 2006). There are, however, emerging problems with expatriation (e.g., high costs, staff shortage, repatriation), and changed circumstances in business and technology (e.g., expansion of the EU, transportation and communication systems, organizational networks and flexible intraorganizational coordination of global units). Also, the increased differentiation within assignments and different forms of more flexible kinds of international work such as short term assignments, frequent flyers, commuters and rotational assignments, business travellers, flexpatriate-assignments as well as virtual assignments (e.g., Demel, 2010; Demel and Mayrhofer, 2008; Dowling et al., 2008; Harris et al., 2001; Mayerhofer et al., 2004a, b; Welch and Worm, 2006) have not led to a more strategic use of different assignments in organizations (Collings et al., 2007; Erten et al., 2006). In this chapter we focus on flexpatriate assignments, a term encompassing short-term, unaccompanied and business travel assignments by people who are still in the job at home office but have a work-load in other countries while their places of residence don't change. The characteristics of a flexpatriate assignment include:

- Frequent alternation between different locations of work including different national and regional cultural contexts
- A flexible schedule of time and/or time zones and an unclear separation between leisure and work time and few daily routines
- Changing social relations and contacts with co-workers and connecting and disconnecting relations with family and friends (Demel, 2010; Dowling and Welch, 2004; Mayerhofer et al., 2004a, b).

An international workload for managers and professionals means opportunities to develop skills and competences in different cultural contexts as well as the likelihood of new physical and psychological demands along with new types of work-family tensions (Adler, 2002; Harvey et al., 2010; Suutari, 2003). Work-life issues are amongst the more frequently cited problems associated with international working patterns regardless of whether the employee is single, in a relationship, or whether he or she has children (Harzing and Ruysseveldt, 2004; Kappelhoff et al., 2006).

The work-life balance debate in general has failed to capture more diversified employee attitudes towards work and their engagement with work (Eikhof et al., 2007). Moreover, Warhurst et al. (2008) argue that the debate on work-life balance should bring the workplace and work experience back into the frame of analysis and the debate in order to progress to a more holistic understanding of life. Therefore, to assess the work-life balance, it is important to understand the underlying assumptions about the relationship between work and life for employees in general, and flexpatriates, in particular. Also, in articulating the relationship between work and life as work-life patterns it is important to consider the characteristics of work practices, structural constraints, lifestyles and logics that shape work-life patterns (Warhurst et al., 2008).

Recognizing these conditions, we aim to offer deeper insights into various lifestyle dimensions and to draw a more differentiated picture of flexpatriates-lifestyle. Finally we identify flexi-lifestyle types in order to better understand special requirements of this group of employees.

16.3 Flexpatriates' Lifestyles

The interface between private and professional spheres is a subject of great concern to flexpatriates (Poster, 2005). Flexpatriates are often not able to discuss their work-life-balance without discomfort, which can be linked to massive tension and conflict brought about by personal and professional pressure.

International business travellers' experience negative effects on work-life-balance, especially regarding family separation, travel stress, and health issues (DeFrank et al., 2000; Welch and Worm, 2006). Evidently, some job characteristics such as time-pressure, time-flexibility, and job security significantly influence interactions within the family (Hughes and Galinsky, 1994). In the research findings about flexpatriates challenges from international work, new circumstances and the lack of "boring" routines are often mentioned (Mayerhofer et al., 2004a, b). According to Schulze (1992) lifestyle determines what is perceived as necessary or beautiful. However, there are also negative outcomes associated with lifestyle like health problems and stress, which can be seen as outcomes of flexi-lifestyle. Lifestyle is one of the eight career anchors of Schein's (1996) typology. According to Schein, people who are focused foremost on lifestyle, look at their whole pattern of living, they try to integrate work and life, instead of balancing it, and they may even take off long periods work to indulge in their passions. This anchor has shown the most change since the original research of the 1960s and 1970s (Schein, 1996) with autonomy and lifestyle having become more important as anchors. Schein (1996) argued that the occupational structure is moving increasingly toward a different concept of employment contract in which the definition of work is gradually changing as the boundaries between jobs, between organizations and between work and family become more fluid and ambiguous. These observations seem to be particularly true for flexpatriates (see also Demel, 2010; Suutari, 2003).

Definitions of lifestyles have been blurred, fuzzy and to some extend contradictory. "Lifestyles are, apart from age (Schulze, 1992), determined by the important socio-structural dimensions education (Schulze, 1992), and/or economic and cultural resource (Bourdieu, 1984)" (von Normann, 2009, p. 383). In general the focus has been on social characteristics that are associated with behaviour patterns. Sociological definitions have their foundation in the work of Max Weber who distinguished between social class based on economic factors and social status indexed by association and symbols. Lifestyles now are seen as complex and changing despite being embedded in the social and market system (Eikhof and Haunschild, 2006; Kelly, 2006). "Lifestyles represent collectively shared patterns of perception, taste and behaviour, which are usually seen as rooted in social class and milieu (Bourdieu, 1984; Weber, 1972)" (Eikhof and Haunschild, 2006, p. 236). People increasingly have to make choices as part of their construction of a "narrative of the self" (Giddens, 1991, p. 76) instead of being able to refer to tradition and custom. Individuals have to produce, stage, and cobble together their own do-it-yourself identities.

The concept of lifestyle is common in sociological analysis (see Bourdieu, 1984; Weber, 1972). However, HRM studies have rarely considered lifestyle although it has proved to be a powerful concept regarding explanations of employment policies and rules and "the co-evolution of organizational HR practices and workers' preferences and attitudes" (Haunschild and Eikhof, 2009, p. 108).

"Lifestyle represents the aims of social norms relevant for the everyday life activities of human beings" (von Normann 2009, p. 385). It is derived from socially useful knowledge that is acquired through one's formal education and training as well as experiences, interests and contacts (Bourdieu, 1984). This understanding of lifestyle and its influences rests on the widely accepted principle that behaviour is a function of the person and the environment and that both variables in this formula are interdependent (Lewin, 1948/1997, p. 188). Therefore, we use the concept of lifestyle as visible and observable contextualised behaviour in everyday life to explore how characteristics of flexpatriate-work influence the daily routines of employees. This is consistent with the views of Richter (2006) that behaviour and actions moulding lifestyles und these actions can be identified and observed. The dimensions of time and space and cultural-social elements of everyday life are useful to describe international flexible work in more detail, and to portray challenges this group of employees faces in their jobs.

Focusing on lifestyle is an appropriate way to explore the characteristics of flexpatriates' work and the influences on their daily routines. This is consistent with the views of Richter (2006) that behaviour and actions, which can be identified and observed, constitute lifestyles. Richter (2006) developed a model of lifestyle consisting of five dimensions:

- *Temporal* dimension refers to working hours in general as well as to the time spent travelling.
- *Spatial* dimension questions the utility and usage of private and public places. Additionally, this dimension encompasses different geographical areas of work.
- *Personal-cultural* dimension refers to differences in an individual's behaviour. Work patterns influence the daily patterns of a person from sleeping habits to times for getting out of bed.
- *Factual-cultural* dimension refers to the material aspects of life, like particular items of IT, equipment and gadgets, clothing or luggage to mention a few. These elements greatly influences people's lives and are an important aspect in differentiating lifestyles.
- *Social everyday occurrence* dimension refers to the social interactions from the business side (to colleagues away and at the home base office) to private life (with partner or spouse, family and friends).

According to Richter, these dimensions determine people's lifestyles. We were interested in whether they would differ within a group of people working as flexpatriates. Differences would then become the basis for our developing a way to describe the lifestyles of flexpatriates.

16.4 Austrian Flexpatriates – Devising a Typology

Since our study focused on gaining deeper insight into the various lifestyle dimensions of flexpatriates and analysing how these dimensions influence different lifestyles an exploratory qualitative research approach was most suited. Interviewees were chosen through purposive sampling (Flick, 2007) according to the following criteria: They are employed in an Austrian multinational enterprise (MNE) or an

Austrian subsidiary of a MNE. They have a graduate degree and have had at least three international assignments within a year. The purpose of the study was to gain a broad picture of flexpatriates' lifestyle; therefore, interview-partners working at different hierarchical levels and different branches were chosen. We interviewed men (60) and women (39) whose ages ranged from 23 to 55.

In the course of the interviews, we asked semi-structured questions about each of the lifestyle dimensions. Some questions were quantified to get an overview of the entire sample and to be able to identify specific patterns in the sample. Answers to open-ended questions were transcribed and encoded according to the derived categories. We applied a qualitative content analysis (Mayring, 2003) to interpret the results and thematically coded the questions concerning the lifestyle dimensions. The results were summarized to gain an overall picture of flexpatriate lifestyles and their various characteristics. The team of three researchers met and discussed the data and processes at each of the steps in order to produce a consensual view (Bortz and Döring, 2003; Schmidt, 2005).

To gain a more diversified picture of the broad group of flexpatriates we additionally asked our interviewees "If you could wish for anything regarding your international travel assignment, for what would you wish?" This helped us to identify key points of reference (what is most important to them) and enabled us to derive an empirically grounded typology of flexpatriates. In a subsequent step we compared the lifestyle dimensions and the flexpatriate-types to identify differences.

Through the question "If you could wish for anything . . ." we developed empirically grounded types of flexpatriates. We identified four different types according to the topics and themes, which are most important to them. The types are internally homogeneous according to the various attributes to which the interview-partners referred (Table 16.1).

Tough travellers' wishes referred to their task and the organization while enjoyers frequently referred to themselves; especially, their health. Cosmopolitans pointed out the travel and the cultural benefits they gained on travelling assignments. Contactors frequently referred to home, their social environment at home and their family and friends. In the following section we will discuss the study results in more detail.

16.5 Results: Flexi-types and Their Lifestyles

A polycontextual environment is the typical world of work for flexpatriates. Although flexpatriates' working circumstances seem to be similar, individual's reactions and the interpretations about important topics and themes are quite different. In the next step we compare the observed behaviour in terms of similarities and differences among the flexi-types along the lifestyle dimensions.

A comparison of the flexi-types with the results of the different lifestyle dimensions showed similarities in reactions towards structures of time and places of work. At the same time, we found distinctive patterns especially on the personal-cultural dimension and the social everyday occurrence dimension and to some extent on the

Table 16.1 Types of flexi-lifestyles

	Description	Main topics and themes according to interview quotes
Type I Tough travellers	Universalism with reference to his/her task and the organization, tough, pragmatic, works everywhere and every time	Desire for organizational support, interviewees would like to have more support for planning and structuring their travels to be able to plan their tasks more effectively
Type II Enjoyers	Individualism and coenaesthesia, with reference to him/herself and his/her body and health	Desire more comfortable and convenient travels (flying business class, 5-star hotels, good meals, time and opportunities to relax)
Type III Cosmopolitans	Intercultural interests, task are seen and fulfilled within the cultural context s/he works in	Desire time and opportunities to see and get to know foreign cultures, to gain new cultural experiences; travels for a change
Type IV Contactors	Strong relationship with family and friends, with reference to his/her social environment	Desire to be able to integrate private life with the travels, worry about family, spouse, kids and/or friends while travelling, want to spend time with their social contacts

factual-cultural dimension. It seems that the organization is the predominant influence on the temporal and spatial dimensions and limits the scope of action for the individuals.

In the following sections we describe the five dimensions (Richter, 2006) determining the lifestyles of our sample. The dimensions are organized into two clusters: one, those dimensions dominated by assignments and their arrangements and two, those dimensions shaped by individual preferences and preconditions.

16.5.1 Commonalities: Structural Effects of the Temporal and Spatial Dimensions

16.5.1.1 In the Tug-of-War Between Speeding Up and Taking Time: The Temporal Dimension

More than two-thirds of the interviewees had completed six or more international business trips in 2006 and 2007. Some 70% of these international trips lasted between 1 and 8 days and most of them were limited to one work-week (5 days). Looking at the total amount of workdays abroad during 1 year, almost one-third of the interviewees were away for 1 month (20 days) and another third was away up to two working months (40 working days). Moreover, nearly 60% of the interviewees stated that their business travel had been increasing over the last 3 years.

How do flexpatriates interpret and react to these work-time conditions? The temporal dimension refers to differences in behaviour in work and leisure time and to differences in the view of time structure (working time and leisure time every day; daily routines and non routines; weekdays and weekends).

An absence from the company as long as 2 months is likely to result in an employee facing a heavy workload in the company-home after returning from an international travel assignment. The mixing up of different time structures (private and working time) like finishing work tasks during the free weekend is supported by so called "all-in-wages" and leads to fluid boundaries of work time.

The interviewees see their international activities as a regular or normal part of their work (and working contract). It is, however, less clear if travel-time is work-time. Contrary to an employee's perception of travelling as an integral part of the job, companies do not regularly compensate travelling hours as working hours:

> I would prefer that time for travelling would be accepted, at least parts of it, as working hours. Since in our company the time for travelling is considered as time for travelling and not as working hours and this time does not flow anywhere.... and this sometimes is very unfair, since sometimes one spends up to twenty hours a week in transit which is just ignored. (I 36, 43 year old, male, tough traveller)

In companies using flexi-time arrangements flexpatriates may be compensated for at least some of their travel time. However, interview-partners did not report any special offers from the corporations for flexpatriates to deal in a more flexible way with working and travelling hours, except in one case. This means that in order to keep up with home office's working demands flexpatriates must work additional hours. Travel time also increases the number of weekly working hours and further reduces leisure time.

We also asked our interviewees what they did if they had spare time before departure. At the airport before departure they check their E-Mails and use their time for preparation or "some work, for that usually is not enough time to do" (I 52, 38 year old, male, tough traveller). So it seems that flexpatriates interpret the timeframe as a given structure and not something to be negotiated with the employer. They just try to keep up with their work and be more effective. This understanding is also reflected in the next dimension.

16.5.1.2 Hotel Room as Working Place: The Spatial Dimension

The spatial dimension investigates the utility and usage of both private and public places (e.g., expanding working routines to different places). Additionally, this dimension describes differences among geographical areas. To operationalize this dimension we asked questions about where flexpatriates were working and how different geographical locations influenced their lifestyles. Our interviewees stated that they were working almost everywhere. The increasing flexibility of work enabled with communication and transport technology (e-mail, notebook and wireless internet access) has expanded working routines to different places (private, public and geographical) as the following statement points out:

> I often have to work in the evening in the hotel or late at night after I have returned home to get all the administrative stuff done. (I 29, 37 year old, male, enjoyer) or It would be good if you had internet access and if it would be possible to telephone everywhere. You lose a lot of time, for example, if you are not contactable while sitting in the plane (I 23, 34 year old, female, tough traveller)

Our findings show that flexpatriates work within polycontextual environments. They work, co-operate and communicate across multiple organizational and national cultures; at times simultaneously. Almost three-quarters of our interviewees work mainly among different regions of Europe (within and outside the European Union), one-quarter worked mainly in Northern America (Canada and US-American cities) and/or in economies such as the United Arab Emerites, Iraq, Egypt, China and India. An interesting relationship emerged between the work destination and the various flexi-types. Tough travellers worked mainly within Europe, while many cosmopolitans had more distant travel-destinations.

Most of the interviewees reported to work at home if they were not in the office or travelling. Also, travel time is used as working time, so they work during their commute, in cars, trains and planes as well as in hotel rooms.

To sum up, all flexpatriates react similarly to structural aspects of time and places of work. The professional work of flexpatriates seems to expand in time as well as in work places. Working time and working places are somewhat fluid. Flexpatriates deliver (and are expected to deliver) results which are not moderated by factors like the number of working hours in the working contract or compensated time for being away from the office. All flexpatriate employees interviewed had to work hard, but there are differences in how, on the individual level, they find positive outcomes from their work. This aspect will be developed further in describing the differences of lifestyle dimensions.

16.5.2 Flexi-Types and Differences in Their Lifestyle-Dimensions

16.5.2.1 Buffet Breakfast: The Personal-Cultural Dimension

Daily routines ranging from time to wake up to sleeping habits are directly influenced by working patterns. In our sample we found that patterns of behaviour were influenced by travel aspects such as being delegated to foreign destinations at short notice, changes in eating and sleeping-working habits as well as leisure and sport activities. The personal-cultural dimension of lifestyle discusses changes in personal habits. Interviewees were asked about their daily routines and whether they modified their habits because of international travel. The interview-partners reflected on their experiences of planning and organizing trips but also on the perceived influence of their experiences on their physical well-being.

Enjoyers in particular pointed out that their travel assignments influence their personal feeling and health: “I have gained weight due to the irregular eating habits” (I 76, 30 years old, female, enjoyer). They also said that due to their travel assignments they have less time to take care of themselves: "I have fewer opportunities to do my physical work out" (I 81, 33 year old, male, enjoyer).

Contactors also perceived travel assignments as having negative consequences on their habits: "A great disadvantage is that you don't have regular times and procedures and those impacts on your health" (I 60, 42 year old, female, contactor).

Interviewees also noted that their travel assignment had positive effects. In particular, enjoyers and cosmopolitans stated their intercultural competence had improved; they enjoyed working in and with other cultures and seeing and meeting new people. Enjoyer and cosmopolitan flexpatriates also reported positive changes of gaining new perspectives and staying mentally flexible as a result of cross-cultural experiences. Travelling was seen as enriching and broadening one's horizon and attitudes towards different cultures: "Long term assignments abroad have improved my tolerance toward other cultures in all respects. My openness to other styles of clothing and meals for example was improved very much. . . .I enjoy it very much." (I 49, 50 year old, female, enjoyer) All in all, new cultural environments provided pay offs to cosmopolitans.

The outlooks of other flexpatriate groups to their temporary travel assignments can be summed up with the statement, if you cannot change it, enjoy it. Some interviewees who felt very strongly about negative aspects of travel assignments had already changed jobs to ones demanding less travel. Others changed their behaviour and put their own personal interests first like the enjoyers. Especially the group of enjoyers referred to health problems. Sometimes they already have experienced or fear illness due to unhealthy working habits.

At the same time, some of the interviewees did not recognize any changed habits caused by their travel assignments. There are two explanations for this. Some employees already had experiences with changing environments like growing up in different nations and studying in different countries; thus, their job did not differ from their normal lifestyle. On the other hand it could be argued that some employees absorb these requirements into their life and do not even realize a change in their behaviour. "Actually my lifestyle didn't change. Travelling is simply part of my job." (I 51, 49 year old, male, enjoyer).

16.5.2.2 Affordable? . . . The World is Our Oyster! The Factual-Cultural Dimension

People's lives are greatly influenced by factual-cultural items. The kinds of elements people use and how they use them is important in differentiating lifestyles. This dimension includes the material items in life, like technical standards and items, clothing and attire and luggage. Companies place these resources at the disposal of their employees so they are accessible nearly anywhere and at any time and can complete their tasks more efficiently. At the same time, these factual resources also facilitate private communication.

Our analysis shows that mobile phones and laptops are standard equipment for all interviewees. "The most important things are my credit card and the notebook. I also need my mobile. Without these I cannot go on a business trip." (I 73, 51 year old, male, tough traveller) Despite these essential material items another important factor is connectivity: "So I would say connectivity is most important, to be able to be connected to home, and therefore the mobile, blackberry and notebook are essential." (I 75, 29 year old, male, tough traveller) Some of the interviewees also mentioned that personal resources like photos of their partner or children, talisman, spa and beauty cases are essential equipment on their journeys; regardless of the length of the trip. Enjoyers mentioned sportswear, cross-trainers, and MP3 players as the personal equipment most important to them: "Business and casual wear, music, MP3-player, pictures, sports wear." (I 76, 30 year old, female, enjoyer); "A good fragrance to relax, casual wear, I always try to take a bit of home with me." (I 95, 28 year old, female, enjoyer) Cosmopolitans focused more on the special features of the cultural destination including climate as well as socio-economic conditions and how to bring back memories: "Cash in appropriate currency, a camera and clothes adjusted to the climate." (I 77, 44 year old, female, cosmopolitan).

In summary, this dimension relatively clearly differentiates material goods along the different lifestyle types. While all interviewees identified professional-material things (mobile phone and notebook) as the most important equipment for international activity other items interviewees considered important distinguished the various lifestyle types. Contactors saw the connection to home as very important whereas enjoyers preferred individual subjects like sports equipment, jewellery or a good book. Cosmopolitans differed with respect to their orientation towards the respective destination and what it had to offer them from an interest or tourist perspective.

16.5.2.3 Long-Lasting Close? Flexible? Business? Private? Social Everyday Occurrence

To explore social everyday occurrence, we analyzed the social interactions of flexpatriates from a business perspective (contact with colleagues and the home base office while away) as well as in terms of private life with partner/spouse/family and friends, both, away and at home. We found that interviewees seldom commented on the everyday dimensions of private life related to family aspects (family status, children and responsibilities for housework) and family-work conflicts. However, our sample characteristics may have influenced this by there being few single breadwinner partnership constellations. Most of the partnerships were dual-earner couples (DEC) or dual-career couples (DCC) and the latter are known to have a high potential to cope with conflict and stress (Sekaran, 1986). In terms of care and family conflicts, one interviewee found his own solution and applied for another job with less international travel: I have changed to this position in order to be on the road less . . . well, this did not work out properly, but I could to some extent reduce travel. (I 31, 48 year old, male, enjoyer).

In our sample there are several examples for the interface between private and professional spheres. Thus, an interviewee stated that a consequence of such intensive travel activity is, "that you cannot maintain private contacts at home if you are travelling a lot. Yes, either you are not here, or you look... primarily after family while being at home or you simply want to rest. This causes you to have fewer social contacts at home." (I 54, 32 year old, male, tough traveller).

It appears that only little time and energy remains for social relations. This is illustrated by way of example with the following statement: "Social contacts change and the only time you see each other is on the weekend." (I 58, 36 year old, male, tough traveller). In some life constellations responsibilities are shifted with the other partner becoming the fulcrum and pivot for the care of social relations: "On the whole nothing has changed. Friendships have been maintained – thanks to my wife. The only one thing to suffer is the activity in the association, in the case that I am on the way for a long time" (I 59, 29 year old, male, cosmopolitan).

It can, however, be emphasized that travelling increases new social contacts, "additional social contacts on site" (I 62, 38 year old, male, tough traveller) or "the travel activity has no influence on the contact and family at home. In addition you have social contacts on site" (I 62, 38 year old, male, tough traveller).

In addition to the challenge of polycontextuality in cross-cultural communication there are positive outcomes like the possibility to become involved in an international network; a point mentioned by more than half of the interviewees. Participants reported few problems regarding interaction with colleagues, especially at the home base office. Some individuals noted that some colleagues at the home office are envious because they do not have the glamour of travel.

16.5.3 Summary of the Facets of Flexpatriates' Lifestyles

Our findings indicate that international assignments affect flexpatriates' employment and living conditions in both similar and different ways. Fluid boundaries between private and work time as well as public and private places are typical aspects of international assignments in a flexible format.

Mobile and virtual work takes place in many more spaces than the traditionally designated workspace. It transforms places; especially public places like cafés or mediums of transportation like taxis into offices (Towers et al., 2006). This third form of rooms or places seems to be another source of evidence of mixing private and work life. These places between home and work offices become legitimised workplaces. Flexpatriates use airplanes, trains and hotel rooms as places of work. In view of time Virilio (1997) declares that these technologies steal "real time" because it shifts "here" and "now" in the background.

Table 16.2 shows how these arrangements lead to a growing temporal workload and intensifying the blur of work-life balance arrangement of flexpatriates and how it affects the four types differently according to Richter's (2006) lifestyle dimensions.

To sum up the most important type-specific differences: Tough travellers focus on time, economy of time and efficiently organized business travels. They rarely make a distinction between places of work. Also enjoyers, who judge personal needs as important, contactors, who focus on maintaining social contacts and cosmopolitans, multiculturally interested persons are all confronted with a reduction of clear time markings and an increase in blending between the boundaries of work and time. But they focus on other points to manage their working life as flexpatriates.

Table 16.2 Overall results according to Richter's lifestyle dimensions (2006)

Dimension		Main characteristics	Main results
Structured by assignments	Temporal	Working hours, travelling frequency and duration	Overall Fluid boundaries of work time (irregular schedule) WLB more imbalanced to work (travelling hours add to regular hours, reduced leisure time)
	Spatial	Work space and geographical destinations	Overall Changing destinations (different regions e.g., Europe, Asia) Fluid boundaries of work places: working everywhere and expanding working routines to different places
Individuals/type specific impacts	Personal-cultural	Habits, change of habits through flexpatriate-job	Overall Changes in habits Individual habits are influenced differently by travel aspects, changes in eating and sleeping habits, trying to find new experiences and wishing to stay in touch with home (office and/or family) Type-specific Tough Traveller: minor awareness of changes Enjoyer: focus on individual interests (health and wellbeing) Cosmopolitan: have new experiences and gain intercultural competence Contactor: stress impacts the wellbeing

Table 16.2 (continued)

Dimension	Main characteristics	Main results
Factual-cultural	Resources, support, compensation and symbols	Overall Perceived standard equipment: the notebook and mobile to communicate and to be reachable everywhere No additional financial rewards for travelling Type-specific Tough traveller: reasons of practicality and usefulness of resources dominates Enjoyer: sport and personal items Cosmopolitan: cameras, local currency and language guide Contactor: communication instruments to maintain contact with relevant reference groups
Social everyday occurrence	Social life, social contacts and social involvement	Overall Contacts established with different reference groups Work-family conflicts and work-private life conflicts Type-specific Tough traveller: reduction of social contact with those who are seen as relevant and having high priority Enjoyer: reflects the dark sides of travelling and its consequences to social life Cosmopolitan: does not seem to be affected Contactor: maintains social contacts with family and the private life sphere

16.6 Conclusions

Our glimpse into the everyday lives of flexpatriates illuminates various facets of their lifestyles. For an increasing number of employees the division between their work and non-work life are no longer bounded by clear time markers. Modern communication serves as a link between time and space (Harvey, 1989) and is also a basic precondition for the increase of short-term business travels. Our results show that companies are using a mix of frequent international work of somewhat short duration. These kinds of work arrangements contribute to a burden of increasing working hours and workload in the home offices and reinforcing the blurring boundaries between professional and private life.

In a nutshell, interviewees' statements differ along personal dimensions like the personal-cultural and to some extent along the factual-cultural as well as the social everyday occurrence, while they have less individual scope of action concerning temporal and spatial issues. Flexpatriate assignments are no longer optional assignments for employees. For most of our interviewees travelling is simply part of their job. Some of them cope with it very easily but others find flexpatriate assignments adversely affect their lifestyle and consequence their working performance.

Welch and Worm (2006) identify perceived positive and negative factors of international assignments. For example, they see the thrill of the deal and personal development as positive factors whereas family separation or health issues are seen as negative ones. The results of our investigation show that, in general, there are no perceived negative or positive aspects applicable to all flexpatriates. Family separation was perceived as a negative factor by the group of contactors but it was not by tough travellers or enjoyers. Thus, indicating that each flexi-type perceives international assignments differently.

Furthermore, our results have shown that flexpatriate assignments have various affects on lifestyle arrangements. Therefore, adaptations of existing methods and the development of differentiated HR measures are necessary. It is important to incorporate flexpatriate considerations into human resource management planning and arrangement processes for organizations. One possibility is to consider different lifestyle types and their particular characteristics as early as in the recruiting process.

The differentiation of lifestyles in this empirical investigation offers a conceptual frame to identify the different actions and sensory patterns of flexpatriates. The identification of these diverse patterns provides a starting point for organisations to develop suitable measures to support actual and efficient working activities. Findings in IHRM show that supporting systems are highly important in order to enhance retention and commitment of employees involved in international assignments (Lazarova and Cerdin, 2007; Mayerhofer et al., 2004a, b; Stahl et al., 2009). The key issue is that "the international assignee's perception about the availability or usefulness of the company's support" influences employee's intention to turnover (Stahl et al., 2009, p. 95). HR management often provides and selects support measures along the type of assignment and the hierarchical level of the workers and less along her/his individual needs and whished lifestyle. Each lifestyle type considers certain aspects important. Thus, it is logical to incorporate these aspects into organisational incentive and supporting systems. Whereas tough travellers are satisfied if the organisation plans their travel and they end up with additional free time or frequent flyer miles, enjoyers and cosmopolitans seem to find that far less attractive. Moreover, cosmopolitans might be interested in challenging destinations because of their individual attraction to cultural experiences. Contactors could be supported with state of the art communication devices.

Our qualitative methods were chosen for their capability to develop an early complex and holistic understanding of a topic about which we have limited empirical data. Thus, we recognize that it is not appropriate to generalize from our results. The emergent typology of tough travellers, enjoyers, cosmopolitans, and contactors, however, shows that within the group of the flexpatriates there are a variety of lifestyles. They also call for the assignment and design of human resource management responses, which we think can be furthered through the use of our flexi-lifestyles types frame for organizations to more effectively understand and work with flexpatriates.

References

Adler NJ (2002) International dimensions of organizational behaviour. South-West, Cincinnati, OH

Bartlett CA, Ghoshal S (1989) Managing across borders: the transnational solution. Harvard Business School Press, Boston, MA

Bortz J, Döring N (2003) Forschungsmethoden und Evaluation für Human- und Sozialwissenschaftler. Springer, Berlin

Bourdieu P (1984) Distinction. Routledge, London

Collings DG, Scullion H, Morley MJ (2007) Changing patterns of global staffing in the multinational enterprise: challenges to the conventional expatriate assignment and emerging alternatives. J World Bus 42:198–203

DeFrank RS, Konopaske R, Ivancevich JM (2000) Executive travel stress: perils of the road warrior. Acad Manage Exec 14:58–71

Demel B (2010) Karriere und Karriereerfolg von österreichischen Expatriates und Flexpatriates. Eine qualitative Studie europaweit tätiger Manager und Managerinnen in Dual-Career Beziehungen. Rainer Hampp Verlag, Mering

Demel, B, Mayrhofer, W (2008) Flexpatriate careers in Europe: aspirations, motives, goals, and consequences. Academy of Management, Annual Meeting, Anaheim, 8–13 August 2008

Dowling PJ, Festing M, Engle AD (2008) International human resource management. Managing people in a multinational context. Thomson, London

Dowling PJ, Welch DE (2004) International human resource management: managing people in a multinational context. Thomson, London

Edström A, Galbraith JR (1977) Transfer of managers as a coordination and control strategy in multinational organizations. Adm Sci Q 22:248–263

Eikhof DR, Haunschild A (2006) Lifestyle meets market: Bohemian entreprenuers in creative industries. Creat Innovat Manage 15:234–241

Eikhof DR, Warhurst C, Haunschild A (2007) Introduction: what work? What life? What balance?. Empl Relat 29:325–333

Erten C, Schiffinger M, Mayrhofer W, Dunkel A (2006) The use of foreign assignments and their relationship with economic success and business strategy: a comparative analysis of Northern, Southern and Eastern European countries. In Morley MJ, Harety N, Collings DG (eds) New directions in expatriate research. Palgrave Macmillan, Hampshire, pp 39–63

Flick U (2007) Qualitative Sozialforschung. Eine Einführung. Rowohlt, Reinbek bei Hamburg
Giddens A (1991) Modernity and self-identity. The Policy Press, Cambridge
Harris H, Brewster C, Erten C (2005) Auslandseinsatz, aber wie? Klassisch oder alternative Formen: neueste empirische Erkenntnisse aus Europa und den USA. In Stahl G, Mayrhofer W, Kühlmann TM eds) Internationales Personalmanagement. Neue Aufgaben, neue Lösungen. Hampp, München, pp 275–292
Harris, H,, Petrovic, J,, Brewster, C (2001) New forms of international working: the panacea to expatriation ills? ESADE Business School, Global Human Resource Management Conference, Barcelona, 20–22 June 2001. Unpublished conference paper. http://www.som.cranfield.ac.uk/som/research/centres/creme/downloads/cronermay2000. Accessed on 5 May 2002
Harvey D (1989) The condition of postmodernity. Blackwell, Cambridge
Harvey M, Mayerhofer H, Hartmann L, Moeller, M (2010) Corralling the "horses" to staff the global organization of 21st century. Organ Dyn 38:258–269
Harvey M, Speier C, Novicevic M (1999) The role of inpatriation in global staffing. Int J Hum Resour Manage 10:459–476
Harzing AW, Ruysseveldt J (2004) International human resource management. Sage, London
Haunschild A, Eikhof DR (2009) From HRM to employment rules and lifestyles. Theory development through qualitative case study research into the creative industries. Zeitschrift für Personalforschung 23:107–124
Hughes D, Galinsky E (1994) Work eExperiences and marital interactions: elaboration the complexity of work. J Organ Behav 15:423–438
Ilies R, Schwind Wilson K, Wagner DT (2009) The spillover of daily job satisfaction onto employee's family lifes. The facilitating role of work-family integration. Acad Manage J 52:87–102
Johnson JP, Lenartowicz T, Apud S (2006) Cross-cultural competence in international business: toward a definition and a model. J Int Bus Stud 37:525–543
Kappelhoff V, Heidemann K, Völker S, Rietz C (2006) Partnerschaft und Auslandseinsatz: Beziehungsdimensionen und ihre Bedeutung für die Qualität des Auslandsaufenthalts bei Expatriates. Zeitschrift für Personalforschung 20:318–342
Kelly JR (2006) Lifestyles. In Beckert J, Zafirovski M (eds) International Encyclopedia of economic sociology. Routledge, London, pp 407–409
Lazarova MB, Cerdin JL (2007) Revisiting repatriation concerns: organizational support vs. career and contextual influences. J Int Bus Stud 38:404–429
Lewin, K (1948/1997) Resolving social conflicts. Filed theory in social science. Harper & Row, Chicago, IL
Mayerhofer H, Hartmann L, Herbert A (2004a) Career management issues for flexpatriate international staff. Thunderbird Int Bus Rev 6:647–666
Mayerhofer H, Hartmann L, Michelitsch-Riedl G, Kollinger I (2004b) Flexpatriate assignments: a neglected issue in global staffing. Int J Hum Resour Manage 15:1371–1389
Mayring P (2003) Qualitative Inhaltsanalyse. Grundlagen und Techniken. Beltz, Weinheim
Poster WK (2005) Three reasons for a transnational approach to work-life policy. In Kossek EE, Lambert SJ eds) Work and life integration. Organizational, cultural, and individual perspectives. Lawrence Erlbaum, Mahwah, NJ, pp 375–400
Prahalad CK, Hamel G (1990) The core competence of the corporation. Harv Bus Rev 68: 79–93
Richter R (2006) Österreichische Lebenstile. Lit-Verlag, Wien
Schein EH (1996) Career anchors revisited: implications for career development in the 21st century. Acad Manage Exec 10:80–88
Schmidt C (2005) Analyse von Leitfadeninterviews. In Flick U, von Kardorff E, Steinke I (eds) Qualitative forschung. Ein Handbuch. Rowohlt, Reinbek bei Hamburg, pp 447–456
Schulze G (1992) Die Erlebnisgesellschaft. Kultursoziologie der Gegenwart. Campus, Frankfurt
Sekaran U (1986) Dual-career families: contemporary organizational and counseling issues. Jossey-Bass, San Francisco, CA

Stahl GK, Chua CH, Caligiuri P, Cerdin JL, Taniguchi M (2009) Predictors of turnover intentions in learning-driven and demand-driven international assignments: the role of repatriation concerns, satisfaction with company support, and perceived career advancement opportunities. Hum Resour Manage 48:89–109

Suutari V (2003) Global managers: career orientation, career tracks, life-style implications and career commitment. J Managerial Psychol 18:185–207

Towers I, Duxbury L, Higgens C, Thomas J (2006) Time thieves and space invaders: technology, work and the organization. J Organ Change Manage 19:593–618

Virilio P (1997) Open sky. Verso, London

Von Normann K (2009) The impact of lifestyles and food knowledge on the food pattern of German children. Int J Cons Stud 33:382–391

Warhurst C, Eikhof DR, Haunschild A (2008) Out of balance or just out of bounds? Analysing the relationship between work and life. In Warhurst C, Eikhof DR, Haunschild A (eds) Work less, live more?. Palgrave Macmillan, London, pp 1–21

Weber M (1972) Wirtschaft und Gesellschaft. J.C.B. Mohr, Tübingen

Welch DE, Worm V (2006) International business travellers: a challenge for IHRM. In Stahl GK, Bjorkman I (Eds.) Handbook of research in international human resource management. Edward Elgar, Cheltham, pp 279–298

Chapter 17
Spatial Aspects of Professionals' Work-Life Integration

Deirdre Anderson and Clare Kelliher

17.1 Introduction

Work-life integration is often associated with workplace flexibility and choice – the ability of individuals to choose when, where and how much to work (Hill et al., 2008). Organizations offer a range of options, including flexible working arrangements, under the umbrella of work-life policies aimed at supporting individuals in their efforts to achieve a satisfactory integration of their work and non-work domains (Poelmans and Caligiuri, 2008). When employees take advantage of flexible working arrangements that change the time and place of work, this may result in the blurring of boundaries between work and non-work activities (Warhurst et al., 2008). For organizations there may be a dual agenda (Bailyn et al., 2001) as business needs may also be major drivers for the instigation of such policies.

This chapter presents findings from a case study of a large UK utilities company. At the time of the research this company had recently implemented a change in working arrangements, whereby existing jobs were re-classified as being office-based, mobile, or home-based. This change meant that significant numbers of staff no longer had office space allocated to them on company premises. The primary aim of this change was to reduce costs by decreasing the requirement for office space. However it was also the company's aim to facilitate greater work-life integration for staff, by giving them greater choice over how they carried out their work under these new arrangements. For example, mobile and home-based staff could exercise some choice over when they did their work and those who were office-based could choose to work from home on some occasions. Thus, in this case, the decision to change working arrangements was not initiated by the employee, as is often the case with flexible working arrangements. Instead, many staff members were asked to agree to changes to their existing arrangements, in terms of place of work, but at the same time these changes brought some freedom for employees to exercise control over when and where they did their work.

D. Anderson (✉)
Cranfield School of Management, Bedford, UK
e-mail: deirdre.anderson@cranfield.ac.uk

S. Kaiser et al. (eds.), *Creating Balance?*, DOI 10.1007/978-3-642-16199-5_17,

In this chapter data will be presented from 19 interviews with professional women and men, addressing the implementation of different working arrangements and the individuals' experiences of them. We examine how employees responded to the changes to spatial and temporal boundaries and what the implications were for their home life and for managing work-life integration. In doing so we aim to further understanding of the (sometimes unanticipated) consequences of shifting the traditional boundaries of time and space associated with work. Although our findings show that work-life integration was facilitated for many employees, there were also examples where the impact was negative. Changes to the location of work meant that the boundaries between the work and non-work domains were no longer clearly delineated and some individuals experienced greater intrusion of work into their home lives, sometimes in unanticipated ways. These findings will be discussed through the lens of boundary theory and considering the discourse of choice within the work-life integration literature.

17.2 Boundary Theory

Different ways of working such as home-based and mobile working as described above mean that the normal spatial and temporal markers of an organization's boundaries no longer exist in the same way (Scott, 2003). The absence of familiar and obvious boundary markers such as place of work and standard working hours can cause difficulties with adjustment as shown in a study of enforced teleworking where home was the full-time base (Harris, 2003). The study found that workers experienced tensions resulting from unclear boundaries between work time and family time.

New working arrangements result in shifting and dynamic boundaries when work is done at places outside the workplace and at times which vary to a greater degree than previous working hours, but this flexibility, or shift in time or space, only refers to one way in which the boundaries can change. Hall and Richter (1988) also identify boundary permeability, or the degree to which a person physically situated in one domain can be psychologically engaged in activity associated with the other.

Permeable boundaries may facilitate greater flexibility in responding to issues which arise in either domain, but equally they can result in the expectation of an "ever-available" response to demands from either the work or home: "a protective barrier between work and home has been removed – protective against the spillover of problems from one domain to the other" (Baruch and Nicholson, 1997, p. 20). Similarly, a study which focused on individuals in managerial and/or professional roles who worked from home found that although they had apparent autonomy, internal control and high responsibilities could lead to ever-availability (Tietze and Musson, 2003, p. 444).

Boundary theory (Ashforth et al., 2000; Nippert-Eng, 1996) and border theory (Campbell-Clark, 2000), building on Kanter's (1977) work, identify different types of relationships between work and non-work domains. The theories conceptualise segmentation (referred to by Kanter as "separation") and integration of the work and non-work domains, and the transitions which take place from one to the other. For instance, an individual segments the domains when s/he chooses to stay late in the office to complete a piece of work, rather than completing it later when at home. Integration occurs when an individual does not have clear boundaries between the two domains, such as an early start to work in the home with a break to take children to school; collecting them and providing an evening meal, before emails and conference calls in the evening with clients in a different time zone.

The Work of Ashforth et al. (2000) described these two concepts on a continuum, with segmented roles, which have inflexible and impermeable boundaries at one end of the continuum, while integrated roles are at the other end, with boundaries which are both flexible and permeable. So working in a designated home office during normal working hours, with no interactions with other members of the household, sits towards the segmentation end of the spectrum despite undertaking work in the physical location of the home. Campbell-Clark (2000) used the term "blending" to describe the process by which individuals experience flexibility and permeability. The expectations of significant others may affect the levels of segmentation and integration which can be achieved. Managers and colleagues may expect an individual to be available throughout standard work hours, limiting the ability to attend to non-work issues during that time. While a person can theoretically plan to have a three hour break from work in the middle of the day, this may not be acceptable to work colleagues who continue to place demands on him/her. Similarly, when working from home, simple physical presence may result in interruptions from family members.

The focus on psychological experiences of flexibility has been highlighted by Kossek et al. (2006) who define boundary management strategy and psychological job control. The former refers to the degree to which an individual attempts to separate work and home roles. Their study found that greater integration of the work and home domains could remove the protective buffer and result in an ongoing refocusing between work and family roles, switching back and forth between the two, which could be a demanding process both mentally and physically. Psychological job control describes the level of perceived control an individual has over where, when and how s/he works, thus incorporating both spatial and temporal aspects as well as introducing a behavioural element. Perceived control was found to be a predictor of individual well-being. Control links to the second area of literature we wish to refer to, that of individual choice.

17.3 Choice

There is a major emphasis within the literature examining the interface between work and non-work domains on the issue of individual choice. For instance, flexible working initiatives often enable the individual to play a major part in determining the amount of hours worked, the arrangement of those hours and the place of work (Hill et al., 2008). Such initiatives may be formally agreed and involve a contractual change, or may be informal, as in the case of an individual having the freedom to vary his/her working hours on an ad hoc basis. An informal arrangement may also be of a more permanent nature where it has been agreed with the line manager but without involvement of the HR function or more senior levels of management (Hall and Atkinson, 2006).

However, the level of real choice open to individuals has been the subject of some debate. For instance, the aforementioned study by Harris (2003) illustrates the emphasis on organizational objectives with a corresponding absence of individual choice. Caproni (2004) points out that the notion of working towards "balance" relies on greater control over life decisions than may actually exist. In practice the demands of, for example, childcare or eldercare may be unanticipated and/or unpredictable. The ability to prioritise work, career, family and other aspects of life also pays little attention to the gendered assumptions about commitment and competence which underpin the concept of the ideal worker (Rapoport et al., 2002). The ideal worker has historically been someone who is able to devote their time to their employing organization with no conflicting demands, because of the presence of another adult within the family unit who is based in the home on a full time basis to attend to domestic and caring responsibilities (Rodgers and Rodgers, 1989). However, households in the 21st century consist of different mixes of number of adults and presence or absence of both children and elderly dependents (Ransome, 2007) and therefore employees can no longer be assumed to be "ideal workers", but instead need to be seen as individuals with varying commitments and priorities outside of work.

To summarize then, when the location and timing of work are subject to change, there are implications for the spatial and temporal boundaries between the work and non-work domains. Theory suggests that the nature of the boundaries between these domains may be flexible and permeable or inflexible and impermeable; segmented or integrated. In the context of work which takes place in the home, the potential for a high permeability and integration is likely to exist, although it may also be possible to maintain segmentation by the erection of barriers and the enactment of rituals to allow for transition between work and non-work. The level of choice or control which individuals experience regarding integration and segmentation may be limited by the expectations of others; colleagues, managers and clients from the work domain, family members from the non-work domain, and societal pressures of gendered role expectations.

17.4 Case Study Background

The study reported here was undertaken in a major UK utilities company which had recently implemented a re-classification of working arrangements, driven by the dual agendas of providing work-life policies for the benefit of employees and cost savings through reduction of office space. The stated intentions were that mobile workers would work away from their base office for at least 60% of the working week, home-based workers would not come to the workplace apart from for occasional meetings and that office workers would be physically present in the office for most of the working week.

Fifteen face-to-face interviews were held on company premises and a further four interviews were conducted via the telephone. The age of the nine women and ten men interviewed ranged from 26 to 50 years. Fourteen of the participants were parents, with eight having at least one child under the age of five and six having school age children. All interviewees worked flexibly, except for one who managed a team which included staff working flexibly. Twelve participants were classified as mobile workers, six as office workers and one person was home-based. However within these classifications there was a further range of flexible working practices being used. In particular, several of the office workers spent part of their contracted hours working from home and/or varied their start and finishing times.

Thus, despite the designation of mobile, office and home-based worker, the actual enactment of working patterns was more variable. For instance, one person had an agreement to work in the office for 3 days a week and at home for 2 days – but the agreement was made to suit personal circumstances rather than because the job required a mobile worker. This meant that in practice, whilst some change was enforced, employees also had some degree of the choice which is usually associated with flexible working arrangements. These changes were implemented company-wide over a period of about two years, with a number of pilot studies taking place. One of the main learning points to emerge from the pilots was the importance of a "team charter" drawn up with agreement from all team members regarding the maintenance of good intra-team relationships, as well as logistical details such as the location and contact details of team members at any given point during the working week. Individual classifications were finalized by the manager after discussion with the individual and the team.

17.5 Findings

Three main themes emerged from the analysis of the interview transcripts and these are presented below. We begin with the changes which individuals experienced as a consequence of the re-classification of jobs/roles. Second, we present data which relates to the intrusion of work into the home. Finally, the ways in which boundaries are managed between the work and non-work domains are reported.

17.5.1 Changed Working Arrangements

For many of the respondents their working arrangement had changed with the implementation of the re-classification of their jobs as mobile, office or home-based working. Although the accompanying opportunity to work more flexibly was largely welcomed, interviewees acknowledged that there was a transition period as they adjusted to a different way of working. This adjustment included recognition of the greater freedom available to them in integrating demands from the work and non-work domains, as explained below by Daniel.

> It was a very slow transition into taking advantage of the flexible working. It took probably four to five months, maybe six months until I got the balance right and where I was taking advantage of it for my benefit I would say. (Daniel – office-based)

We found little evidence of guidance or training available to help employees adapt to the challenges of working flexibly or on how to reap the benefits. However, overall we found that respondents were content with the new arrangements and some reported increased job satisfaction as a direct outcome of the changed ways of working, for example,

> I've got the empowerment to deliver in the way in which I want to deliver. . . . I do get a lot of job satisfaction out of the way in which my role is. . . I'm not sure if it's designed or not, but the way in which I'm allowed to deliver. (Vince – mobile)

Several respondents expressed such satisfaction in terms of greater ownership, explaining that their new working arrangement allowed them greater autonomy in carrying out their responsibilities and achieving their objectives. One interviewee reflected,

> I think that it gives you ownership of what you need to deliver. And if you are responsible towards meeting your targets and your objectives and are motivated to do so then I think that it has certainly contributed to my performance in that I'm taking ownership of it and I'm managing it myself. (Geraldine – mobile)

However, in addition to the benefits which many workers discussed, there were also some less welcome issues which were often unanticipated. For instance, a key driver of the change in working arrangements had been the desire to reduce operating costs through a decreased need for office space and the subsequent closure of a number of offices. Reduced office space meant that the number of available desks had been significantly reduced which resulted in ongoing difficulties for mobile or home workers in actually finding a desk at which to work when they did need to visit a particular office.

> The offices, well, they're more crowded basically. We were moved to a smaller building and sometimes it's extremely difficult if it's a day when lots of people go to the office. Like I said we are in a very small place. (Belinda – mobile)

Additionally, those who were home-based or mobile workers no longer had storage facilities available to them in any office. Despite a push for minimizing the amount of paper used, respondents pointed out that a certain amount of documentation was inevitably needed in hard copy which therefore had to be carried around with them. Mobile workers frequently stayed away from home for work purposes, and so accumulated paperwork over a number of days, as reflected in the comment below,

> There is a tendency to carry all your work around with you, so your bag becomes heavier and heavier as the week goes on. I think it's because everything is in your bag. So I think we need to ensure that we have adequate storage at sites, someone's nominal site if they truly are flexible. (Karen – mobile)

A further unwelcome consequence of the changed arrangements was the reduced opportunity for social interaction, which for many had been part of the normal working day. The more formal work related side of such interaction was often dealt with through the use of technology with many people talking about an increased use of teleconferencing. The more intangible elements such as informal chats and sharing of ideas which had reportedly often led to productive suggestions and actions were missing and individuals had to adapt and make arrangements to replace those previously spontaneous interactions.

> It can be quite isolated. In the old world, if you like, or if you come into this office, you're there with your colleagues, you're bouncing things off them, you pick up on what's going on, you pick up on ideas and all these kinds of things that, when you're home alone you miss. And you do need to find ways to make sure you're talking to your colleagues quite a lot. You know, meetings and even if you're out and about and saying when you're on the phone, 'Where are you? I'm only round the corner. Let's go and grab a cup of tea'. So you do need to plan that in more. (Neil – home-based)

Thus whilst there had been some initial opposition to the changes imposed upon staff in this organization, the resulting opportunity to exercise greater control of when work took place was generally welcomed by employees. A number of interviewees indicated that it had taken time to adjust to the new working arrangements and in particular to gain benefits for themselves, but that overall, giving them greater control had had a positive impact on their levels of job satisfaction. Notwithstanding such positive effects, we also found evidence of a number of practical difficulties which arose, such as the difficulties of obtaining desk space when they did need to be in the workplace, the logistics of not having an office base for storage and the loss of regular social interaction with colleagues.

17.5.2 Intrusion of Work in the Home

A second major theme to emerge from the analysis of the interview transcripts was that of intrusion when the home becomes a regular and frequent place of work. Many instances were given of work intruding into home life in a more pervasive way than simply taking place there. For instance, Vince observed that the equipment provided was designed for an office environment, and hence was not always suitable for a home.

> The reality is that people's homes aren't offices and thus some of the pieces of equipment that we've been given are very big. So having a full size telephone with all of the different options, it's big. I've got a fax machine, I've then got a laser jet printer, and then I've got a colour printer. These are all big pieces of kit. I'm fortunate, I've got a dedicated room, but there would have been no way in which we could have put that into our previous property. (Vince – mobile)

Space for working in the home was a particular concern as not everyone was in Vince's situation and able to allocate a room as a study or office dedicated to work. For example some interviewees talked of working at the kitchen or dining room table, while others mentioned a desk in the bedroom. Working in such areas meant that those individuals were unable to create the physical barrier between work and home by closing a door. Furthermore, having office equipment in the home could be a source of intrusion in other ways when designated space was not available; one manager gave an example of team members being woken in the night by incoming faxes when fax machines were located in bedrooms.

Mobile workers, whose home was their base, received their mail at home, which could be inconvenient especially if there was a large volume. One person reported overcoming the problem of mail accumulating by setting up a P.O box so that he could collect it at his convenience. Another manager talked of a further intrusion of work into the home when needing to store large deliveries even if only for a short time.

> I've just ordered lots of diaries for my engineers, because they have diaries that they put their meetings in and rosters etc. So I've just had a box of 80 diaries turn up plus ones for me and other things. So there's a few boxes of diaries or your papers and all of that sort of stuff. (Neil – home-based)

In addition to the space taken up by transient paperwork, a further challenge was the issue of the build-up of work related post while an individual was away on holiday. In some cases arrangements were then made for a colleague to enter the home to collect mail. This presents a further intrusion into home life as other members of the household may not be comfortable with allowing a stranger to enter their home when they are absent.

Family members were impacted in other ways due to work being carried out in the home. A number of interviewees talked of telling children to be quiet "because I'm on the phone" and that this had generated feelings of guilt because "it's their home too". Similarly, conversations which take place in the course of work may be robust and unpleasant, and some individuals reported experiencing discomfort with work interactions of this nature taking place within their home and sometimes in front of their families. However, we found that some respondents had developed ways of dealing with such situations. Neil explained how he avoided this conflict,

> Because your business face is quite often different to your personal face, isn't it? And how you have to be, act and react to some situations would probably be quite alien for your family to see you being like that. And I do find myself walking down into the garden. I've got my business line in an alcove in the kitchen whilst we're having work done in the house. So if I've got an office, as I've had before, it makes no difference, you close the door, but there are quite a few that have got the dining room, an alcove in the kitchen, the spare room or whatever and people around all the time. And it can be.... we walk outside. You do have some difficult conversations. (Neil – home-based)

17.5.3 Control Over Work/Home Boundaries

However, despite the challenges presented by work taking place in the home, the third theme to be identified in the data was an increased sense of control over the boundaries between work and home. Many participants expressed satisfaction as a result of the perceived greater flexibility which had resulted from their changed working arrangements. The notion of freedom was frequently mentioned as part of the greater integration of work and life, with regard to the choice available to them to attend to demands from the work and non-work domains in a way that suited the individual, rather than being dictated by a traditional working pattern of being present in the workplace between certain designated hours. This greater level of flexibility was experienced not just by those with mobile or home-based roles, but also by those whose jobs had been re-classified as office-based. For instance, Robert observed,

> It does help you balance your work-life situation when you've got the freedom and flexibility to [plan your own working pattern]. (Robert – office-based)

Linked to this expression of freedom and control was the ability to organize days spent working from home in a different way from those in the office. Of particular value to participants was the chance to attend to domestic chores in the course of the day. This was not seen as a distraction from work activities, but simply as a means of dealing with minor, but ongoing demands from the non-work domain.

> It gives you the flexibility to do things that don't impact on your actual work. (Vicky – office-based)

Instead, we found evidence of employees interweaving demands from the non-work domain into natural pauses and small gaps in work activity, as in Geraldine's example below.

> It makes a huge difference, just being able to do things like prepare some food for the evening. I can put a stew in the oven in ten minutes and three hours later it's done and ready in the evening. So I do try and fit things like that in my day when I'm working from home. I don't think that that makes any difference to my performance and what I can deliver for the business. (Geraldine – mobile)

Another perceived benefit came from the increased level of control over one's time spent with family. Work commitments were adjusted to take account of other needs, while maintaining an overall sense of fulfilling responsibilities in a conscientious way. Interviewees talked freely of such an approach, with a clear indication that the organization was aware of, and encouraged, such behaviour. Achievement of work goals and targets were seen to be of primary importance, but it was also seen as acceptable for individuals to meet the demands of the non-work domain.

> I'm a family person and so there are times when there may be children's events at school or I need to go and do a child pickup or something like that. So yesterday morning I had to take children to school so I didn't actually arrive in the office 'til 10 am But then I'll pick that up, if not that same day, it may get picked up another day. So yes, I think for me the flexibility of being able to do some of those things is really valuable. (Richard – mobile)

Importantly it was not just parents who appreciated the positive effect on their overall lives of the enhanced control over their time. For instance the youngest participant (aged 26) explained:

> I don't have responsibilities like children, so for me I suppose, it does mean I am able to socialize in the evenings, and if I'm doing something like going to the theatre then at least

> that day I'll come in early or I'll work a bit longer the day before just to make sure I've done stuff, so I have that freedom. (Janice – mobile)

Similarly, Caroline explained how she was able to structure her week and therefore allocate time to differing priorities, from both domains.

> I can plan my week and I can balance my life, so have the ability to say, well, if I'm going up to Leeds that day, I'll make sure I'm perhaps working from home the next. (Caroline – mobile)

The second aspect of time management involved self discipline when working from home, with many participants commenting on the tendency to work longer hours and to return to their laptops or computers later in the evenings and at weekends. Moving work from the workplace into the home had an impact on how they managed the temporal boundaries, in some cases resulting in employees working extended hours. Daniel explained,

> It probably started with me being able to work from home, which started with working on a Sunday evening to catch up and so I was getting no benefit but the business was getting a benefit because I was working more hours. (Daniel – office-based)

The need for self discipline when transitioning to the new working arrangement is explained by Vince.

> You have to be good at logging off though. You have to be really dedicated to say "Okay I've done enough today", because you've got access to the company systems, you could go on and on and on, and I think part of my development plan probably this time last year was around identifying what is my natural logging off time. (Vince – mobile)

Similarly Richard commented favourably on having greater choice over working hours but pointed out the drawback of work always being in the background when at home.

> Being able to be more in control of your working day, your working week and being able to work flexibly is a great help. The risk is of course that the technology enables you to work at times when you wouldn't normally work and so you've got to, again, have a discipline to switch off and make that concerted switch-off moment happen. The risk is that a lot of us, at times, will leave laptops linked up to the internet at home during the evening or whatever and that means that you can dip in and dip out. And there's sometimes value in that and sometimes it's very distracting. (Richard – mobile)

The need to be strict about maintaining a boundary between the domains of work and home when actually at home was evident to participants. The challenge came not just in dealing with the seductive pull of emails ("I'll just check to see..."), but also with the expectations of others. Most interviewees reported a good relationship with their manager and a shared understanding of what constituted a satisfactory standard of performance. However a minority talked of the expectations of other stakeholders, such as clients or senior managers, of always being available, often beyond even extended working hours such as seven am to eight pm, five days a week.

> There are key contacts in the business that know that I'm going to be available on Sunday evening normally to take emails and things like that and so they take advantage of that to some extent which, being the kind of person I am and I see myself very much supporting

> the business, I don't mind to some extent. But it becomes an expectation and there are some weekends when I'm not going to be back home until late on a Sunday and I don't really fancy logging in. But it's almost ... I do feel, a little bit of a guilt feeling that I'm not available when I normally would be. (Daniel – office-based)

This blurring of boundaries between the two domains clearly has advantages in terms of the ability to adjust working hours in response to non-work priorities, but also has disadvantages beyond the ever-present nature of work within the home.

17.6 Discussion and Conclusions

This chapter has presented findings from a case study where an organization, in an attempt to reduce costs, reclassified jobs as mobile, home or office-based, resulting in a significant number of staff no longer having office space allocated to them or a designated workplace. At the same time as this imposed change, the organization offered greater flexibility to the employees, in order to facilitate the co-ordination of their work and non-work activities. In many ways we found that this increased flexibility to exercise some degree of choice over the timing of work served to temper the potentially negative reactions to the imposed change, in contrast to Harris' (2003) findings of increased stress and tension and little benefit to the individual. This deviation may be due in part to the large scale introduction of the new working arrangements in this case, rather than Harris' small group who felt that they had been singled out.

Being away from the workplace for some or all of their working time opened up a degree of control for the mobile and home-based workers in this case study, to make decisions about how to meet the sometimes conflicting demands placed upon them by the work and non-work domains. Caproni's (2004) concerns about the unpredictability of demands from the non-work domain were not supported. Instead we found evidence of some employees choosing to integrate work and non-work activities during the course of a, sometimes extended, working day. In contrast, there was also evidence of some employees choosing to segment their work and non-work roles, in some cases adopting working time patterns similar to those in office-based jobs. This increased control and autonomy experienced by the interviewees does however confirm that the ideal worker was not typical in this company, with the blurred boundaries between work and home allowing individuals to attend to domestic and caring responsibilities.

When work is moved into the home, the spatial boundaries of the workplace are eroded (Scott, 2003) and we found employees adopting a number of different strategies in response to this. It may be assumed that if an employee requests to work from home, then there will be a dedicated space available. However in reality, such office space is not always exclusively work space, and it may be unrealistic to expect it to be so. Some of our respondents were able to create spatial boundaries in the home with a room designated as a work space and a door as a physical marker of the divide between work/non-work. However, for those unable to do so, work and home activities had to share space, sometimes creating tensions in the home environment as a result. We also found evidence of work intruding on the home in a number of unanticipated ways. In some cases the presence of work activity and work-related equipment in the home permeated the work-non-work boundary, largely outside the control of the employee. This study therefore provides support for Baruch and Nicholson's (1997) argument that working from home removes the protective barrier between the work and non-work domains.

In this case moving work out of the workplace was coupled with increased freedom of time for employees, which was highly valued by our respondents. Professional workers are likely to experience some degree of control over the work they do; moving their work into the home environment opened up the opportunity to integrate work and non-work activities more easily. Being at home meant that time available through natural breaks in workflow could be utilized for non-work activity. Other categories of employees who experience less control over their work may have fewer opportunities to do this.

In this chapter we have examined how employees have experienced the removal of the spatial and temporal boundaries traditionally associated with work. Our findings show that whilst employees have generally responded positively to the freedom that accompanied these changes, there were a number of unanticipated consequences. These findings shed light on a number of issues which have tended not to be taken into consideration by organizations when developing policies which change the location and sometimes the timing of work. We found evidence that the transition to working in a different way was not always straightforward for employees and may at least in the short-term have impacted on their performance. The provision of some form of support for this transition could aid the process and enable employees to gain the benefits more effectively, which we found to be highly valued by them. Equally, the type of intrusions into home life reported here may well have longer-term consequences for employee well-being which could mitigate the positive outcomes reported. Some of these intrusions could however be ameliorated relatively easily. For example where work moves from the workplace to the home, consideration could be given to the nature of the equipment provided to employees, so that it fits with the home environment, rather than simply replicating what was provided in the workplace.

References

Ashforth BE, Kreiner GE, Fugate M (2000) All in a day's work: boundaries and micro role transitions. Acad Manage Rev 25:472–491

Bailyn L, Rayman P, Bengtsen D, Carre F, Tierney M (2001) Fleet Financial and Radcliffe explore paths of work/life integration. J Organ Excell 20:49–64

Baruch Y, Nicholson N (1997) Home, sweet work: requirements for effective home working. J Gen Manage 23:15–30

Campbell-Clark S (2000) Work/family border theory: a new theory of work/family balance. Hum Relat 53:747–770

Caproni PJ (2004) Work/life balance: you can't get there from here. J Appl Behav Sci 40:208–218

Hall L, Atkinson C (2006) Improving working lives: flexible working and the role of employee control. Empl Relat 28:374–386

Hall DT, Richter J (1988) Balancing work-life and home-life: what can organizations do to help?. Acad Manage Exec 3:213–223

Harris L (2003) Home-based teleworking and the employment relationship: managerial challenges and dilemmas. Pers Rev 32:422–437

Hill EJ, Grzywacz J, Allen S, Blanchard V, Matz-Costa C, Shulkin S, Pitt-Catsouphes M (2008) Defining and conceptualizing workplace flexibility. Commun Work Fam 11:149–163

Kanter RM (1977) Men and women of the corporation. Basic Books, New York, NY

Kossek EE, Lautsch B, Eaton S (2006) Telecommuting, control, and boundary management: correlates of policy use and practice, job control, and work-family effectiveness. J Vocat Behav 68:347–367

Nippert-Eng C (1996) Home and work: negotiating boundaries through everyday life. University of Chicago Press, Chicago, IL

Poelmans S, Caligiuri P (2008) Harmonizing work, family and personal life: from policy to practice. Cambridge University Press, Cambridge, MA

Ransome P (2007) Conceptualizing boundaries between 'life' and 'work'. Int J Hum Resour Manage 18:374–386

Rapoport R, Bailyn L, Fletcher J, Pruitt B (2002) Beyond work-family balance: advancing gender equity and workplace performance. Jossey-Bass, San Francisco, CA

Rodgers FS, Rodgers C (1989) Business and the facts of family life. Harv Bus Rev 67:121–129

Scott WR (2003) Organizations: rational, natural and open systems. Prentice Hall, New Jersey

Tietze S, Musson G (2003) The times and temporalities of home-based telework. Pers Rev 32: 438–455

Warhurst C, Eikhof DR, Haunschild A (2008) Out of balance or just out of bounds? Analsing the relationship between work and life. In Warhurst C, Eikhof DR, Haunschild A (eds) Work less, live more? Critical Analysis of the work-life boundary. Palgrave Macmillan, Basingstoke, pp 1–21

About the Authors

Andrea E. Abele-Brehm (in publications "Abele" only) is professor at the Chair of Social and Applied Social Psychology at the University of Erlangen-Nuremberg. She studied psychology, history, and sociology at the University of Konstanz, where she received her doctoral degree in 1976. She had several stays in the US and in Great Britain, and passed her Habilitation at the University of Bielefeld in 1982. She is editor and on board of several scientific journals, she is head of the Interdisciplinary Institute on Labor Market Research, and she is the university's chairperson for good scientific practice. She has published more than 150 papers, and her current research interests are career research, gender research, and self-concept research.

Deirdre Anderson is Lecturer in Organizational Behaviour at Cranfield School of Management. She is a member of the International Centre for Women Leaders and is a Senior Fellow of the Foundation for Management Education. In all of her work, Deirdre seeks to increase awareness of diversity and inclusion, challenging the often unconscious processes which can limit individual actions and organizational practices. Deirdre frequently presents her research at academic and practitioner conferences both within the UK and internationally. She has published journal articles and book chapters on flexible working and work life balance, women's career choices and women only leadership development.

Chris Baldry is Emeritus Professor of Human Resource Management at the University of Stirling. His recent research interests have focused on the changing world of work and employment, and the role of the built working environment, and office buildings in particular, on the way that work is experienced. His work on occupational health and safety has led to publications looking at ill-health in offices and call centres, and safety trends for trackside rail workers.

T. Alexandra Beauregard is Lecturer in Employment Relations and Organisational Behaviour at the London School of Economics, and a member of the editorial board for *Equality, Diversity and Inclusion: An International Journal*. Her research

S. Kaiser et al. (eds.), *Creating Balance?*, DOI 10.1007/978-3-642-16199-5,

interests focus on the work-life interface, and include the implications of flexible working practices for workplace relationships and career progression.

Ronald J. Burke is Professor of Organization Studies, Schulich School of Business, York University, in Toronto. His research interests include corruption in organizations, corporate reputation, and occupational health and safety.

Simon Burnett is Senior Research Associate at Lancaster University Management School, responsible for a 2 year project investigating fathers and the effects of flexible working practices intended to elicit a positive work-life balance. In addition to these topics he is interested in the nature and spread of "new" capitalism, organisational well-being and work-life boundaries. He completed his PhD in 3 years, an in-depth analysis of the current societal obsession with and contemporary manifestations of the personal, organisational and cultural pursuit of happiness; and is currently in the process of preparing a number of articles for publication.

Lydia A. Buswell is a graduate student at Brigham Young University. She is studying Youth and Family Recreation with an emphasis on fathers. She has done previous research in the areas of marriage, working with the National Healthy Marriage Resource Center, and families and religion. She is married and is excitedly expecting her first baby at the beginning of 2010.

Sarah June Carroll is Research Assistant in the Family Studies Center where she has aided the research of Dr E. Jeffrey Hill for more than 4 years. As well, she is a graduate student in Marriage and Family Therapy at Brigham Young University.

Uracha Chatrakul Na Ayudhya, PhD, is Lecturer in Human Resource Management and Organizational Behaviour at Middlesex University Business School (UK). Her research focuses on exploring the experiences of work-life balance and transitions in and out of employment of individuals from a life course perspective. She is a member of the Editorial Advisory Board for the journal *Community, Work and Family*. She is a co-editor of *Critical Issues in Human Resource Management* (CIPD, 2011).

Judith A. Clair (Ph.D. University of Southern California) is Associate Professor at Boston College. She studies social identities at work and impacts of critical events, such as crises and downsizing, on individuals and organizations. Her work appears in journals such as *Academy of Management Review*, *AMLE*, *Organizational Dynamics* and *Human Relations*.

Malissa A. Clark (M.A., Wayne State University) is a PhD student in the Industrial and Organizational Psychology program at Wayne State University, Detroit Michigan. Her research focuses on emotions in the workplace and work-family conflict.

Cary Cooper (CBE) is Professor of Organizational Psychology and Health at Lancaster University Management School at Lancaster University and Chair of the Academy of Social Sciences. He is the author of over 100 books (on occupational stress, women at work and industrial and organizational psychology), has written over 400 scholarly articles, and is a frequent contributor to national newspapers, TV and radio. He is a Fellow of the British Academy of Management and the Academy of Management (having also won the 1998 Distinguished Service Award), and also the lead scientist to the UK Government Office for Science on their Foresight programme on Mental Capital and Well Being (2007–2008). He was appointed a member of the expert group on establishing guidance for the National Institute for Health and Clinical Excellence on "promoting mental wellbeing through productive and healthy working conditions", 2009. HR Magazine named him the 6th Most Influential Thinker in HR in 2009.

Josje Dikkers is Assistant Professor at the Department of Management & Organization of the VU University Amsterdam. Her PhD project (2008, Work- & Organizational Psychology, Radboud University Nijmegen) was titled "work-home interference in relation to work, organizational, and home characteristics". She also works at Qidos as a research consultant. Since September 2005, she coordinates courses on Organizational Behavior and Research Methodology, as well as Bachelor and Master Theses within the field of Human Resources. Her research interests primarily focus on the interaction between people's work and private lives and work-home culture. With the Amsterdam Center for Career Research (www.accr.nl), Josje has coordinated a large data collection project on the careers of 9000 lawyers in the Netherlands.

Nikos Dimotakis is a PhD student in the Eli Broad Graduate School of Business at Michigan State University. His research interests include employee health and well-being, discretionary behaviors in the workplace, emotions, motivation, and interpersonal interactions.

Linda Duxbury is Professor at the Sprott School of Business, Carleton University. Within the past decade she has completed major studies on Balancing Work and Family in the public, private sectors and not for profit sectors. Dr Duxbury has published widely in both the academic and practitioner literatures in the area of work-family conflict, change management and the use and impact of office technology. Within the business school at Carleton, Dr Duxbury teaches masters and PhD courses in Managing Change as well as the master course in Organizational Behaviour and the PhD course in Organizational Theory. Dr Duxbury has received numerous awards for her both research and teaching.

Doris Ruth Eikhof (PhD Hamburg) is Lecturer in Organisation Studies at the Stirling Institute for Socio-Management, Stirling Management School, University of Stirling. Her research focuses on creative work and enterprise, social theories in organisation studies, work life-boundaries and women's work. Doris has published in international books and journals, including Journal of Organizational Behavior. She is co-editor of *Work less, live more? Critical Analyses of the Work-Life Relationship* (Palgrave 2008) and of an Employee Relations special issue on work-life balance (2007, Vol. 29:4).

Ghada El-Kot is Professor and administrator at the College of Management and Technology, Arab academy for Science and Technology and Maritime Transport in

Alexandria. Her research interests focus on human resource management practices in Egypt, total quality management, and the Islamic work ethic.

Tamara A. Fackrell is a practicing attorney in Utah focusing on family law since 1998. She is also part-time faculty at the J. Reuben Clark Law School. She graduated cum laude from BYU Law School and magna cum laude for undergraduate in Psychology and Communication. Tamara is a Master Mediator and Primary Trainer for the State of Utah and performs certifications in mediation and divorce mediation for professionals.

Lisa Fiksenbaum is currently completing her PhD in the Department of Psychology, York University, in Toronto. Her research interests include stress and health, work and family conflict, and research methodology.

Adam M. Galovan received his undergraduate degree in Family, Consumer, and Human Development with an emphasis in Family Finance from Utah State University. He graduated from USU as the valedictorian of the Emma Eccles Jones College of Education and Human Services. He received his master's degree in Marriage, Family, and Human Development at Brigham Young University and is currently working of his doctoral degree in Human Development and Family Studies at the University of Missouri. His research interests focus around marriage, coparenting, fatherhood, and household role interactions.

Timothy Galpin is Associate Professor at the University of Dallas, College of Business. Tim has over 20 years of experience as a management consultant and business manager working with boards and senior management around the world on strategic planning, strategy execution, merger and acquisition integration, divestitures, restructurings, human capital management, business productivity improvement, and organizational culture change. He has also authored three management books: *The Human Side of Change*, *Making Strategy Work*, and *The Complete Guide to Mergers and Acquisitions*.

Caroline Gatrell is Lecturer at Lancaster University Management School. Her research focuses on motherhood and the maternal body, examining relationships between family practices, motherhood, fatherhood and paid/unpaid work. Caroline explores the notion of maternal "work" in the context both of reproductive and productive labour. She has recently published two books: *Embodying Women's Work*, Open University Press and *Gender and Diversity in Management: A Concise Introduction* (with Elaine Swan), Sage. Caroline's work has been published in a range of journals including *The Sociological Review, The International Journal of Human Resource Management, Journal of Gender Studies, European Journal of International Management* and *Gender, Work and Organization.*

Danna Greenberg (PhD Boston College) is Associate Professor at Babson College. Her primary research focuses on the relationship between individuals and organizational change particularly as it relates to the work-life interface. She also does research on pedagogy pertaining to business education. Her work has appeared in journals such as *Administrative Sciences Quarterly*, *Academy of Management*

Learning and Education, Journal of Applied Behavioral Science and *Group and Organization Management.*

E. Jeffrey Hill is Associate Professor in the School of Family Life at Brigham Young University. His research examines finding harmony between paid work and family life. Before coming to BYU he was a work and family subject matter expert at IBM where he pioneered many flexible work options including paternity leave, professional part-time employment, and telecommuting. Dr Hill recently developed and analyzed the 2007 IBM Global Work and Life Issues survey with more than 25,000 respondents from 79 countries. In 2006 he conducted the Singapore National Study of Work-Life Harmony. Dr. Hill has published more than 50 scholarly articles and book chapters. He and his wife Tammy are the parents of 12 children and 11 grandchildren.

Jeff Hyman is Emeritus Professor of Management at the University of Aberdeen, Scotland. His main research interests include employee participation in organisational decision-making and more recently he has recently been involved with colleagues in an examination of the future of work, with specific reference to the experience of employment in new sectors such as software. Jeff Hyman is the author or co-author of several books and articles which derive from these research interests.

Remus Ilies is the Valade Research Fellow and a Professor of Management in the Eli Broad Graduate School of Business at Michigan State University. His research interests include individual differences, employee health and well-being, work-family processes, emotions, attitudes, leadership, and motivation.

Weng Jing is Professor in the school of Hospitality and Tourism, Beijing Union University. Her research interests include the teaching of hospitality and tourism, and improving tourism and hospitality in China

Blake L. Jones is a postdoctoral Research Associate at the University of Illinois at Urbana-Champaign. He works in the Family Resiliency Center, examining how family routines (e.g., parental work characteristics, family meals) relate to health outcomes (e.g., obesity, sleep problems, emotional regulation). He also studies these and other factors within work and family life in relation to differences in poverty, gender, and culture. Blake completed his M.S. degree from Purdue University, and his PhD from Brigham Young University.

Stephan Kaiser is Professor for Human Resources Management and Organizational Theory at the Universität der Bundeswehr München. He received his PhD from the Catholic University of Eichstätt-Ingolstadt. He published several books, and his work appeared in journals like *Organization, Industry and Innovation, Int. Journal of Work Organization and Emotion, Int. Journal of Knowledge Management Studies*, etc. His research focuses on human resource and organizational issues in knowledge-intensive work.

Clare Kelliher, PhD, is Reader in Work and Organisation at Cranfield School of Management, Cranfield University. She has a longstanding interest in flexible working and has published widely in this field. She is co-chair the IIRA Study Group on

Flexible Work Practices and has co-convened sub-themes on New Ways of Working at EGOS Conferences.

Ulla Kinnunen (PhD in psychology from University of Jyväskylä in 1989) is Professor at the Department of Psychology, University of Tampere, Finland. At the present her research activities focus on three areas: work-family interaction among individual employees and marital couples, well-being at work from the viewpoints of stress, burnout and recovery, and job insecurity and temporary work in organizations.

Mustafa Koyuncu is Associate Professor of Management, Faculty of Commerce and Tourism Education, Nevsehir University in Turkey. His research interests include tourism education, management in the tourism and hospitality sector, and cross-cultural research in tourism.

Jamie J. Ladge (Ph.D. Boston College) is Assistant Professor at Northeastern University. Her primary research focuses on the intersection of identity, work-family and careers. She also does research on stigmatized social identities in the workplace. Her work has appeared in the *Journal of Vocational Behavior*, *Organizational Dynamics* and the journal of *Negotiations and Conflict Management Research.*

Suzan Lewis, Ph.D., is Professor of Organisational Psychology at Middlesex University Business School (UK). Her research focuses on work-personal life issues and workplace practice, culture and change in different workplace and national social policy contexts. She has led many national and international research projects and has advised governments and worked with employers and policy makers on these issues in Europe and Japan. She is currently a member of a European Union Expert Group on Women in Science, Engineering and Technology. She has published widely in the field and is a founding co-editor of the international journal *Community, Work and Family*. Her most recent book is *Work, Families and Organisations in Transitions: European Perspectives* (Policy Press, 2009).

Rosemary Maellaro is Assistant Professor and Academic Director for the Organization Development MBA at the University of Dallas College of Business. She has over 25 years of management experience and has held senior HR positions in the high tech, hospitality, and retail industries. Her research interests lie in the areas of adult learning, leadership interpersonal skill development, and team building. Rosemary has served clients such as Alcon Laboratories, Baylor Health Care Systems, and Starbucks in her consulting practice. She was recently recognized as the 2009 *HR Southwest Educator of the Year.*

Saija Mauno (PhD in work psychology from the University of Jyväskylä in 1999) is currently working as Academy Research Fellow at the University of Jyväskylä, Finland. Her research interests cover work-family interface, job insecurity and temporary work as well as well-being at work (e.g., work engagement, job stress).

Helene Mayerhofer is Assistant Professor at Institute of Human Resource Management, Department of Management, at the WU Vienna (Vienna University of Economics and Business). International assignments (expatriation and alternatives), comparative IHRM and different aspects of human resource development are fields of research and practical work.

Jesse S. Michel (PhD, Wayne State University; Post-Doctoral, Michigan State University) is Assistant Professor of Industrial and Organizational Psychology at Florida international University, Miami Florida. His research focuses on the dynamics between work and family, personality and behavioral stability, and leadership and abusive supervision.

Barbara Müller is Assistant Professor at the Institute of Human Resource and Change Management at the JKU Johannes Kepler University Linz. She received her master and doctoral degree at the WU Vienna where she was on faculty from 2004-2010. Her current research topics focus on social systems theory, organizational learning processes and qualitative methods of research. Her current research focuses on social systems theory, organizational learning and knowledge management.

Ann Chunyan Peng is a PhD student at Department of Management, Eli Broad College of Business, Michigan State University. She received a research master degree from Lingnan University of Hong Kong. Her current research interests include stress coping and well-being, self-identity, work-family interface, and leadership.

Bram Peper is Assistant Professor of sociology at the Erasmus University Rotterdam, the Netherlands. His main area of expertise is cross-national research regarding work–family arrangements in organizations in different welfare state regimes. Presently he is working several research projects on utilization of work-life policies, and on well being of employees. His latest co-edited books are one on flexible work, and a Dutch introduction to sociology.

Miguel Pina e Cunha is a professor at NOVA – School of Business and Economics. He has a PhD from Tilburg University. One of the topics he is investigating is positive and toxic organizing. He is especially interested in the emergent nature of toxic organizational contexts, i.e., in the dynamics that make organizational madness possible.

Johanna Rantanen (PhD in psychology from University of Jyväskylä in 2008) is a researcher at the Department of Psychology, University of Jyväskylä, Finland. At present her main research interests are work-family and work-life interaction from individual and life-span perspectives.

Cornelia U. Reindl is Research Assistant and PhD student at the Chair of Organizational Theory and Human Resource Management at Ingolstadt School of Management (Catholic University Eichstaett-Ingolstadt). She studied Educational Science and Psychology of Excellence in Business and Education (M.A.) at the University of Munich. Her research interests include human resource management, leadership, and particularly research on work-life integration.

Max Josef Ringlstetter is Professor of the Chair of Organizational Theory and Human Resource Management at Ingolstadt School of Management (Catholic University Eichstaett-Ingolstadt). His research, teaching and consulting focuses especially on organizational studies, human resource management and corporate strategy. In addition to that he is an expert for of media and professional service firms.

Angelika Schmidt is Assistant Professor at the Institute of Change Management and Management Development at the WU Vienna (Vienna University of Economics and Business). Her research focuses on issues of employment relations, HR-strategies, work-life boundaries and gender issues.

Rob Smart is a post-doctoral Researcher and Lecturer at the Sprott School of Business, Carleton University. His primary research interests include teamwork, organizational change, and role overload. Rob has worked on several research studies in the areas of work-family conflict and role overload, including projects in the hospital, policing and specialized eldercare environments. Rob has also conducted recent research on the impact of BlackBerry adoption on balancing work and family for professional employees.

Paul Sparrow is the Director of the Centre for Performance-led HR and Professor of International Human Resource Management at Lancaster University Management School. His research interests include cross-cultural and international HRM, HR strategy, cognition at work and changes in the employment relationship. He recently researched *Globalizing HRM* for the CIPD and *International Mobility* for Brookfield GRS. He has consulted with major multinationals, public sector organisations and inter-governmental agencies. In 2009 he was voted amongst the Top 15 Most Influential HR Thinkers by Human Resources Magazine.

Martin L. Stolz is a Research Assistant and PhD student at the Chair of Organizational Theory and Human Resource Management at Ingolstadt School of Management (Catholic University Eichstaett-Ingolstadt). He studied management sciences at the Catholic University Eichstätt-Ingolstadt and the Università degli Studi de Modena e Reggio Emilia, Italy. His research interests include strategic development of professional service firms and consultancies, work-life balance and employee professional and organizational commitment.

Kati Tillemann (M.Sc. in Social Psychology from Tallinn University, 2000; PhD student of Psychology at the University of Jyväskylä) is Lecturer of organizational behavior, general and social psychology, and introduction to sociology at the Estonian Business School (since 2002). Her current research activities and interests are work-family issues, occupational health and parenting.

Marloes van Engen is Assistant Professor at the Department of Human Resource Studies, Faculty of Social and Behavioral Sciences at Tilburg University where she lectures in Diversity in Organizations. She studied social psychology with a minor in the psychology of culture and religion at the Radboud University in Nijmegen. Her research interests are in the area of gender in organizations, gender and careers, work-family issues in organizations, diversity in teams and organizations, and methodology such as meta-analysis and multi-level analysis. She earned her PhD in 2001 on gender and leadership. She was a visiting academic at Northwestern University (USA), the University of Queensland (Australia) and Monash University (Australia). She has published in Psychological Bulletin, Journal of Organizational and Occupational Psychology and International Journal of Human Research Management.

Claartje Vinkenburg is Associate Professor of organizational behavior and development at the VU University Amsterdam. After her PhD she worked as a management consultant and was a visiting scholar and adjunct lecturer at Northwestern University (USA). As managing director of the Amsterdam Center for Career Research (www.accr.nl), Claartje's research focuses on gender, leadership and career advancement, including the effects of normative beliefs about parenting on women's career patterns and outcomes. She has published several book chapters and articles on her research, as well as edited a book on "Top potentials" for the Dutch Foundation of Management Development.

Judith Volmer is Assistant Professor of work and organizational psychology at the Department of Psychology, University of Erlangen-Nuremberg, Germany. Before, she worked as a doctoral candidate at the Technical University of Braunschweig, Germany and the University of Konstanz, Germany. She received her PhD for her work on star performers in software design teams. Her current research interests include emotions at work, careers and career development, and performance and learning at work.

J. Lee Whittington is Professor at the University of Dallas College of Business. His industry experience includes over 25 years in manufacturing and distribution. He has held management positions in marketing, logistics, and general management. His consulting experience includes engagements with Nokia, FedEx-Kinko's, RadioShack, Camp Fire Boys and Girls, US Army Corps of Engineers, and Siemens ElectroCom. His research has been published in The Leadership Quarterly, Journal of Management, Academy of Management Review, Journal of Organizational Behavior, Journal of Applied Social Psychology, Journal of Management Spirituality and Religion, and Journal of Managerial Issues.

Index